T0327507

SYSTEM SIMULATION TECHNIQUES WITH MATLAB® AND SIMULINK®

SYSTEM SIMULATION TECHNIQUES WITH MATLAB® AND SIMULINK®

Dingyü Xue
Northeastern University, China

YangQuan Chen
University of California, Merced, USA

This edition first published 2014
© 2014 John Wiley & Sons, Ltd

Registered office
John Wiley & Sons Ltd, The Atrium, Southern Gate, Chichester, West Sussex, PO19 8SQ, United Kingdom

For details of our global editorial offices, for customer services and for information about how to apply for permission to reuse the copyright material in this book please see our website at www.wiley.com.

The right of the author to be identified as the author of this work has been asserted in accordance with the Copyright, Designs and Patents Act 1988.

All rights reserved. No part of this publication may be reproduced, stored in a retrieval system, or transmitted, in any form or by any means, electronic, mechanical, photocopying, recording or otherwise, except as permitted by the UK Copyright, Designs and Patents Act 1988, without the prior permission of the publisher.

Wiley also publishes its books in a variety of electronic formats. Some content that appears in print may not be available in electronic books.

Designations used by companies to distinguish their products are often claimed as trademarks. All brand names and product names used in this book are trade names, service marks, trademarks or registered trademarks of their respective owners. The publisher is not associated with any product or vendor mentioned in this book.

Limit of Liability/Disclaimer of Warranty: While the publisher and author have used their best efforts in preparing this book, they make no representations or warranties with respect to the accuracy or completeness of the contents of this book and specifically disclaim any implied warranties of merchantability or fitness for a particular purpose. It is sold on the understanding that the publisher is not engaged in rendering professional services and neither the publisher nor the author shall be liable for damages arising herefrom. If professional advice or other expert assistance is required, the services of a competent professional should be sought.

MATLAB® is a trademark of MathWorks, Inc. and is used with permission. MathWorks, Inc. does not warrant the accuracy of the text or exercises in this book. This book's use or discussion of MATLAB® software or related products does not constitute endorsement or sponsorship by MathWorks, Inc. of a particular pedagogical approach or particular use of the MATLAB® software.

Library of Congress Cataloging-in-Publication Data

Xue, Dingyü.
  System simulation techniques with MATLAB and Simulink / Dingyü Xue, YangQuan Chen.
    1 online resource.
  Includes bibliographical references and index.
  Description based on print version record and CIP data provided by publisher; resource not viewed.
    ISBN 978-1-118-69435-0 (Adobe PDF) – ISBN 978-1-118-69437-4 (ePub) – ISBN 978-1-118-64792-9 (cloth)
1. System analysis–Data processing.   2. Computer simulation.   3. MATLAB.   4. SIMULINK.   I. Chen, YangQuan, 1966–   II. Title.
  T57.62
  620.00285′53–dc23

                                                                                        2013025348

A catalogue record for this book is available from the British Library.

ISBN: 978-1-118-64792-9

Typeset in 10/12pt Times by Aptara Inc., New Delhi, India

1   2014

Contents

Foreword

It is a pleasure for me to write a foreword for this book by Dingyü Xue and YangQuan Chen. Dingyü came to the University of Sussex in 1988 to study for his DPhil with me. At the time, computing, relating to control engineering, was starting to move from Fortran to MATLAB, first on terminals connected to a central mainframe computer and then to standalone desktop machines. Digital simulation languages, which had replaced analog computers, were also heading in the same direction. The original version of MATLAB used on the mainframe was written in Fortran, followed by the much faster C version a few years later. One great advantage of MATLAB was that its fundamental data type was the matrix, the concept of which I first came across in the now little known language APL. APL was a very efficient coding language, so much so that a fair comment would be that it required as many lines of commenting as coding for a person to understand a program, and it also required a special keyboard. Other major features of MATLAB were the very good graph plotting facilities and the tools available for providing an excellent graphical user interface for a program. The graphical features provided for programming and for the display of results in Simulink were also a major improvement over the features of existing digital simulation languages.

In the early days of MATLAB, I had several general programs on the mainframe computer which used a question and answer interface and gave the output as a printed plot of points. Dingyü, in doing his research, developed a deep understanding of MATLAB and the capabilities of the GUI, one eventual result of which was the program CtrlLAB which is freely available from the MathWorks library. The genesis of this was a program described in my 1962 doctoral dissertation written in Manchester Autocode, which used paper tape to provide the data input and the values of points as output. Intermediate stages had seen its coding in APL and MATLAB using a question and answer format. Dingyü has therefore used MATLAB and Simulink avidly for the past 25 years, including, I suspect, most of the versions issued over that period. He has spent thousands of hours writing new code and modifying existing routines to be compatible, or to take advantage of new features in the changing versions of MATLAB and Simulink. I have known YangQuan Chen – whom Dingyü first met in Singapore about twenty years ago – for the past ten years. Since they first met, they have cooperated a lot with their complementary research interests being united by their use of MATLAB and Simulink.

This book is therefore written by two people who have had a wealth of first-hand experience of using MATLAB/Simulink in control engineering research and teaching its use to students in China and the USA in both mathematical and control-related courses for over two decades. Also, much of the material has been available in earlier versions of the book in Chinese, where it has been extremely well received, and it is used at many universities. Feedback from these publications has provided suggestions for improvements which have been incorporated here.

The coverage of the book is such that it provides a basic introduction to the use of MATLAB/ Simulink before going on to address their usage in many facets of mathematics and engineering. After covering the general aspects of programming and computation in MATLAB, details of applications in many areas of scientific computation are given, covering areas such as differential equations and optimization. Chapters 3–6 are primarily devoted to Simulink, starting from consideration of the functions of the various blocks and continuing to describe a variety of applications covering topics such as linear and nonlinear system simulations, multivariable systems, vectorized blocks, output blocks, the animation of results, linearization of nonlinear systems, S-functions and optimization in simulations. Chapter 7 discusses the more specific engineering application blocks for electronic systems, electrical drive systems and so on, that are available in Simscape, and in chapter 8 some simulation applications for non-engineering systems, image processing and finite state machines are described which show the wide applicability of modeling and simulation techniques.

I'm sure that this book with its many examples and problems will prove a major asset to you, the reader, in learning the simulation capabilities of MATLAB/Simulink, but as Dingyü and YangQuan would no doubt confirm, the only way to really learn is by the hard work of "doing". So attempt the exercises and also design your own to possibly clarify certain points and gain greater understanding.

Derek P Atherton
Professor Emeritus
University of Sussex, UK
March 2013

Preface

As Confucius has said, *"The mechanic, who wishes to do his work well, must first sharpen his tools"*, so MATLAB/Simulink is the right tool to solve problems in the field of systems simulation. It can free the scientist and engineer from tedious, laborious and error-prone work in low-level computer programming, and it is obvious that by the use of MATLAB and Simulink, the efficiencies of researchers can be significantly improved. In communities such as systems simulation and control engineering, MATLAB/Simulink is the de facto international computer language, and the importance of such a tool is being taught in universities worldwide.

Although MATLAB itself was developed and advocated by mathematicians, it was in fact first acknowledged by researchers in the engineering community, and in particular, by the researchers in the field of control engineering. The development of MATLAB and Simulink received a significant amount of innovative contribution from scholars and researchers in the field of control engineering. Already, a significant number of toolboxes and blocksets are oriented to control problems. MATLAB itself has extremely strong capabilities for solving problems in scientific computation and system simulation, with its handy graphical facilities and integrated simulation facilities. It is being used by researchers in more and more engineering and other scientific fields, and it has huge potential and great applications possibilities in related fields.

The authors have been consistently using MATLAB in education and scientific research since 1988, and have had some of their MATLAB packages added to MATLAB Central. A significant amount of first-hand knowledge and experience have been accumulated.

The first author started introducing MATLAB into education more than twenty years ago, and has tried to instruct students in the use such tools. For instance, the book "Computer-aided control systems design — MATLAB languages and applications" published by Tsinghua University Press was regarded as the first of its kind and one of the best in China and has been cited by tens of thousands of journal papers and books. The second author has had more than ten years of experience of scientific research and education in universities in the United States, after his work in industry. He has built up a lot of experience in MATLAB/Simulink based simulation as well as hardware-in-the-loop simulation and real-time design of control systems. Two other books have also been written by the authors and introduced into English world, concentrating on, respectively, the fields of automatic control and scientific computation.

The first edition of this present book was published by Tsinghua University Press in Chinese in 2002, and the second edition was published there in 2011. It has been used as a textbook and reference book by many universities in China. With evolution of MATLAB, Simulink and related products, a lot of new material and innovative work has emerged. It is not possible to cover all the material in one book, so the material here was carefully chosen, and tailored to meet the demands of engineering students and researchers in the relevant disciplines. The current shape of this book was finalized in the course at the Northeastern University, China, and also by offering seminars and series lectures at Utah State University in the USA, at Baosteel Co. Ltd and at Harbin Institute of Technology in China. Based on the programming and educational experiences of over twenty years, the authors have finally debuted the book to the English-speaking world, and we feel sure that this book will be welcomed by readers worldwide.

The educational work in this book, together with other related educational work, was directed and encouraged by the former supervisors, Professors Xingquan Ren and Xinhe Xu of Northeastern University, China, and Professor Derek P Atherton at Sussex University, UK. It was them who guided the first author into the field of system simulation and, in particular, into the paradise of MATLAB/Simulink programming and education.

A lot of suggestions were received during the preparation of related books, and among them, the authors are in particular grateful for the help given by Professor Hengjun Zhu of Beijing Jiaotong University, the late Professor Jingqing Han of the Institute of System Sciences of Academia Sinica, Professor Xiaohua Zhang of Harbin Institute of Technology, China, and Professor Igor Podlubny of the University of Kosice, Slovakia.

Fruitful discussions with colleagues Drs Feng Pan, Dali Chen, Ying Wei, Jianjiang Cui, Liangyong Wang, Zheng Fang resulted in some new ideas and materials in this book, and Zhuo Li in proofreading an early draft of the book. The Chapter 9.6 was based on contribution of MESA LAB Ph.D. students Brandon Stark, Zhuo Li and Brendan Smith of UC Merced.

We wish to thank editorial staffs of Wiley: Tom Carter, Project Editor; Paul Petralia and Anne Hunt. Copyediting by Paul Beverley is particularly appreciated!

This book was supported by the MathWorks Book Program, and MATLAB, Simulink and related products can be acquired from

MathWorks, Inc., 3 Apple Hill Drive, Natick, MA, 01760-2098 USA

Tel:+01-508-647-7000, and FAX: +01-508-647-7101

Email: info@mathworks.com, Webpage: http://www.mathworks.com

Last but not least, the authors are grateful to their family members for the understanding and support during the years of working. Dingyü Xue would like to thank his wife, Jun Yang, and daughter Yang Xue, and YangQuan Chen would like to thank his wife, Huifang Dou, and his sons Duyun, David and Daniel.

Dingyü Xue, *Northeastern University, China*
YangQuan Chen, *University California, Merced, USA*

1

Introduction to System Simulation Techniques and Applications

1.1 Overview of System Simulation Techniques

Systems are the integrated wholes composed of interrelated and interacting entities; these could be engineering systems or non-engineering systems. Engineering systems are the whole composed of interacting components such that certain system objectives can be achieved. For instance, motor drive systems are composed of an actuating component, a power transfer component and a signal measurement component, so as to control the motor speed or position, among other objectives.

The field of non-engineering systems is much wider. From universe to micro world, any integrated whole can also be regarded as a system, since there are interrelated and interacting relationships.

In order to quantitatively study the behavior of a system, the internal characteristics and interacting relationship should be extracted, to construct a model of the system. System models can be classified as physical models and mathematical models. Following the rapid development and utilization of computer technology, the application of mathematical models is more and more popular.

A mathematical model of a system is a mathematical expression describing the dynamical behavior of the system. They can be used to describe the relationship of quantities in the system and they are the basis of system analysis and design. From the viewpoint of the type of mathematical model, systems can be classified as continuous-time systems, discrete-time systems, discrete event systems and hybrid systems. Systems can also be classified as subclasses of linear, nonlinear, time invariant, time varying, lumped parameters, distributed parameters, deterministic and stochastic systems.

System simulation is a subject within which the system behavior can be studied on the basis of the mathematical models of the actual systems. Usually computer simulation of the systems is the main topic of the subject, including the topics of systems, modeling, simulation algorithms, computer programming, display of simulation results, and validation of simulation results.

Of all the topics listed above, simulation algorithms and computer programming are the most important topics, and determine whether the original problems can be solved. The modeling and simulation result display and validation can be solved easily with MATLAB, and Simulink, the most authoritative and practical computer language. Using MATLAB and Simulink is the innovative characteristic of the whole book. In Section 1.2, a brief introduction to the historical development and future expectation of computer mathematical software and simulation languages will be given. In Section 1.3, development of MATLAB/Simulink programming is presented, and practical examples

System Simulation Techniques with MATLAB® and Simulink®, First Edition. Dingyü Xue and YangQuan Chen.
© 2014 John Wiley & Sons, Ltd. Published 2014 by John Wiley & Sons, Ltd.

are explored, such that the reader can start to experience the powerful facilities of MATLAB. In Section 1.4, the main contents and the characteristic behavior of systems are presented.

1.2 Development of Simulation Software

Historically, computer simulation techniques went through the following stages of development: In the 1940s, analog simulation was the major way of simulation. Digital simulation began in 1950s, and in the 1960s, the first simulation languages and packages began to emerge. In the 1980s, development of object oriented simulation techniques was the leading trend. With the popularity and wide availability of digital computers, in the past 30 years, a great many professional computer simulation languages and tools have appeared, such as CSMP, ACSL, SIMNON, MATLAB/Simulink, MatrixX/System Build and CSMP-C. Because MATLAB/Simulink has become more and more popular and powerful, most of the above-mentioned simulation packages are no longer available. MATLAB/Simulink has became the de facto standard computer language and tool for system simulation.

1.2.1 Development of Earlier Mathematics Packages

The rapid development of digital computers and programming languages powered research into numerical computation. In the early stages of the development of scientific computation, a lot of famous packages emerged such as the LINPACK package [1] – linear algebraic equation solver, the eigenvalue-based package EISPACK [2, 3], the NAG package [4] developed by the Numerical Algorithm Group in Oxford, and the subroutines provided in the well-established book *Numerical Recipes* [5]. These packages were very popular and had a very good reputation among the users worldwide.

The well-established EISPACK and LINPACK packages are mainly used to solve eigenvalue problems and singular value decomposition based linear algebra algorithms. These packages were all written in Fortran.

For instance, to find all the eigenvalues of a real square matrix A of size N, and the eigenvalues are represented by W_R and W_I, for real and imaginary parts, and the eigenvector matrix is represented by Z, the following subroutine calls are suggested in the EISPACK package

```
CALL BALANC(NM,N,A,IS1,IS2,FV1)
CALL ELMHES(NM,N,IS1,IS2,A,IV1)
CALL ELTRAN(NM,N,IS1,IS2,A,IV1,Z)
CALL HQR2(NM,N,IS1,IS2,A,WR,WI,Z,IERR)
IF (IERR.EQ.0) GOTO 99999
CALL BALBAK(NM,N,IS1,IS2,FV1,N,Z)
```

Before the above subroutine calls, you should write a piece of code to assign the matrix to the program. Then with the above statements, the main program can be written. After the compiling and linking process, the executable file can be generated. The results can finally be obtained with the executable file.

A large number of numerical subroutines are provided in the NAG package and in the book, *Numerical Recipes* [5]. The NAG package is even more professional since many more subroutines are provided. In *Numerical Recipes*, a large number of high-quality subroutines, written in C, Pascal and Fortran, are provided; these subroutines can be used directly by researchers and engineers. There are more than 200 effective and reliable subroutines, and the subroutines are trusted by researchers worldwide.

Readers with a knowledge of Fortran and C programming might already know that, in those two programming languages, the scientific computation of matrices and graphics are rather complicated. For instance, to solve a linear algebraic equation, the elements in the matrices should be assigned first. Then a subroutine has to be written to implement the solution algorithms, such as the Gaussian elimination algorithm, and finally the result has to be output. If the subroutine written or selected is not reliable, misleading conclusions may be reached. Normally such a low-level subroutine can consist of over 100 statements. A small programming error can result in wrong conclusions.

Writing programs with packages has the following disadvantages

- **Inconvenience**. If the user is not familiar with the package being used, it might be very difficult to write programs with it, and it is always error-prone. If the slightest error is made in the program, erroneous results and misleading conclusions can be obtained.
- **Trivial procedures are involved**. A main program has to be written, and compiling and linking to the program should be made to generate executable files. A lot of effort is needed to debug the program and to validate the program.
- **Too many executables**. To solve a specific problem, a dedicated program has to be prepared. An executable file must be generated for this specific problem. The code reuse is not good, where a lot of similar problems may have to be solved.
- **Not suitable for data transfer between independent programs**. Each program can solve one particular problem. It might be difficult to transfer data from one standalone program to another. And it might not be suitable for solving one common problem by several standalone programs.
- **Difficult to allocate the array size**. In many mathematical computation problems the most important variables can be matrices. In most packages, the dimensions of the matrices might be set very low; for instance, in the package for control systems analysis and design in [6], the dimension is normally set to 10. It cannot be used to solve very high order systems.

Also, most earlier packages were written in Fortran. The plotting facilities of standard Fortran are not very good. Some other packages such as GINO-F [7] have to be used instead. However, on some platforms this package may not be available.

Apart from the above-mentioned shortcomings there is yet another difficult problem. A program written in Fortran or C cannot be easily transported to other platforms, since the source code on different platforms may not be compatible. For instance, a program written for Microsoft Windows cannot be executed at all on Linux without changes. Modifications must be made to the source code, and the source code has to be recompiled to generate executables. This is a rather difficult task, especially when plotting facilities are part of the source code.

Despite this, the development of mathematical packages is still going on. The most advanced numerical algorithms are implemented in mathematical packages, and more effective, more accurate and faster mathematical packages are still being produced. For instance, in the field of numerical linear algebra, the brand new LAPACK is becoming the leading mathematical package [8]. However, the objective of the new packages is no longer to support the average user; they are provided as low-level support to mathematical languages. In new versions of MATLAB, the base packages LINPACK and EISPACK have been abandoned, and LAPACK is used instead to provide support for linear algebra computation.

1.2.2   Development of Simulation Software and Languages

It can be seen from the limitations of these software packages that it might be rather complicated to complete simulation tasks with them. It is not wise to restart everything from low-level programming, and abandon the well-established packages, since the packages carry the experience and effort of scientists in the field. Low-level programming cannot achieve such a goal. Thus high-level packages and languages with good reputation, such as MATLAB/Simulink, should used instead to perform simulation tasks.

Simulation techniques gained the attention of scholars and experts worldwide, and the International Simulation Councils Inc (SCi) was founded in 1967 to formalize simulation language standards. Computer Simulation Modeling Program (CSMP) can be regarded as the earliest simulation language using that standard.

In the early 1980s, Mitchell and Gauthier Associates released a brand new simulation language ACSL (Advanced Continuous Simulation Language) [9], based on SCi's standard. Due to its powerful facilities for simulation and analysis, ACSL dominated simulation languages in relevant research communities.

In ACSL, the user should write a model program file with its dedicated syntaxes. The file was then compiled and linked with the ACSL library, to create an executable file. ACSL commands can then be used to perform simulation and analysis tasks. The main difference between ACSL and Fortran is that ACSL is much easier to program, and the library is more powerful. ACSL can directly call the subroutines written in Fortran. Many ACSL blocks (macros) are provided, such as transfer function block `TRAN`, integrator block `INTEG`, lead-lag block `LEDLAG`, time delay block `DELAY`, dead zone nonlinearity block `DEAD`, hysteresis block `BAKLSH` and rate limited integrator block `LIMINT`. These blocks can be used to describe a simulation model of the system. ACSL commands can then be used to analyze simulation results and to draw curves.

After the ACSL source program has been written, the compiler and linker are used to create an executable file. Running the program will automatically generate a prompt: `ACSL>`. At the prompt, relevant commands can then be issued.

Example 1.1   *Consider the well-known Van der Pol equation described by* $\ddot{y} + \mu(y^2 - 1)\dot{y} + y = 0$. *If* $\mu = 1$, *a set of state variables* $y_1 = y$, $y_2 = \dot{y}$ *can be selected, the Van der Pol equation can be represented as* $\dot{y}_1 = y_1(1 - y_2^2) - y_2$, $\dot{y}_2 = y_1$. *The following statements in ACSL can be written for describing the equation*

```
PROGRAM VAN DER POL EQUATION
CINTERVAL CINT=0.01
CONSTANT  Y1C=3.0, Y2C=2.5, TSTP=15.0
   Y1=INTEG(Y1*(1-Y2**2)-Y2, Y1C)
   Y2=INTEG(Y1, Y2C)
TERMT (T.GE.TSTP)
END
```

where the display step size is specified as `CINT` $= 0.01$. *The initial values are represented by the variables* `Y1C` *and* `Y2C`. *The final simulation time* `TSTP` *is assigned as 15. The time variable* `T` *is assumed. Compiling and linking the source ACSL program, an executable file can be generated. Executing the program, the prompt* `ACSL>` *is given. At this a prompt, the following commands can be used:*

```
ACSL> PREPAR T, Y1, Y2
ACSL> START
ACSL> PLOT Y1, Y2
```

These inform the ACSL model to reserve the variables T, Y1 *and* Y2, *and the phase plane trajectory of* Y1 *and* Y2 *can be obtained. The internal parameters in the system can also be set, with the command*

```
ACSL> SET Y1C=-1, Y2C=-3
```

Other packages and simulation languages similar to ACSL appeared at the same time, such as the SIMNON package [10] and ESL [11]. These packages have similar statement structures, since they were based on the same standard.

The emergence and popularization of MATLAB brought mathematical computation to a completely new level. The Simulink environment equipped researchers with new solution methodologies and schemes. Since the release of MATLAB many other software packages appeared, which imitated the syntaxes and ideas of MATLAB, such as Ctrl-C, Matrix-X, O-Matrix, and the CemTool proposed by Professor Kwan at Seoul National University. Octave [12] and Scilab [13] are still available as free software. In this book, MATLAB is extensively and exclusively used for discussing different kinds of simulation problems.

Computer algebra systems, or symbolic computation systems, brought into the field brand new ideas and solutions. Deriving analytical formulae by using programming languages such as C, even by very experienced programmers, may not be easy; indeed, sometimes it is impossible. High-quality computer algebra systems were developed generation by generation. The earlier muMath was developed by IBM, and the Reduce software introduced new solutions to such problems. The dominating Maple [14] and Mathematica [15] soon took the lead in computer algebra systems, and became very successful.

In earlier versions of Mathematica, there was an interface called MathLink to communicate with MATLAB. To better solve computer algebra problems in MATLAB, a Symbolic Math Toolbox was developed, which used Maple as its symbolic computation engine to combine the two major systems together. Then the engine was replaced by muPad.

Since these software packages and languages are usually too expensive for average users, more users are interested in getting free, open-source languages. The MATLAB-like languages such as Octave and Scilab attracted the attention of software users, but the facilities provided by this software are not powerful enough to compete with the sophisticated MATLAB language.

MATLAB and Simulink are the leading-edge tool in scientific computation and system simulation research. It is also the top selected computer language in research fields such as automatic control. In this book, MATLAB and Simulink will be extensively illustrated.

1.3   Introduction to MATLAB

1.3.1   Brief History of the Development of MATLAB

The creator of MATLAB, Professor Cleve Moler, is an influential scientist in numerical analysis, especially in numerical linear algebra [1, 2, 3, 16, 17, 18]. In the late 1970s, while he was the director of the computer department of the University of New Mexico, he found it inconvenient to solve linear algebra problems numerically with the then popular EISPACK [2] and LINPACK [1] packages. He then conceived and developed an interactive MATLAB (which stands for matrix laboratory); it indeed brought great convenience for users to solve related problems. With the use of MATLAB, matrix computation becomes a very easy problem. Earlier version of MATLAB can only be used to solve matrix computation problems. There were very few functions related to matrix computation. Since its emergence, MATLAB has received a great deal of attention and was welcomed by educators and researchers alike world wide.

Cleve Moler and Jack Little co-founded The MathWorks Inc. to develop MATLAB-related products. Cleve Moler is still the Chief Scientist at MathWorks. In 1984, the first commercial MATLAB was released, with the supporting language changed from Fortran to C. Powerful graphics, multimedia facilities and symbolic computation were gradually introduced into MATLAB. All these make MATLAB more powerful still. Earlier versions on PCs were called PC-MATLAB, and the workstation version is called Pro MATLAB. In 1990, MATLAB 3.5i was released and it was the first version executable on Microsoft Windows, where the command window and graphics windows can be displayed separately. The SimuLAB environment emerged later, introducing block diagram based simulation facilities, and it was renamed Simulink the following year.

In 1992, the epoch-making MATLAB 4.0 was released by MathWorks, and in 1993, a PC version was released. Graphical user interface programming was introduced, and in 1994, version 4.2 had an enhanced interface design with new methods.

In 1997, MATLAB 5.0 was released and more data types such as cells, structured arrays, multi-dimensional arrays, classes and objects were supported. Object oriented programming was possible for the first time. In 2000, MATLAB 6.0 was released with many useful windows such as a history command window and several different graphics windows could be displayed at the same time. The kernel of LAPACK [8] and FFTW [19] were used instead of the original LINPACK and EISPACK. The speed of computation and the numerical accuracy and stability were greatly enhanced. Graphical user interface design methods were more flexible, and the interface with C was greatly improved. In 2004, MATLAB 7.0 was introduced, the innovations and concepts of multi-domain physical modeling and simulation were very attractive to engineers.

In 2012, MATLAB 8.0, also known as MATLAB R2012b, was released. In particular, Simulink modeling and simulation facilities were significantly updated.

MathWorks is currently releasing two versions a year now, named version a and version b. At the of writing, the current one was released on September 2012, named 2012b. This version is used in this book to address simulation facilities with MATLAB/Simulink.

MATLAB has now become the de facto top scientific computation and simulation language. The current MATLAB is no longer merely a "matrix laboratory"; it is now a promising, completely new, high-level computer language. MATLAB has been referred to as a "fourth generation" computer language. It is playing an important role in education, academic research and industry. The MATLAB language becomes ever more powerful, to adapt to ever growing needs. More and more software and languages are now providing interfaces to MATLAB and Simulink, where it is becoming a standard in many fields. It is not difficult to reach the conclusion that, in the fields of scientific computation and system simulation, MATLAB will keep its unique and leading position for a long time to come.

1.3.2   Characteristics of MATLAB

MATLAB can run on almost all computers and operating systems. For instance, in Microsoft Windows, Linux and Mac OS X, MATLAB, source code written on one is completely compatible with the others. It can be claimed that MATLAB is independent of computers and operating systems.

From the viewpoint of the authors, the relationships between MATLAB and other computer languages such as C, are similar to the relationship between C and assembly language. Although the execution efficiency of C is much higher than MATLAB, the readability, programming efficiency and portability of MATLAB is much higher than C. Thus for scientific computation purposes, MATLAB should be adopted. In this way, the efficiency of programming, its reliability and the quality of the programs is much higher. For researchers in the area of scientific computation system simulation, MATLAB can easily reproduce all the functions implementable with C or Fortran. Even if programmers have no knowledge of C or Fortran, they can still design high quality, user-friendly, reliable, high quality programs with high efficiency.

Generally, MATLAB has a very high accuracy of numerical computation, mainly because of the double-precision scheme adopted for the computation. Also, advanced well-tested algorithms with a good reputation are adopted in MATLAB functions. In matrix-related computation, the accuracy can reach the 10^{-15} level. Also, symbolic computation can derive analytical solutions to many problems.

Simulink is another shining point in MATLAB applications. The block diagram based modeling techniques and its leading-edge multi-domain physical modeling technique bring users new solutions to simulation problems. The finite state machine system provided in Stateflow, for example, provides practical new tools for the modeling and simulation of discrete event systems and hybrid systems. The interface to external hardware bridges the gap between pure numerical simulation and hardware-in-the-loop simulation and real-time control.

MATLAB/Simulink is now widely used in automatic control, aerospace engineering, the automobile industry, biomedical engineering, speech and image processing and computer engineering applications. In many fields, MATLAB/Simulink has already become the number one computer language.

1.4 Structure of the Book

1.4.1 Structure of the Book

As in the learning of any computer language, active and repetitive practice are essential to mastering MATLAB and Simulink. Only regular practice will improve your ability in MATLAB programming and your application skills. For the student readers, the statements, programs and models should be used in person to gain more knowledge and skill with MATLAB. First-hand knowledge is very important for mastering computer languages and tools.

In this first chapter, system simulation concepts are briefly discussed. The development of computer packages and simulation languages is also briefly introduced. In Chapter 2, the concentration is on the fundamentals of MATLAB programming. Useful topics in MATLAB programming such as data types, statement structures, function programming, graphical visualization and graphical user interface design are logically presented. Essential knowledge of MATLAB programming are fully covered in this chapter. In Chapter 3, simulation-related scientific computation problem solutions with MATLAB and applications are explained. The topics of numerical linear algebra, differential equations, nonlinear equation solutions and optimization, dynamic programming, data interpolation and statistical analysis are presented. These topics are the essential mathematical fundamentals for solving simulation problems. In Chapter 4, primary knowledge on Simulink modeling is presented. A brief introduction to commonly used Simulink model groups is given first. The use of the Simulink environment is presented, and examples are used to demonstrate Simulink applications in mathematical modeling. Modeling and simulation of linear systems are presented, followed by the simulation studies of stochastic continuous systems. In Chapter 5, intermediate knowledge of Simulink modeling is presented. Application skills of commonly used blocks, nonlinearities modeling, algebraic loop avoidance, zero-crossing detection and solutions of various differential equations are extensively studied. Simulation result visualization via gauges and virtual reality techniques are also illustrated. Subsystem modeling and block masking techniques are presented in this chapter, and the F-14 aircraft control problem is used to demonstrate the use of subsystem model techniques. In Chapter 6, advanced techniques in Simulink modeling are presented. Mainly programming based modeling techniques are introduced. Statement based modeling methods are introduced first, and then linearization and S-function programming are introduced. Optimization based optimal controller design problems are demonstrated. Chapter 7 presents engineering system simulation and multi-domain physical modeling technique. Simulation tools such as Simscape, SimPowerSystems, SimElectronics and SimMechanics are presented, through examples, and the simulation of electrical, electronic and mechanical systems is presented. In Chapter 8, we look at some non-engineering system simulation techniques, including pharmacodynamical modeling and control problems, image and video processing problems and discrete event system modeling problems. In Chapter 9, hardware-in-the-loop real-time simulation and control problems are considered.

1.4.2 Code Download and Internet Resources

The MATLAB functions and models developed for the book can be downloaded directly from the book service website at Wiley (`http://www.wiley.com/go/xue`) or from:

`http://mechatronics.ucmerced.edu/simubook2013wiley`

However, we suggest that you do not use all the downloaded files directly. It would be better to input the functions and models yourself, since this is also a useful stage of learning and practical experience. If the solutions obtained by the users are different from the ones given in the book, the downloaded materials can be used for comparison.

The whole set of PDF and HTML manuals for MATLAB and its related toolboxes can be downloaded directly from the official MathWorks website. There are also a lot more free third-party toolboxes downloadable from internet. Moreover, active and experienced MATLAB users may answer various of your questions. The commonly used websites and forums are:

- MathWorks Website: `http://www.mathworks.com`.
- User forum: `http://www.mathworks.com/matlabcentral/newsreader/`.

Here are two suggestions for the use of forums: first, when a problem is encountered, first try to solve the problem by yourself. Sometimes the answers obtained by one's own effort can be of great benefit. Second, actively contribute to the questions to which you know the answers, or participate in discussion, so as to improve the skills of others.

1.4.3 Fonts Used in this Book

The fonts in the book are illustrated as follows, to help you understand better the materials presented here:

- Times-Roman fonts are for constants in formulae such as e, dx, and x axis.
- Italic Times-Roman font are provided by MATLAB to represent variables such as x t in MATLAB equations, while bold italic Times-Roman font are used to present vectors and matrices, such as A, x and $f(t, x)$.
- Typewriter font is used to represent program listings, as well as function names, such as `eig()`, `tic`, `stateflow`.
- The text in interfaces, Simulink group and block names are denoted by bold Helvetica font, such as **File** menu, **OK** button and **Step** block.

Exercises

1.1 A large number of demonstration programs are provided in MATLAB. To invoke the main demonstration program type the `demo` command in the MATLAB command window. Run the demonstration program and get a feel of the powerful facilities provided in MATLAB.

1.2 Programs and models designed for this book can be downloaded from the website for the book, and all the code is repeatable. It is advisable to input the program and block diagrams yourself, rather than use the downloaded ones directly, so as to understand better the materials presented in the book.

1.3 A powerful on-line help system is provided in MATLAB. Also the command `lookfor` allows you to search for keywords and function names. You can also use the `help` or `doc` commands to search for information, including syntax, of a particular MATLAB command. For example, a Riccati matrix equation is given by

$$PA + A^{\mathrm{T}}P - PBR^{-1}B^{\mathrm{T}}P + Q = 0$$

and

$$A = \begin{bmatrix} -27 & 6 & -3 & 9 \\ 2 & -6 & -2 & -6 \\ -5 & 0 & -5 & -2 \\ 10 & 3 & 4 & -11 \end{bmatrix}, \quad B = \begin{bmatrix} 0 & 3 \\ 16 & 4 \\ -7 & 4 \\ 9 & 6 \end{bmatrix}, \quad Q = \begin{bmatrix} 6 & 5 & 3 & 4 \\ 5 & 6 & 3 & 4 \\ 3 & 3 & 6 & 2 \\ 4 & 4 & 2 & 6 \end{bmatrix}, \quad R = \begin{bmatrix} 4 & 1 \\ 1 & 5 \end{bmatrix}.$$

Try to use the `lookfor riccati` command to find a possible Riccati equation solver, then use the `help` command to find the syntax of the solver and solve P for the above equation.

References

[1] J J Dongarra, J R Bunch, C B Moler, *et al.* LINPACK user's guide. Philadelphia: Society of Industrial and Applied Mathematics (SIAM), 1979

[2] B T Smith, J M Boyle, J J Dongarra. Matrix eigensystem routines – EISPACK guide, Lecture notes in computer sciences, volume 6. New York: Springer-Verlag, (2nd Edition), 1976

[3] B S Garbow, J M Boyle, J J Dongarra, *et al.* Matrix eigensystem routines – EISPACK guide extension, Lecture notes in computer sciences, volume 51. New York: Springer-Verlag, 1977

[4] Numerical Algorithm Group. NAG FORTRAN library manual, 1982

[5] W H Press, B P Flannery, S A Teukolsky, *et al.* Numerical recipes, the art of scientific computing. Cambridge: Cambridge University Press, 1986

[6] J L Melsa, S K Jones. Computer programs for computational assistance in the study of linear control theory. New York: McGraw-Hill, 1973

[7] CAD Center. GINO-F Users' manual, 1976

[8] E Anderson, Z Bai, C Bischof, *et al.* LAPACK users' guide. SIAM Press, 3rd Edition, 1999

[9] E E L Mitchell, J S Gauthier. Advanced continuous simulation language (ACSL) – user's manual. Mitchell & Gauthier Associates, 1987

[10] K J Åström. Computer aided tools for control system design, In: Jamshidi M and Herget C J. (eds.) Computer-aided control systems engineering. Amsterdam: Elsevier Science Publishers B V, 1985, 3–40

[11] R E Crosbie, S Javey, J L Hay, *et al.* ESL – a new continuous system simulation language. Simulation, 1985, 44(5): 242–246

[12] Octave Language Webpage. http://www.octave.org/

[13] SciLAB Language Webpage. http://scilabsoft.inria.fr/

[14] F Garvan. The Maple book. Boca Raton: Chapman & Hall/CRC, 2002

[15] S Wolfram. The Mathematica book. Cambridge: Cambridge University Press, 1988

[16] G E Forsythe, M A Malcolm, C B Moler. Computer methods for mathematical computations. Englewood Cliffs: Prentice-Hall, 1977

[17] G E Forsythe, C B Moler. Computer solution of linear algebraic systems. Englewood Cliffs: Prentice-Hall, 1967

[18] D Kahaner, C B Moler, S Nash. Numerical methods and software. Englewood Cliffs: Prentice Hall, 1989

[19] M Frigo, S G Johnson. The design and implementation of FFTW3. Proceedings of IEEE, 2005, 93(2):215–231

2

Fundamentals of MATLAB Programming

Different kinds of computer languages for system simulations have been summarized in Chapter 1. In this chapter, the top computation and simulation language MATLAB will be systematically introduced. The programming and skills of the MATLAB language will be presented. In Section 2.1, MATLAB windows and on-line help facilities will be presented. In Section 2.2, the fundamentals of MATLAB programming will be illustrated, including data types, statements and matrix representation. Matrix manipulations, such as algebraic computation, logical and relationship expressions and data conversion will be presented in Section 2.3. In Section 2.4, the use of flow charts in programming will be illustrated, including loop structures, conditional structures, switches and trial structures. In Section 2.5, the MATLAB function programming and pseudo code processing will be presented. Two-dimensional and three-dimensional graphics and visualization techniques will be presented in Sections 2.6 and 2.7. In Section 2.8, graphical user interface (GUI) techniques will be explained. Equipped with the new GUI programming skills, user-friendly interfaces can be designed. Section 2.9 will explore the skills of high speed, high efficiency programming, finally, we look at vectorized programming methodology and MEX programming fundamentals.

2.1 MATLAB Environment

2.1.1 MATLAB Interface

At the time of writing, the current version of MATLAB is R2012b (or MATLAB version 8.0), released by MathWorks Inc. in September 2012. Two versions are released each year, in March and September respectively and labeled versions a and b. If MATLAB is installed and invoked, the graphical interface shown in Fig. 2.1 will appear. Apart from the main **Command Window**, there are other windows, such as the **Current Folder** window, **Command History** window and **Workspace** window. The window layout can be rearranged with the **Desktop** → **Desktop Layout** menu item. In MATLAB 8.0, brand new toolbar systems are made available.

2.1.2 MATLAB On-line Help and Documentation

All the manuals for MATLAB and its Toolboxes can be downloaded for free from MathWorks's website http://www.mathworks.com. They are provided both in PDF and HTML formats.

System Simulation Techniques with MATLAB® and Simulink®, First Edition. Dingyü Xue and YangQuan Chen.
© 2014 John Wiley & Sons, Ltd. Published 2014 by John Wiley & Sons, Ltd.

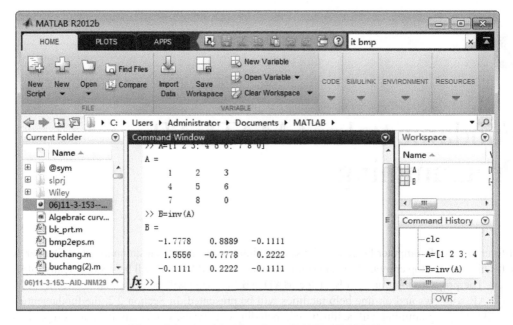

Figure 2.1 Graphical interface of MATLAB R2012b.

In the MATLAB interface, select the **Help** menu in the **RESOURCES** panel in the MATLAB interface (in earlier versions, click the menu **Help** → **MATLAB Help**); the on-line help window can then be opened, as shown in Fig. 2.2, from which different types of information can be retrieved. For a quick reference, `doc` or `help` commands can also be used, and the `lookfor` command can be used for keyword searches.

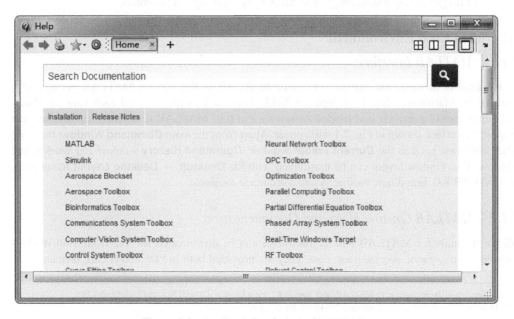

Figure 2.2 On-line help window of MATLAB.

2.2 Data Types in MATLAB

MATLAB has many powerful and accurate numerical facilities. In order to get the highest accuracy in computation, double-precision floating point data type is used as the default data type in MATLAB. Double precision floating point data type is composed of 8 bytes (64 bits) of binary digits, with 11 exponential bits, one sign bit and 52 bits to represent the number; this follows the IEEE standard. The value range is $-1.7 \times 10^{308} \sim 1.7 \times 10^{308}$, and its MATLAB description is `double()`. The fundamental data type in MATLAB is the double precision floating point complex matrix. In certain applications, such as image processing, unsigned 8 bit integer data type, `uint8()`, can be used to save memory and to increase speed in MATLAB. Other data types are also supported in MATLAB. These data types include `int8()`, `int16()`, `int32()`, `uint16()` and `uint32()`, and it is easy to work out the meaning of those data types from their names.

Apart from numerical computation, a symbolic data type is also provided in MATLAB and its Symbolic Math Toolbox, for finding analytical solutions and doing theoretical computation. Symbolic variables can be declared with the `syms` command.

For the need of more advanced programming, other data types such as strings, multi-dimensional arrays, structured arrays, cells, classes and objects are also supported in MATLAB. In this section, constants and variables are presented first, followed by MATLAB statements. Finally matrix representation and other related topics are presented.

2.2.1 Constants and Variables

By convention, the variable name in MATLAB is led by a letter, followed by other letters, digits and underscores. For instance, `MYvar12`, `MY_Var12` and `MyVar12_` are valid variable names, while `12MyVar` and `_MyVar12` are not. In MATLAB, the variable names are case-sensitive, so `Abc` and `ABc` are different variable names.

MATLAB reserves a few special names for specific constants. Although these names can be reassigned, it is suggested that we should be very careful with them, and avoid reassigning them whenever possible.

- `eps`: Machine-specific error tolerance for the floating point operations. For PCs, the default value of `eps` is 2.2204×10^{-16}. If the absolute value of a certain quantity is smaller than `eps`, it can be regarded as 0, from a numerical viewpoint.
- `i` and `j`: The default values of `i` and `j` are $\sqrt{-1}$. In real programming, these names may be rewritten. For instance, they are usually reused as loop variables. So we must be very careful with these constants, instead, `1i` or `1j` can be used instead. We can also restore the variables with $i = $ `sqrt(-1)`.
- `Inf`: MATLAB representation of $+\infty$, and it is also written as `inf`. Similarly $-\infty$ can be expressed as `-Inf`. In the execution of a MATLAB program, the quantity divided by 0 can be accepted, and the result is assigned to `Inf`. A warning message of "Divided by zero" is given.
- `NaN`: This is not a number, always obtained by 0/0, `Inf/Inf`, and other possible computations. The product of `NaN` and `Inf` is `NaN`.
- `pi`: A double-precision representation of the circumference ratio π.

2.2.2 Structure of MATLAB Statements

Two types of statements are supported in MATLAB:

1) Direct assignment: The basic structure of direct assignment is

```
variable = expression
```

and the expression on the right hand side is executed and the result is assigned to `variable` on the left hand side. If there is a semicolon at the end of the statement, the variable will not be displayed. If the left hand side variable name is not given, the result is returned to the reserved variable `ans`.

2) Function call statement: The basic structure is

[returned variable list] = fun_name (input variable list)

where the function name `fun_name` should be the same as the filename, and it should be a *.m file under a MATLAB search path. For instance, the function `my_fun` should correspond to my_fun.m file. MATLAB also provides quite a lot of built-in functions, such as `inv()` function.

The input and output variables can be composed of many arguments, and they can be separated with commas. For instance, $[U, S, V] = $ `svd`(X). In such a function, singular value decomposition is made to the input matrix X, and three matrices U, S and V are returned from the function. The same function may be called by different syntaxes, for example the `svd()` function can also be called $D = $ `svd`(X).

2.2.3   Matrix Representation in MATLAB

The complex double-precision matrix is the basic MATLAB variable type. In MATLAB, it is quite easy and straightforward to express a matrix. For instance, the matrix

$$A = \begin{bmatrix} 1 & 2 & 3 \\ 4 & 5 & 6 \\ 7 & 8 & 0 \end{bmatrix}$$

can be entered into the MATLAB workspace with the following command

```
>> A=[1,2,3; 4 5,6; 7,8 0]
```

where `>>` is the MATLAB prompt, displayed automatically. At the prompt, a MATLAB command can be issued, and the execution results can be obtained directly. Therefore, MATLAB is often regarded as an interactive language. In the above statement, commas and spaces inside the brackets are used to separate the matrix elements in the same row, while semicolons are used to indicate another row. Once the command is executed, a matrix variable A can be established in the MATLAB workspace and other commands can also be used to analyze such a matrix. For instance, command $B = $ `inv`(A) can be used to get the inverse matrix of A, and assign it to the variable B. Since A is a double-precision variable, numerical methods can be used to find the inverse matrix. If the exact inverse of matrix A is expected, we should convert A to symbolic variable, such that $B = $ `inv(sym`(A)`)`, the analytical inverse, can be obtained.

In MATLAB programming, it is important to know that when a semicolon is used at the end of a statement, the results are not displayed. Thus the following command assigns the matrix A in the MATLAB workspace, but the result is not displayed.

```
>> A=[1,2,3; 4 5,6; 7,8 0];
```

In MATLAB, vectors and scalars can also be entered into MATLAB in the same way. Also, it is easy to understand the following MATLAB statements and the results

```
>> A=[[A; [1 3 5]] [1;2;3;4]]
```

It can be seen that a 4×4 matrix A can be constructed. Thus MATLAB makes it possible to dynamically change the size of an existing matrix arbitrarily. It might be very difficult to do so in other languages such as C.

MATLAB also provides a colon expression $a = s_1 : s_2 : s_3$, such that a row vector a can be established easily, where s_1 is the starting value, s_2 is the increment and s_3 is the terminating value. If the increment s_2 is negative, then s_1 should be larger than s_3, otherwise an empty vector a will be created. If the increment s_2 is omitted, the default increment of one will be used.

Consider the command a $= 0:0.1:1.16$, and the a vector created is $a = [0, 0.1, 0.2, 0.3, 0.4, 0.5, 0.6, 0.7, 0.8, 0.9, 1, 1.1]^{\mathrm{T}}$. It should be noted that the last entry in the vector is 1.1, rather than 1.16 (the value of s_3).

Complex matrices can also be entered into MATLAB in a similar way. The notations i and j can be used directly in entering a complex matrix. For example, to enter the following matrix into MATLAB

$$B = \begin{bmatrix} 1 + 9j & 2 + 8j & 3 + 7j \\ 4 + 6j & 5 + 5j & 6 + 4j \\ 7 + 3j & 8 + 2j & 0 + j \end{bmatrix},$$

the following statements can be used

```
>> B=[1+9i,2+8i,3+7j; 4+6j 5+5i,6+4i; 7+3i,8+2j 1i]
```

The command $B = \text{sym}(A)$ can be used to convert a double-precision matrix A into symbolic form.

2.2.4 Multi-dimensional Arrays

Apart from the matrix expression (which can, in fact, be considered as a two-dimensional array), three-dimensional and multi-dimensional arrays are also defined in MATLAB. It is relatively easy to understand. Assume that there are several matrices $A_1, A_2, \cdots, A_m$ of the same size, then the matrices can be piled up to form a three-dimensional array, as shown in Fig. 2.3. Three-dimensional arrays are important in color image representation, since for an RGB color picture, the red, green and blue components in pixels can be expressed in a three-dimensional array.

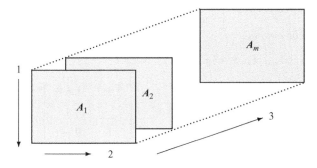

Figure 2.3 Illustration of a three-dimensional array.

Assume that the matrices A_1, A_2 and A_3 can be entered into MATLAB, and the three-dimensional array A_4 can be constructed.

```
>> A1=[1,2,3; 4 5 6; 7 8,9]; A2=A1'; A3=A1-A2;
   A4(:,:,1)=A1; A4(:,:,2)=A2; A4(:,:,3)=A3
```

The three-dimensional array can be constructed and displayed as

$$A_4(:,:,1) = \begin{bmatrix} 1 & 2 & 3 \\ 4 & 5 & 6 \\ 7 & 8 & 9 \end{bmatrix}, \quad A_4(:,:,2) = \begin{bmatrix} 1 & 4 & 7 \\ 2 & 5 & 8 \\ 3 & 6 & 9 \end{bmatrix}, \quad A_4(:,:,3) = \begin{bmatrix} 0 & -2 & -4 \\ 2 & 0 & -2 \\ 4 & 2 & 0 \end{bmatrix}.$$

A function `cat()` is provided in MATLAB to construct multi-dimensional arrays in MATLAB, $A = \text{cat}(n, A_1, A_2, \cdots, A_m)$, where when $n = 1$ and 2, the matrices $[A_1; A_2; \cdots; A_m]$ and $[A_1, A_2, \cdots, A_m]$ can be constructed, and the result is still a matrix. If $n = 3$, a three-dimensional array can be established. It can be seen that the three-dimensional array in A_4 can also be used with the following statements:

```
>> A5=cat(3,A1,A2,A3)
```

2.3 Matrix Computations in MATLAB

2.3.1 Algebraic Computation

Assume that a matrix A has n rows and m columns. The matrix A is referred to as an $n \times m$ matrix. If $n = m$, matrix A is called a square matrix. The following algebraic manipulations of matrices are processed in MATLAB.

1) Matrix transpose: In mathematics, the transpose of matrix A is denoted as A^{T}. Assume that matrix A is $n \times m$, the transpose matrix B, whose elements are defined as $b_{ji} = a_{ij}, i = 1, \cdots, n, j = 1, \cdots, m$, is an $m \times n$ matrix. When matrix A contains complex elements, its Hermit transpose matrix B is defined as $b_{ji} = a_{ij}^*, i = 1, \cdots, n, j = 1, \cdots, m$, that is the transpose of the complex conjugate of each element is performed. The Hermit transpose is mathematically denoted by A^*. In MATLAB, the Hermit transpose of matrix A can be obtained with A', while the direct transpose can be obtained with A'. They can also be obtained respectively with the MATLAB functions `ctranspose()` and `transpose()`.

Example 2.1 *The matrix* $A = \begin{bmatrix} 5+\text{j} & 2-\text{j} & 1 \\ \text{j}6 & 4 & 9-\text{j} \end{bmatrix}$ *can be entered into MATLAB and different transposes can be obtained*

```
>> A=[5+i, 2-i, 1; 6*i, 4, 9-i]; B=A', C=A.'
```

$$B = \begin{bmatrix} 5-\text{j} & 0-6\text{j} \\ 2+\text{j} & 4 \\ 1 & 9+\text{j} \end{bmatrix}, \quad C = \begin{bmatrix} 5+\text{j} & 0+6\text{j} \\ 2-\text{j} & 4 \\ 1 & 9-\text{j} \end{bmatrix}.$$

2) Addition and subtraction of matrices: The sum and difference of two matrices A and B can be obtained directly with the commands $C = A + B$ and $C = A - B$. If the two matrices A and B are compatible, that is the sizes of them are the same, the corresponding elements of matrix A can be added up or subtracted from those of matrix B, and the result is assigned to matrix C. If they are not, an error message will be given to prompt that the two matrices are not compatible. The addition and subtraction of two matrices can also be obtained respectively with `plus(A, B)` and `minus(A, B)`.

3) Matrix multiplication: For the two matrices A and B, where the number of columns in A must be the same as the number of rows in B, or one of them is a scalar, the two matrices are compatible for multiplication. For an $n \times m$ matrix A, and an $m \times r$ matrix B, then $C = AB$ matrix is an $n \times r$ matrix, whose elements can be mathematically obtained as

$$c_{ij} = \sum_{k=1}^{m} a_{ik}b_{kj}, \quad \text{where } i = 1, 2, \cdots, n, j = 1, 2, \cdots, r \qquad (2.1)$$

In MATLAB, the multiplication of two matrices A and B can easily be obtained with the command $C = A * B$. If the two matrices are compatible, their product can be obtained and assigned to variable C. If they are not compatible, an error message will be given. The product of matrices A and B can also be evaluated with the MATLAB function $C = $ `mtimes(A, B)`.

4) Matrix divisions: Matrix divisions are closely related to matrix inverse or linear simultaneous equation solutions. In MATLAB, left division and right division are defined.

Left division of two matrices can be evaluated from $X = A \backslash B$, and it is the solution of the matrix equation $AX = B$. For equations where no solution exists, the least squares solution X can be obtained. Left division of matrices can also be obtained with $X = $ `mldivide(A, B)`.

Right division of two matrices can be evaluated from $X = B/A$, and it is the solution of the matrix equation $XA = B$. Right division can also be obtained with MATLAB function $X = $ `mrdivide(A, B)`.

5) Matrix flipping and rotation: Flipping of a matrix A can be obtained easily by existing MATLAB functions. Function $B = $ `fliplr(A)` flips a matrix A from left to right, while $C = $ `flipud(A)` function flips matrix A from up to down. Matrix rotation can also be implemented with MATLAB function $D = $ `rot90(A)`, which rotates matrix A in clockwise direction $90°$. The function $D = $ `rot90(A, k)` rotates a matrix A in clockwise direction $90k°$, where k is an integer.

6) Power of a matrix: Mathematically the power of a matrix A is denoted by A^x. It is also important to know that A must be a square matrix. In MATLAB, the power can be evaluated with $F = A\hat{\ }x$, or $F_1 = $ `mpower(A, x)`.

■ Example 2.2 *For the matrix* $A = \begin{bmatrix} 1 & 2 & 3 \\ 4 & 5 & 6 \\ 7 & 8 & 0 \end{bmatrix}$ *that we discussed earlier, the cube and cubic root of it can easily be obtained with the following statements, where the* `norm()` *function is used to find the norm of an error matrix.*

$$B = \begin{bmatrix} 279 & 360 & 306 \\ 684 & 873 & 684 \\ 738 & 900 & 441 \end{bmatrix}, \quad C = \begin{bmatrix} 0.7718 + 0.6538j & 0.4869 - 0.0159j & 0.1764 - 0.2887j \\ 0.8885 - 0.0726j & 1.4473 + 0.4794j & 0.5233 - 0.4959j \\ 0.4685 - 0.6465j & 0.6693 - 0.6748j & 1.3379 + 1.0488j \end{bmatrix}.$$

```
>> A=[1,2,3; 4,5,6; 7,8,0]; B=A^3, C=A^(1/3), norm(A-C^3)
```

The obtained cubic root can be validated by taking the norm of the error matrix, which is 1.0682×10^{-14}. *In fact, there should be three cubic root matrices, and the other two can be obtained by rotating the obtained* $\boldsymbol{C}$ *matrix by* $120°$ *and* $240°$, *that is* $\boldsymbol{C}e^{2j\pi/3}$ *and* $\boldsymbol{C}e^{4j\pi/3}$. *The following statements can be used*

```
>> r=exp(sqrt(-1)*2*pi/3); A1=C*r, A2=C*r^2,
 e1=norm(A1^3-A), e2=norm(A2^3-A)
```

It is found that

$$A_1 = \begin{bmatrix} -0.9521 + 0.3415j & -0.2297 + 0.4296j & 0.1618 + 0.2971j \\ -0.3814 + 0.8058j & -1.1388 + 1.0137j & 0.1678 + 0.7011j \\ 0.3256 + 0.7289j & 0.2497 + 0.9170j & -1.5772 + 0.6342j \end{bmatrix},$$

$$A_2 = \begin{bmatrix} 0.1803 - 0.9953j & -0.2572 - 0.4137j & -0.3382 - 0.0084j \\ -0.5071 - 0.7332j & -0.3085 - 1.4930j & -0.6911 - 0.2052j \\ -0.7941 - 0.0825j & -0.919 - 0.2422j & 0.2393 - 1.6830j \end{bmatrix}.$$

Validation shows that the norms of the error matrices are 1.1847×10^{-14}, *and* 1.8817×10^{-14}.

7) Dot operation: In MATLAB, a special matrix operation, called the dot operation, is defined, such that the corresponding elements are processed directly. For instance, dot multiplication of two matrices can be obtained with the command $\boldsymbol{C} = \boldsymbol{A}.*\boldsymbol{B}$, where the elements in the resulting $\boldsymbol{C}$ matrix are $c_{ij} = a_{ij}b_{ij}$. Dot multiplication of two matrices is also known as the Hadamard product. Dot operation of two matrices requires that the sizes of them must be the same, or either of them can be a scalar.

■ **Example 2.3** *For the following two simple matrices*

$$A = \begin{bmatrix} 1 & 2 & 3 \\ 4 & 5 & 6 \\ 7 & 8 & 0 \end{bmatrix}, \quad B = \begin{bmatrix} 2 & 3 & 4 \\ 5 & 6 & 7 \\ 8 & 9 & 0 \end{bmatrix},$$

the following statements can be used to calculate their product and dot product

$$C = \begin{bmatrix} 36 & 42 & 18 \\ 81 & 96 & 51 \\ 54 & 69 & 84 \end{bmatrix}, \quad D = \begin{bmatrix} 2 & 6 & 12 \\ 20 & 30 & 42 \\ 56 & 72 & 0 \end{bmatrix}.$$

```
>> A=[1 2 3; 4 5 6; 7 8 0]; B=[2 3 4; 5 6 7; 8 9 0]; C=A*B, D=A.*B
```

It can be seen that the two results are different. The first is the matrix product, and the second is the product of corresponding terms of the two matrices.

We can also specify the **A.^A** *command to find the following results*

$$
A.\hat{\ }A = \begin{bmatrix} 1 & 4 & 27 \\ 256 & 3125 & 46656 \\ 823543 & 16777216 & 1 \end{bmatrix}, \quad i.e., \quad \begin{bmatrix} 1^1 & 2^2 & 3^3 \\ 4^4 & 5^5 & 6^6 \\ 7^7 & 8^8 & 0^0 \end{bmatrix}.
$$

Dot operation is very important in MATLAB, especially for drawing plots. For instance, to draw the curve of x^2, a vector $x = [x_1, x_2, \cdots, x_m]$ is generated first. Then the square value of each term in x vector is evaluated, that is $x.\hat{\ }2 = [x_1^2, x_2^2, \cdots, x_m^2]$, and it cannot be written as $x\hat{\ }2$. In fact, some of the mathematical functions such as `sin()` and `log()` are performed as dot operations. Dot multiplication of two matrices A and B can also be obtained with `times(A, B)`.

8) Kronecker product: For two matrices A and B, where A is an $n \times m$ matrix, and B is a $p \times q$ matrix, the Kronecker product of matrices A and B is defined as

$$
C = A \otimes B = \begin{bmatrix} a_{11}B & a_{12}B & \cdots & a_{1m}B \\ a_{21}B & a_{22}B & \cdots & a_{2m}B \\ \vdots & \vdots & \ddots & \vdots \\ a_{n1}B & a_{n2}B & \cdots & a_{nm}B \end{bmatrix}. \tag{2.2}
$$

It can be seen from the above expression that the Kronecker products $A \otimes B$ and $B \otimes A$ are all $np \times mq$ matrices, but normally $A \otimes B \neq B \otimes A$. There are no compatibility problems in the Kronecker product. In MATLAB, the Kronecker product of two matrices can be obtained with $C = $ `kron(A, B)`.

Example 2.4 *For the following two matrices*

$$
A = \begin{bmatrix} 1 & 2 \\ 3 & 4 \end{bmatrix}, \quad B = \begin{bmatrix} 1 & 3 & 2 \\ 2 & 4 & 6 \end{bmatrix},
$$

the Kronecker products of them can be evaluated with the following statements

$$
C = \begin{bmatrix} 1 & 3 & 2 & 2 & 6 & 4 \\ 2 & 4 & 6 & 4 & 8 & 12 \\ 3 & 9 & 6 & 4 & 12 & 8 \\ 6 & 12 & 18 & 8 & 16 & 24 \end{bmatrix}, \quad D = \begin{bmatrix} 1 & 2 & 3 & 6 & 2 & 4 \\ 3 & 4 & 9 & 12 & 6 & 8 \\ 2 & 4 & 4 & 8 & 6 & 12 \\ 6 & 8 & 12 & 16 & 18 & 24 \end{bmatrix}.
$$

```
>> A=[1 2; 3 4]; B=[1 3 2; 2 4 6]; C=kron(A,B), D=kron(B,A)
```

2.3.2 Logical Operations

In earlier versions of MATLAB, there were no logical data types defined. Any nonzero value can be regarded as logical "1", and zero is logical "0". In new versions, this definition is still used. But the `logical` data type is also supported.

Assume that matrices A and B are of the same size. Logical operations $A \& B$, $A \mid B$, $\sim A$ and $\texttt{xor}(A, B)$ respectively express the "and", "or", "not" and "exclusive or". These operations return a logical matrix of size A. The above operation is also valid if either of the matrices is a scalar.

Example 2.5   *Assume that*

$$A = \begin{bmatrix} 0 & 2 & 3 & 4 \\ 1 & 3 & 5 & 0 \end{bmatrix}, \quad B = \begin{bmatrix} 1 & 0 & 5 & 3 \\ 1 & 5 & 0 & 5 \end{bmatrix}.$$

The following statements can be used to perform various logical operations

$$A_1 = \begin{bmatrix} 0 & 0 & 1 & 1 \\ 1 & 1 & 0 & 0 \end{bmatrix}, \quad A_2 = \begin{bmatrix} 1 & 1 & 1 & 1 \\ 1 & 1 & 1 & 1 \end{bmatrix}, \quad A_3 = \begin{bmatrix} 1 & 0 & 0 & 0 \\ 0 & 0 & 0 & 1 \end{bmatrix}, \quad A_4 = \begin{bmatrix} 1 & 1 & 0 & 0 \\ 0 & 0 & 1 & 1 \end{bmatrix}.$$

```
>> A=[0 2 3 4;1 3 5 0]; B=[1 0 5 3;1 5 0 5];
A1=A&B, A2=A|B, A3=A, A4=xor(A,B)
```

2.3.3   Comparisons and Relationships

Different comparison and relationship operations are defined in MATLAB. For instance, $>$ and $<$ are defined for larger than and smaller than, while $==$ and $\sim=$ are for equal and not equal, respectively. Other comparison operators such as $>=$ and $<=$ are also provided. These relationship operators are defined as in the dot operations.

Example 2.6   *For the two matrices defined in Example 2.5, the following comparisons are made*

$$A_1 = \begin{bmatrix} 0 & 0 & 0 & 0 \\ 1 & 0 & 0 & 0 \end{bmatrix}, \quad A_2 = \begin{bmatrix} 0 & 1 & 0 & 1 \\ 1 & 0 & 1 & 0 \end{bmatrix}, \quad A_3 = \begin{bmatrix} 1 & 1 & 1 & 1 \\ 0 & 1 & 1 & 1 \end{bmatrix}.$$

```
>> A=[0 2 3 4;1 3 5 0]; B=[1 0 5 3;1 5 0 5]; A1=A==B, A2=A>=B, A3=B~=A
```

MATLAB also provides other special functions such as $\texttt{find()}$, $\texttt{all()}$ *and* $\texttt{any()}$, *and they are very useful in programming. For instance,* $\texttt{find()}$ *function can be used to get the indexes in the array whose corresponding elements satisfy a certain criterion. The following statements can be issued where it is found that* $k = [2, 3, 6, 7]^{\mathrm{T}}$, $i = [2, 1, 2, 1]^{\mathrm{T}}$, $j = [1, 2, 3, 4]^{\mathrm{T}}$.

```
>> k=find(A2==1), [i,j]=find(A2==1); [i,j]
```

2.3.4   Data Type Conversion

For a non-integer matrix A, a corresponding integer of it can be defined as follows:

* $\texttt{floor}(A)$ finds the next integer of the elements in A towards $-\infty$.

* $\texttt{ceil}(A)$ finds the next integer of the elements in A towards $+\infty$.

* $\texttt{round}(A)$ finds the nearest integer of the elements in A.

* $\texttt{fix}(A)$ finds the next integer of the elements in A towards 0.

Also, the rat() function finds the rational approximation to the elements in a matrix, with the syntax of $[N, D] = $ rat(A), where N and D are integer matrices such that $A = N./D$.

Example 2.7 *Consider a 4×4 Hilbert matrix. The* rat() *function can be used to find the rational representation of matrix A*

```
>> A=hilb(4); [n,d]=rat(A)
```

$$
n = \begin{bmatrix} 1 & 1 & 1 & 1 \\ 1 & 1 & 1 & 1 \\ 1 & 1 & 1 & 1 \\ 1 & 1 & 1 & 1 \end{bmatrix}, \quad d = \begin{bmatrix} 1 & 2 & 3 & 4 \\ 2 & 3 & 4 & 5 \\ 3 & 4 & 5 & 6 \\ 4 & 5 & 6 & 7 \end{bmatrix}.
$$

2.4 Flow Structures

As a programming language, MATLAB supports different flow structures such as loops, conditional structures, switches and the new trial structure. In this section, these structures are briefly presented.

2.4.1 Loop Structures

There are two kinds of loop structures in MATLAB – the for loop and the while loop. The two kinds of loops are different, but each has different advantages. The basic structure of for loop is defined as

for $i = v$,   loop statements,   end

where v is a given vector. The for statement takes each entity in vector v, and assigns it to i, then the loop statements are executed once, and it goes back to the for statement, until all the entities in vector v are extracted. In special cases, if vector v is defined as a colon expression, that is $v = [s_1 : s_2 : s_3]$, and $s_2 > 0$, the MATLAB loop structure is the same as the loop in C, such that for $(i = s_1; i <= s_3; u += s_2)$. It can be seen that the for structure is much more flexible than the counterpart in C, since v can be any vector. Noted too that for each loop and other flow structure, the structure should be terminated with an end statement.

Example 2.8 *To evaluate* $\sum\limits_{i=1}^{100} i$ *in MATLAB, loop structure can be used. The following statements can be issued to find that the sum is $s = 5050$.*

```
>> s=0; for i=1:1:100, s=s+i; end; s
```

Please note, this example is used only in demonstrating the loop structure. In real applications, a much simpler command can be used to evaluate the sum: sum$(1:100)$. *This is much more effective than the* for *loop structure.*

Another type of loop structure, the while loop, is also provided in MATLAB, with the syntax

while k,   loop statements,   end

where k is a logical variable or logical expression. The while loop structure can be executed in the following way. The condition k is examined first to see whether or not k is true (more specifically, nonzero). If it is true, the loop statements are executed once, then it goes back to the `while` statement to check the new k again, until k is no longer true.

Now consider again the problem in Example 2.8. If the `while` structure is used, the program can be rewritten as follows. The same results, of course, can be obtained.

```
>> s=0; i=0; while (i<=100), i=i+1; s=s+i; end, s
```

It can be seen that for this particular example, the `while` loop structure is much more complicated than the `for` loop structure.

In MATLAB, the loop structures, `for` or `while` loops, can be nested, and the `break` statement can be used to terminate the current loop command. This is the same as in other languages such as C.

Example 2.9 *If the problem in Example 2.8 is changed to find the minimum number of m such that*

$$\sum_{i=1}^{m} i > 10000,$$

the `sum()` *command cannot be used, nor can the* `for` *loop, since the vector v cannot be established. In this case, the* `while` *loop can be used instead, and the following statements can be issued such that $m = 141$, $s = 10011$.*

```
>> s=0; m=0; while s<=10000, m=m+1; s=s+m; end, m, s
```

2.4.2 Conditional Structures

MATLAB supports various conditional structures led by the keyword `if`. The simplest if conditional structure is

`if` k, conditional statements, `end`

where again k is a logical variable or a logical expression. If k is true (or k is nonzero), the conditional statements are executed, otherwise, the conditional statements are bypassed. Note again that the `if` block should be completed with `end`.

Example 2.10 *Consider again the problem in Example 2.9. The* `for` *loop structure can be used with the aid of an* `if` *block. The problem can be solved with the following statements and the same results can be obtained. Here the* `break` *statement is used to terminate the last* `for` *or* `while` *loop.*

```
>> s=0; for m=1:1000, s=s+m; if (s>10000), break; end, end
```

More complicated `if` conditional structures are also supported, with the syntaxes

```
if condition
    conditional statements  1
else
    conditional statements  2
end
```

```
if condition 1,    conditional statements  1
elseif condition 2, conditional statements  2
    :
else, conditional statements  n + 1
end
```

These structures are the same as in other programming languages, such as C or Fortran. Examples will be given later to further demonstrate these structures.

2.4.3   Switches

Switch structures are supported in MATLAB, with the syntax of

```
switch k
case k₁, statements 1
case {k₂, k₃, ···, kₘ}, statements 2
    ⋮
otherwise, statements n
end
```

where k and k_i are expressions. If expression k is the same as k_1, statements 1 will be executed. Then, unlike in the case of C, the switch structure is completed. If expression k is the same as one of the expressions $k_2, \cdots, k_m$, where they are composed of a cell data type (bounded by { and }), the statements 2 will be completed. If none of the expressions are met, the statements under the otherwise keyword will be executed. The switch-case structure is similar to the structure in C, but there are differences, as listed below:

- When the expression k is the same as k_1, statements 1 will be executed. After the execution, the structure is completed. However, in C, the other cases will be tried after execution. To get C to produce the same results, a break statement should be inserted before the next case clause.
- To describe the case "one of the several expressions $k_2, \cdots, k_m$ are satisfied", they should be expressed in cells.
- The keyword otherwise is equivalent to the default keyword in C.
- If the same expression k_n is used in two different cases, the latter case will never be executed. Thus the expressions must be exclusive.

2.4.4   Trial Structure

A new trial structure is supported in MATLAB, with the syntax

```
try, statements 1, catch, statements 2, end
```

In this structure, the statements 1 are executed first. If no error occurs during the execution, the structure is then completed. If an error occurs, the statements 2 are executed instead, and the error message can be extracted with the lasterr command.

The trial structure is very useful in programming practice. For instance, to solve a particular problem there are two algorithms, one is very fast, but it may have problems sometimes. Another algorithm may be extremely slow, but it is reliable. We can try the first algorithm; if it works then the problem is successfully solved. If the first algorithm fails, the second algorithm can then be tried, to guarantee a solution. In this way, the solution to certain problems may be made more effective. Also the trial structure is useful in graphical user interface design and will be illustrated later.

2.5   Programming and Tactics of MATLAB Functions

Two types of MATLAB programs are allowed, both of them written in ASCII files. The first type is composed of some sequential MATLAB statements. For instance, the following statements

```
>> s=0; m=0; while s<=10000, m=m+1; s=s+m; end, m, s
```

can be saved to a *.m file (e.g., save it to test.m file). This type of program is referred to as an M-script file, or simply an M-file. The M-files can be executed directly to solve certain problems. For instance, the above-mentioned M-file can be invoked by typing `test` under the MATLAB prompt, and the results of the program can be returned to the MATLAB workspace. M-files are only suitable for solving small-scale problems. If there is a slight change in the original problem, the program itself needs modification. This will make the use of it limited. For instance, to solve a new problem similar to the previous one: to find the minimum m which makes the sum exceed 20,000, the source code of the M-file has to be manually modified. Thus a new kind of program is needed.

Another program type is the so-called M-function. It is the mainstream programming style in MATLAB. In this section, M-function programming and skills in writing M-functions will be presented.

2.5.1 Structures of MATLAB Functions

MATLAB functions are led by the keyword `function`, and the basic structure of MATLAB functions are organized as follows

```
function [returned variable list] =fun name (input argument list)
comment part, led by %, explains the use of the function and syntaxes
input and argument checking, the numbers and the contents
the main body of the function – this is essential in M-function programming
```

The numbers of inputs and returned arguments can be obtained with the commands `nargin` and `nargout`, while the input and returned variables can be extracted from the cell arrays `varargin` and `varargout`, respectively. If more than one returned variable is involved, square brackets are necessary. In the variable list, the arguments should be separated by commas. The comment statements are useful for high-quality programming. The `help` command can be used to display information of the file prepared in the comments. Examples will be given in the section for illustrating MATLAB programming skills.

Example 2.11   *Consider the problem in Example 2.9. It has been indicated that the M-script has certain disadvantages. One solution to the problem is to rewrite it in an M-function, where one input argument M and two output arguments, m and s can be assigned. The following M-function can be written*

```
function [m,s]=findsum(M)
s=0; m=0; while s<=M, m=m+1; s=s+m; end
```

Example 2.12   *Suppose we want to write an M-function to generate an n × m Hilbert matrix, whose elements are described by $1/(i + j - 1)$. Two additional requests are*

1) If only one input argument n is given, an n × n square matrix is generated.

2) Help information must be well written, to include use, syntax and argument descriptions.

When learning programming, it is always a good habit to write comments regularly, because they are helpful for both programmers and maintainers of the program. A MATLAB function `myhilb.m` *listed below can be written*

```
function A=myhilb(n,m)
%MYHILB  generates a rectangular Hilbert matrix
%   A=MYHILB(N, M) generates NxM rectangular Hilbert matrix A
```

```
%    A=MYHILB(N) generates NxN square Hilbert matrix A
%See also: HILB.

%  Designed by Professor Dingyü XUE, Northeastern University, PRC
%  5 April, 1995, Last modified by DYX at 22 March, 2011
if nargin==1, m=n; end
for i=1:n, for j=1:m, A(i,j)=1/(i+j-1); end, end
```

In the function, most of the statements are prefixed by %; *these are the comments where the use, syntaxes and argument descriptions are presented. A blank line follows, and then come other comments indicating the author, modification date and so on. Because of the blank line, this information will not be displayed in the on-line help system, via the* help *command.*

To implement the requirement 1), the number of input arguments is measured. If it is 1, the column number m is assigned to row number, such that the matrix generated is a square matrix. The double for *loops generate each element of the expected Hilbert matrix.*

Using help myhilb *command, the following on-line help information is displayed*

```
MYHILB  generates a rectangular Hilbert matrix
   A=MYHILB(N, M) generates NxM rectangular Hilbert matrix A
   A=MYHILB(N) generates NxN square Hilbert matrix A
See also: HILB.
```

The following statements can be used to generate Hilbert matrices of different sizes.

$$
A_1 = \begin{bmatrix} 1 & 0.5 & 0.3333 \\ 0.5 & 0.3333 & 0.25 \\ 0.3333 & 0.25 & 0.2 \\ 0.25 & 0.2 & 0.1667 \end{bmatrix}, \quad A_2 = \begin{bmatrix} 1 & 0.5 & 0.3333 & 0.25 \\ 0.5 & 0.3333 & 0.25 & 0.2 \\ 0.3333 & 0.25 & 0.2 & 0.1667 \\ 0.25 & 0.2 & 0.1667 & 0.1429 \end{bmatrix}.
$$

```
>> A1=myhilb(4,3), A2=myhilb(4)
```

Example 2.13 *Recursive calls can be used in MATLAB functions; that is, the function may call itself from within itself. Let us consider implementing factorial n! with a recursive structure. It is known that* $n! = n(n-1)!$. *Exits must be assigned in recursive functions, otherwise an infinite loop may be executed. The exits of the recursive function for this problem are* $1! = 0! = 1$. *The following M-function can be written.*

```
function k=my_fact(n)
if abs(n-floor(n))>eps | n<0
   error('n must be a non-negative integer');
end
if n>1, k=n*my_fact(n-1);      % for n > 1, recursive structure is used
elseif any([0 1]==n), k=1;     % the exits 0! = 1! = 1 are assigned
end
```

It can be seen that the function will judge first whether n is a nonnegative integer. If not, an error message is given and the function is terminated. If n is a nonnegative integer, recursive function calls can be made, until the exit is met. With MATLAB, factorial of n can easily be obtained from the command prod(1:n)*. If n is large, symbolic manipulation must be used with* prod(sym(1):n)*.*

2.5.2   Handling Variable Numbers of Arguments

It has been indicated that the numbers of input and output arguments can be measured with nargin and nargout, and the input and output arguments can be extracted from the cell arrays varargin and varargout, respectively. More specifically, the input arguments are directly accessible with commands varargin{1}, ···, varargin{n}.

Example 2.14   *In MATLAB, the function* conv() *can be used to perform multiplication of two polynomials, such that* $p = \text{conv}(p_1, p_2)$, *and it cannot handle multiple polynomials, unless nested calls can be used. Based on the function, the following function can be written as*

```
function a=convs(varargin)
a=1; for i=1:length(varargin), a=conv(a,varargin{i}); end
```

By the use of such a function, multiple polynomial multiplication problems can be handled and, theoretically speaking, this function can be used to handle multiplication of arbitrary numbers of polynomials. For instance, the following statements can be used

```
>> P=[1 2 4 0 5]; Q=[1 2]; F=[1 2 3]; D=convs(P,Q,F),
   G=convs(P,Q,F,[1,1],[1,3],[1,1])
```

and the following results can be obtained

$$D = [1, 6, 19, 36, 45, 44, 35, 30], G = [1, 11, 56, 176, 376, 578, 678, 648, 527, 315, 90].$$

2.5.3   Debugging of MATLAB Functions

MATLAB provides a program editing and debugging interface, within which the internal local variables can also be debugged and monitored. A simple example will be given to demonstrate the program debugging facilities. The myhilb.m can be opened with the edit myhilb command. An editing window will open the program. Breakpoints can be set in the program editing window. If you click the 🔲 button, a red dot will appear before the sentence to indicate that there is a breakpoint. When running the program, it will stop at the breakpoint. When the user moves the cursor to a variable name in the editing window, the contents of the variable can be displayed. Meanwhile, in the command window, the MATLAB prompt is changed to K>>. We can enter an internal local variable name at the prompt so that it can be displayed in the command window. In this way, internal local variables can be manipulated directly.

Multiple breakpoints can be set with the editing interface. To cancel a breakpoint, click again on the red dot. To cancel all the breakpoints, the button 🔲 can be clicked.

The MATLAB debugging interface supports the facilities of single step execution by clicking the 🔲 button. Clicking the 🔲 button moves the single step execution mode into the subfunction. Click the 🔲 button, and single step execution is performed. Clicks the 🔲 button, and single step execution process leaves the subfunction. Clicking the 🔲 button allows the user to continue to execute the program to the next breakpoint, if any. If there is no further breakpoint, the program completes and the debugging facilities are terminated. Clicking the ⬚ button will allow the program to leave debugging mode, and the whole program will be executed. The MATLAB prompt in the command window will be changed back to >>.

2.5.4   Pseudo Codes

There are two major applications of MATLAB pseudo code techniques, one of which is to speed up the program execution, since the *.m file can be converted to executable code. Another application of pseudo code techniques is that the ASCII code in the source MATLAB function can be transformed into binary code. This prevents other people from reading the source code.

A pcode command is provided in the MATLAB environment, and it can be used to transform the .m functions into pseudo code files, with a suffix of p. To convert file mytest.m into pseudo code, the command pcode mytest should be issued. To generate p files in the same folder as the original .m files, the command pcode mytest -inplace should be used. To convert all the *.m files in a folder to *.p files, use the cd command in that folder, and issue the command pcode *.m. If there is a grammar error in the files, the conversion process will be aborted and error messages will be given. Users can find potential errors in the code in this way. If both *.m and *.p files with the same name exist, the *.p files have the priority in execution.

It is very important to note that the source *.m files should be saved in a safe place, so that they cannot be deleted by mistake, since the *.p file cannot be converted back.

2.6   Two-dimensional Graphics in MATLAB

Apart from the powerful facilities in scientific computations, another important factor that is widely accepted by scientists and engineers is its straightforward graphics capabilities. Before the advent of MATLAB and other similar software and languages, it was rather difficult to generate two-dimensional plots, not to mention three-dimensional graphics. For instance, to draw a curve in a Fortran program, you have to preprocess the data, and find the minimum and maximum values within the data, and then compute the possible range of axes. Library or package subroutines, such as the well-known GINO-F library [1], can be called to display graphics on the screen. This is extremely complicated, and the effect and results are heavily dependent upon the experience of the programmers.

Also, if the programmer wants to port the program to another platform or system – for instance, if the programmer wants to transfer it to C – the whole graphics part has to be completely rewritten. This will be a great burden on the programmers. Even if the programmer wants to transfer the program from the PC to a Sun workstation under the same programming language, again the graphics part of the program has to be rewritten.

MATLAB provides powerful graphics facilities. Flexible ways of creating and modifying are provided, so the graphics generated by MATLAB can look elegant and neat.

So-called "handle graphics" concepts have been introduced into MATLAB [2], and they are very useful in object oriented programming for graphics processing. Compared with earlier versions of MATLAB, the main difference is that each element of the graphics, such as axes, curves and text on the graph, is an independent object. The properties of these objects can be assigned independently without affecting other objects. Each object has a handle, and the properties of the object can be accessed by its handles.

2.6.1   Basic Two-dimensional Graphics

MATLAB plotting facilities are very straightforward and flexible. Suppose that a set of data was obtained either by experiment or by computation. For instance, at time instances, $t_1, t_2, \cdots, t_n$, the values of the function are measured as $y = y_1, y_2, \cdots, y_n$. The data can be entered into MATLAB, and two vectors can be generated with the commands $t = [t_1, t_2, \cdots, t_n]$ and $y = [y_1, y_2, \cdots, y_n]$. The relationship between the two variables can be established graphically in a two-dimensional curve with `plot(t, y)`. It can be seen that the plotting command is quite simple and straightforward. The "curve" obtained is in fact a set of polylines joining the adjacent points. If the points are densely distributed, the polylines can look like curves. Thus later we shall call the plotting curves. In practical applications, the syntax of the `plot()` function can further be extended as

1) If t is still a vector, and y is a matrix, i.e.,

$$y = \begin{bmatrix} y_{11} & y_{12} & \cdots & y_{1n} \\ y_{21} & y_{22} & \cdots & y_{2n} \\ \vdots & \vdots & \ddots & \vdots \\ y_{m1} & y_{m2} & \cdots & y_{mn} \end{bmatrix},$$

then the same command can be used, and t curves are drawn. Note that, the column numbers in matrix y should be the same as the length of vector t.

2) The arguments t and y are matrices of the same size. The curve between each of the pairs of rows in t and y can be drawn.

3) If there are many pairs of such vectors or matrices, $(t_1, y_1), (t_2, y_2), \cdots, (t_m, y_m)$, the syntax `plot(t_1, y_1, t_2, y_2, \cdots, t_m, y_m)` can be used.

Example 2.15   *If we want to draw a sinusoidal curve for one complete cycle, a vector t of the independent variable should be generated first, with a colon expression. The function* `sin()` *can be used to evaluate the sinusoidal value for each element in vector t. Then the command* `plot()` *can be used to draw the curve, as shown in Fig. 2.4(a).*

```
>> t=0:.1:2*pi; y=sin(t); plot(t,y)
```

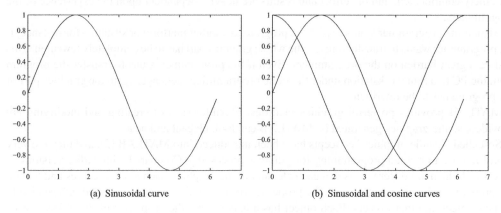

(a) Sinusoidal curve   (b) Sinusoidal and cosine curves

Figure 2.4   Examples of MATLAB plotting.

In MATLAB, a number of curves can be drawn simultaneously with the following commands, where the sinusoidal and cosine curves can be obtained as shown in Fig. 2.4(b).

```
>> t=0:.1:2*pi; y=[sin(t); cos(t)]; plot(t,y)
```

*In the statements, a vector **t** is generated first, the two vectors* $\sin(t)$ *and* $\cos(t)$ *are evaluated, and a matrix **y** with two rows can be constructed. The two curves can be drawn simultaneously with the* plot() *function. From the MATLAB graphical window obtained, it can be seen that the two curves are in solid lines, but the colors are different, one in blue and the other in green. However, in the book, they are in black and white form, so the two colors in grayscale cannot be distinguished. We can use different line styles to distinguish them.*

MATLAB function plotyy() can be used to draw special two-dimensional plots with scales on both side of the y axis. The syntax of the function is plotyy(t_1, y_1, t_2, y_2). For instance, the curves of $\sin t$ and $0.01 \cos t$ can be drawn with this function, and the results obtained are as shown in Fig. 2.5. It can be seen that since there are significant differences in the scales of the functions, the plot() may be used, but the curve of the cosine curve cannot be displayed at all. By using the plotyy() function, the two curves can be displayed satisfactorily.

```
>> t=0:.1:2*pi; plotyy(t,sin(t),t,0.01*cos(t))
```

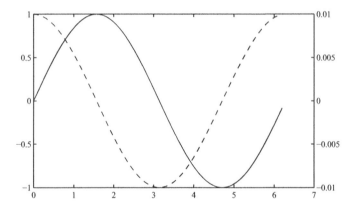

Figure 2.5   Double y axis plots with the plotyy() function.

2.6.2   *Plotting Functions with Other Options*

Properties of curves such as line styles, line widths and colors can be assigned in the following plotting function

plot(t_1, y_1, option 1, t_2, y_2, option 2, $\cdots, t_m, y_m$, option m)

where options can be assigned in the forms given in Table 2.1, and proper combinations of them can also be specified. For instance, to draw red dash dot lines for a curve, and indicate the data points by pentagrams, the option string 'r-.pentagram' can be used, meaning red dash dot lines, with sample points marked by pentagrams.

Table 2.1 Options in MATLAB graphics.

line styles		color specification				mark specification			
options	for	options	for	options	for	options	for	options	for
'-'	solid line	'b'	blue	'c'	cyan	'*'	*	'pentagram'	☆
'--'	dash line	'g'	green	'k'	black	'.'	·	'o'	○
':'	dotted line	'm'	magenta	'r'	red	'x'	×	'square'	□
'-.'	dash dot	'w'	white	'y'	yellow	'v'	▽	'diamond'	◇
'none'	no line					'^'	△	'hexagram'	✿
						'>'	▷	'<'	◁

2.6.3 Labeling MATLAB Graphics

When plotting commands are completed, further modifications are allowed and supported in MAT-LAB. For instance, one can use the following commands to decorate the plots:

```
>> grid,                          % adding grids in the plots
 xlabel('This is my X axis'),     % add a label to the x axis
 ylabel('My Y axis'),             % add a label to the y axis
 title('My Own Plot')            % add a title to the plot
```

Here `grid` commands automatically add a grid in dotted lines to the axes to increase readability of the plots. The commands `xlabel()` and `ylabel()` will add labels to the x and y axes, and the labels added to the y axis will automatically be rotated by 90°. The `title()` command adds a title to the plot.

MATLAB automatically selects suitable x and y axis ranges according to the data to be drawn, so that the graphs obtained are informative and neat. Normally speaking, users need not worry about the automatically assigned range of axes. However, to change the range of an axis, the function `axis()` can be used to set the range manually using the syntax `axis([`$x_{\mathrm{m}}, x_{\mathrm{M}}, y_{\mathrm{m}}, y_{\mathrm{M}}, z_{\mathrm{m}}, z_{\mathrm{M}}$`])`. With this command, the ranges of x, y and z axes can be assigned. For a two-dimensional plot, only the four values x_{m}, x_{M}, y_{m} and y_{M} are needed. Also, the functions `xlim()`, `ylim()` and `zlim()` can be used to independently specify the ranges of x, y and z axes.

2.6.4 Adding Texts and Other Objects to Plots

Various tools are provided in the toolbar of the MATLAB graphics window, allowing the users to add decorations to the plots. For instance, the user can add text, arrows and lines to the graph. A subset of the well-established LaTeX commands[3] can be used to represent formatted text or mathematical formulae on the plots. For instance, the commands `\bf`, `\it` and `\rm` can be used to specify bold, italic and normal font. For example, `The {\bf word is bold}` will give display of The **word is bold** in the graphics window. Superscripts and subscripts can be assigned. For instance, `y = x^{abc}` will give a display of $y = x^{abc}$, while `y = x^abc` will produce $y = x^a bc$. Similarly, underscores can be used to describe subscripts.

Unfortunately, the qualities of the mathematical formulae thus generated are not very high. In order to embed such a graph into a LaTeX documentation system, the package **overpic** should be used to superimpose the text on the graphs.

2.6.5   Other Graphics Functions with Applications

Apart from the standard two-dimensional plotting functions, special two-dimensional functions are also supported in MATLAB. The commonly used functions and syntaxes are given in Table 2.2, where the arguments x and y store the data for the x and y axes, respectively. The argument c is the color option, while u and l are the upper and lower bounds in the error plots. The following examples can be used to demonstrate the plotting functions.

Table 2.2   Special two-dimensional plotting functions provided in MATLAB.

function names	plotting facilities	commonly used syntaxes
bar()	bar plot	bar(x, y)
comet()	comet plots to show trajectories	comet(x, y)
compass()	arrow plots from origin	compass(x, y)
errorbar()	curves with error bars	errorbar(x, y, l, u)
feather()	velocity vector plots	feather(x, y)
fill()	filled polygons	fill$(x, y,$c$)$
hist()	histogram	hist(y, n)
loglog()	both axes are in logarithmic scales	loglog(x, y)
polar()	polar plots	polar(x, y)
quiver()	quiver or velocity plots	quiver(x, y)
stairs()	stairstep graphs	stairs(x, y)
stem()	discrete sequence stem plots	stem(x, y)
semilogx()	one of the axis in logarithmic scale	semilogx(x, y), semilogy(x, y)

Example 2.16   *The command* subplot() *can be used to divide the graphics window into* 2×2 *portions, and in each portion, plotting functions can be issued independently, so that the following commands draw the plots shown in Fig. 2.6.*

```
>> t=-pi:0.3:pi; y=1./(1+exp(-t));
   subplot(221), plot(t,y); title('plot(t,y)')
```

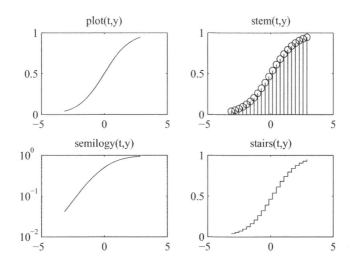

Figure 2.6   Plotting of different types of curves.

```
subplot(222), stem(t,y); title('stem(t,y)')
subplot(223), semilogy(t,y); title('semilogy(t,y)')
subplot(224), stairs(t,y); title('stairs(t,y)')
```

2.6.6   Plotting Implicit Functions

If an explicit function $y = f(t)$ is given, we can generate a t vector first. Then we can use the dot operation to compute, for each value in the t vector, the function value, to generate the vector y. The function plot() can then be used to draw two-dimensional curves. However, if an implicit function is given, such as $x^2 + 3y^2 = 5$, the above method cannot be used. The MATLAB function ezplot() has to be used instead to draw implicit functions. Several examples are given here to illustrate implicit function facilities.

Example 2.17   *The implicit function $x^2 + 3y^2 = 5$ can be drawn with the following statements, and the ellipse obtained is shown in Fig. 2.7(a).*

```
>> ezplot('x^2+3*y^2=5'), axis([-4,4,-4,4])
```

If a portion of the ellipse is expected, for instance the portion in the range $x \in (-\pi/4, \pi)$, $y \in (-1, 3)$ is expected, the following statements can be given; the result obtained is shown in Fig. 2.7(b).

```
>> ezplot('x^2+3*y^2-5',[-pi/4,pi,-1,3]), axis([-4,4,-4,4])
```

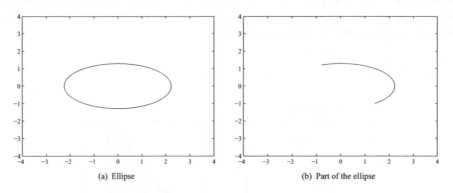

(a) Ellipse   (b) Part of the ellipse

Figure 2.7   The ellipse plots obtained with ezplot() functions.

Example 2.18   *For the more complicated implicit function*

$$x^2 \sin(x + y^2) + y^2 e^{x+y} + 5 \cos(x^2 + y) = 0,$$

we can use the ezplot() function to draw its curves, as shown in Fig. 2.8. It can be seen that although the curves of the multi-valued implicit function look very complicated, it can still be drawn directly with a single function call.

```
>> ezplot('x^2 *sin(x+y^2) + y^2*exp(x+y)+5*cos(x^2+y)')
```

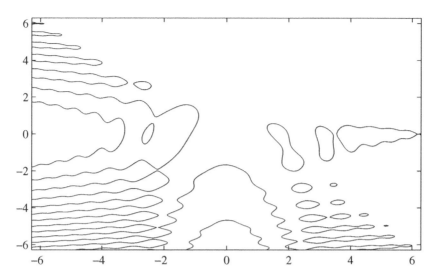

Figure 2.8 Complicated implicit function.

Example 2.19 *If the parametric equation for a spatial point* (x, y) *satisfies* $x = \sin 3t \cos t$, $y = \sin 3t \sin t$, $t \in (0, \pi)$. *The following MATLAB statements can be used, and the three-dimensional trajectory can be obtained as shown in Fig. 2.9.*

```
>> ezplot('sin(3*t)*cos(t)','sin(3*t)*sin(t)',[0,pi])
```

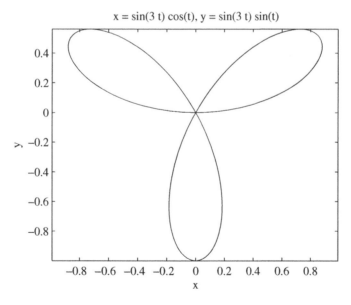

Figure 2.9 Parametric equation.

2.7 Three-dimensional Graphics

2.7.1 Three-dimensional Curves

Similar to two-dimensional plots, a function `plot3()` can be used to drawn three-dimensional plots, with the syntax `plot3(x, y, z, options)`, where x, y and z are vectors of the same length, storing the coordinates, and the options available are exactly the same, as shown in Table 2.1.

▣ Example 2.20   *For the time variable t, the spatial coordinates can be computed from* $x = \sin t$, $y = \cos t$, $z = t$, *and we want to show the trajectory in green, the following statements can be used, and the three-dimensional curve is as shown in Fig. 2.10(a).*

```
>> t=0: pi/50: 2*pi; x=sin(t); y=cos(t); z=t; h=plot3(x,y,z,'g-');
   set(h,'LineWidth',4*get(h,'LineWidth'))
```

Alternatively the three-dimensional curve can be drawn with the following commands, and that gives the new curve drawn in Fig. 2.10(b).

```
>> ezplot3('sin(t)','cos(t)','t',[0,2*pi])
```

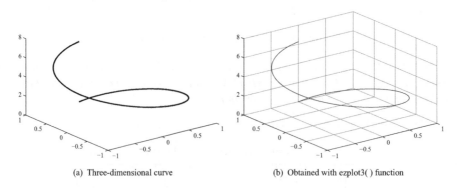

(a) Three-dimensional curve   (b) Obtained with ezplot3( ) function

Figure 2.10   Three-dimensional curve.

Similar to the two-dimensional functions stem() *and* fill(), *stem plots and filled plots can be obtained respectively with* stem3() *and* fill3(), *as shown in Figs 2.11(a) and (b), with the following statements:*

```
>> t=0: pi/50: 2*pi; x=sin(t); y=cos(t); z=t; stem3(x,y,z), grid on
   figure; fill3(x,y,z,'g'), grid off
```

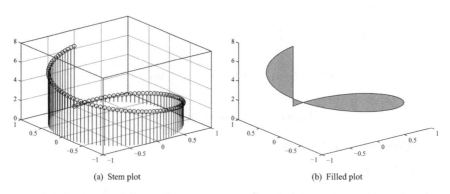

(a) Stem plot   (b) Filled plot

Figure 2.11   Other three-dimensional plots.

2.7.2   Surface Plots

Three-dimensional surfaces and mesh grid plots can be drawn with the functions `surf()` and `mesh()`, with the same syntax of `mesh(x, y, z, c)`, where x and y are matrices of the mesh grid coordinates on the x-y plane. The variable z is the height matrix. The argument c is the color matrix, indicating the range of colors. If the c argument is omitted, MATLAB will assign it to $c = z$, that is the setting of the color is proportional to the value of z.

Example 2.21   *Consider a two-dimensional function $z = f(x, y) = (x^2 - 2x)e^{-x^2-y^2-xy}$, and we can select a rectangular area on the x-y plane, and draw the three-dimensional surface for the function. The function `meshgrid()` can be used first to generate mesh grids on the x-y plane, and the function can be evaluated with dot operations to the coordinates of x and y to form matrix z. The function `mesh()` can be used to draw the mesh grid plot, as shown in Fig. 2.12(a).*

```
>> [x,y] = meshgrid(-3:0.1:3,-2:0.1:2);
   z=(x.^2-2*x).*exp(-x.^2-y.^2-x.*y); mesh(x,y,z)
   axis([-3,3,-2,2,-0.7,1.5])
```

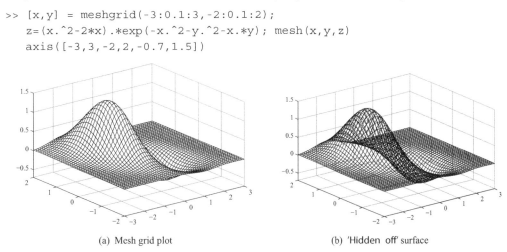

(a) Mesh grid plot   (b) 'Hidden off' surface

Figure 2.12   Three-dimensional mesh grid plot with `mesh()` function.

It can be seen from Fig. 2.12(a) that part of it has been hidden and to show the hidden parts, the command `hidden off` can be used, as shown in Fig. 2.12(b).

If the `mesh()` function is replaced by `surf(x, y, z)`, the surface plot is obtained as shown in Fig. 2.13(a). The command `colorbar` can be used to display a color bar beside the plot, as shown in Fig. 2.13(b).

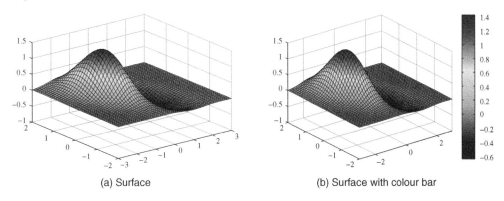

(a) Surface   (b) Surface with colour bar

Figure 2.13   Three-dimensional surface with `surf()` function.

With the functions `ezmesh()` *and* `ezsurf()`, *the same plots can also be obtained.*

```
>> ezmesh('(x^2-2*x)*exp(-x^2-y^2-x*y)')
   figure; ezsurf('(x^2-2*x)*exp(-x^2-y^2-x*y)')
```

It should be noted that in the above functions, dot operation is not used, since x and y are simply symbolic variables, not the mesh grid matrices.

The MATLAB command `shading` can be used to specify the coloring scheme. For instance, if the option `interp` is used, the surface of the plot is interpolated, with a much smoother surface obtained as shown in Fig. 2.14(a).

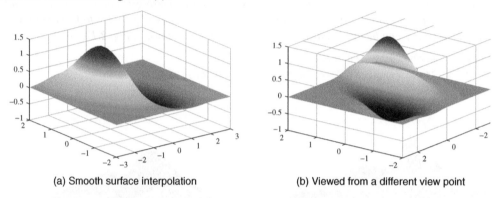

(a) Smooth surface interpolation   (b) Viewed from a different view point

Figure 2.14   Further processing of three-dimensional surfaces.

The options `faceted` (the default one) and `flat` can also be used, and the latter removes the mesh grid lines in the default surface plot.

The button ⟲ in the toolbar of the graphics window can be used to rotate the three-dimensional plot. For instance, after rotation, the plot is as shown in Fig. 2.14(b). The button ⟲ allows the user to get the coordinates on the surface with a mouse click. The toolbar in the graphics window also provides other facilities.

2.7.3   Local Processing of Graphics

The constant NaN is useful in the processing of graphics. To hide or cut off a particular part of the plots, the function values in this part can deliberately be set to NaN. This method can also be used in two-dimensional graphics.

■ **Example 2.22**   *If in the three-dimensional surface obtained in Example 2.21, the areas x ≤ 0 and y ≤ 0 are to be cut off. The following statements can be provided, and the effect is as shown in Fig. 2.15.*

```
>> [x,y]=meshgrid(-3:0.1:3,-2:0.1:2);
   z=(x.^2-2*x).*exp(-x.^2-y.^2-x.*y);
   ii=(x<=0)&(y<=0); z1=z; z1(ii)=NaN;
   surf(x,y,z1), shading flat; axis([-3 3 -2 2 -0.7 1.5])
```

2.8   Graphical User Interface Design in MATLAB

For successful software, of course, its contents and functions are the most important factors. The graphical user interface is also very important, since it determines the quality and level of the

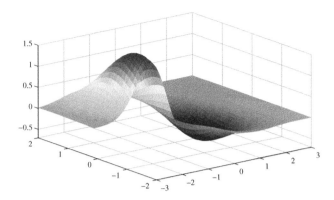

Figure 2.15   Cutting and effects in three-dimensional surfaces.

software. The graphical user interface in this case acts like the outward appearance of the product. Thus, by mastering the skills of graphical user interface techniques, it is possible to design high-quality software for general purposes.

The GUI design interface, named "Guide" (which stands for GUI development environment), is a very powerful tool for designing GUIs, in a visual manner. Equipped with MATLAB programming experiences and skills, high-standard GUI programming can be achieved. Illustrations are presented through examples of MATLAB implementations of the GUI and its applications.

2.8.1   Graphical User Interface Tool – Guide

Type `guide` in the MATLAB command window, and the initial interface shown in Fig. 2.16 is displayed to prompt the user to select a suitable template for the GUI to be designed. It can be seen that the existing GUI templates available are a default **Blank GUI**, a **GUI with Uicontrols**, a **GUI with Axes and Menu** and a **Modal Question Dialog**. Also we can **Open Existing GUI**.

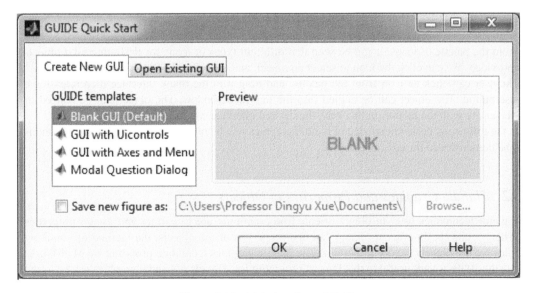

Figure 2.16   Main interface of Guide.

Here we shall only discuss the blank GUI design method. Selecting the **Blank GUI** template from the list, and clicking the **OK** button, the GUI design interface shown in Fig. 2.17 is displayed. The prototype blank interface is also shown in the figure.

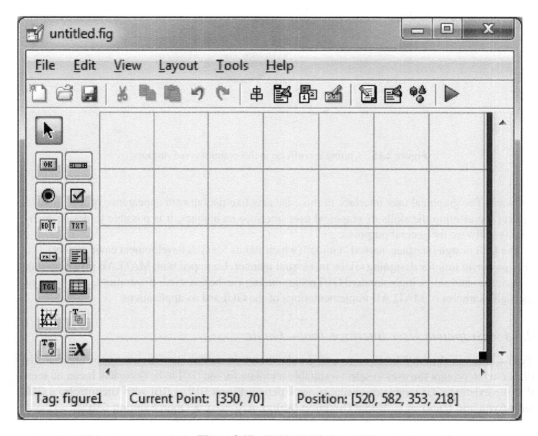

Figure 2.17   Guide interface.

From the palette of icons, it can be seen that different types of icons for designing GUI objects are provided. We can click the icon to select an object, and drag it to the blank window.

The user can click an icon from the palette, and drag it to the blank interface then release the mouse button. The object can be copied into the interface window. The object can be resized or moved easily with the mouse. In this way, the desired interface can then be drawn.

In the following, basic knowledge of handle graphics will be presented, and GUI design methods will be demonstrated through an example.

2.8.2   *Handle Graphics and Properties of Objects*

The major techniques used in object oriented programming are to extract and assign properties to different objects. In order to manipulate the properties of the objects, the method of obtaining handles for the objects is very important. The handle graphics technique proposed by MathWorks Inc. in the 1990s is very useful in this kind of programming. Details of handle graphics can be found in [2].

In MATLAB interface design, object oriented methodology is extensively used. The window is an object, and the controls, such as buttons and edit boxes, on the window are also objects. In object oriented programming, each object has its handle and properties.

Double click the prototype window, and a **Property Inspector** shown in Fig. 2.18(a) will be displayed. The inspector shows the properties, and their values, of the selected object – in this case the window object – and the user can set or change the properties easily by using the mouse. For instance, if we want to change the color of the window, the box to the right of the **Color** property can be clicked, and a dialog box shown in Fig. 2.18(b) is opened. More colors can be selected when the button **More Colors** is clicked. The standard color dialog box is shown in Fig. 2.18(c), from which more colors can be selected. The new color will take effect immediately in the prototype window.

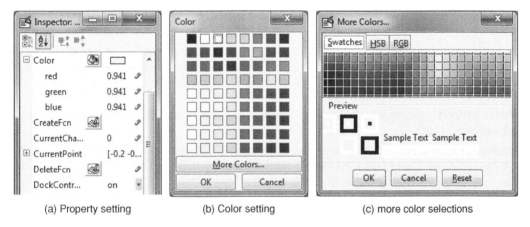

| (a) Property setting | (b) Color setting | (c) more color selections |

Figure 2.18   Parameter modification of interfaces.

It can be seen from the property browser that many properties are provided for the window object. It is not necessary to modify all of them. The commonly used window properties are:

- **MenuBar property** describes the form of the menu bar of the proposed window, and the available options are **figure** (for standard figure window menu – the default) and **none** (a window with no menu bar, for the time being). If necessary, the new menu system can be added, or appended to the existing ones, with the menu editor to be discussed later.

- **Name property** sets the title of the current window with its string property value. It should be used in conjunction with the **NumberTitle** property.

- **NumberTitle property** determines whether or not to prefix the window title with "Figure No *:". The property values are **on** (default) and **off**.

- **Units property** specifies the length unit used in the window property. The possible options are **pixels** (default), **inches**, **centimeters**, **normalized** (between 0, 1) and so on. The **Units** property can also be set with the property editor. When the property value box is clicked in the **Units** property, a listbox will appear, as shown in Fig. 2.19(a), and we can choose from it the desired property value.

- **Position property** determines the size and position of the current window. The property value is a 1×4 vector, with the first two elements specifying the coordinates of the lower left corner of the window, while the latter two values specify the width and height of the window, as shown in

Fig. 2.19(b). The property should be used in conjunction with the **Unit** property. The best way to set the **Position** property is to directly move and resize the prototype window by using the mouse. The current position and size will be filled to the property value automatically.

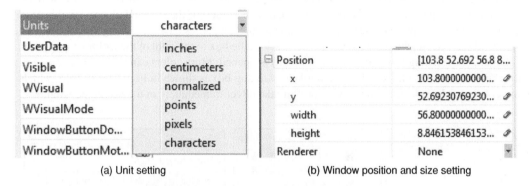

(a) Unit setting   (b) Window position and size setting

Figure 2.19   Setting of window position and size.

- **Resize property** determines whether the window is resizable or not. Two options are available, **off** and **on** (default).

- **Toolbar property** indicates whether or not to add visible toolbars to the window. The available options are **none**, **figure** (standard toolbar for the figure window) and **auto** (default). Toolbars can be designed visually by the Toolbar Editor.

- **Visible property** determines whether the window is visible or not, initially, with the options **on** (default) and **off**.

Object property extraction and setting in handle graphics can be implemented with two functions: set() and get(), with the syntaxes

```
v = get(h,property name)   % e.g., v = get(gcf,'Color')
set(h,property name 1,property value 1,property name 2,property value 2,...)
```

where h is the handle of the object. The command gcf can be used to get the handle of the current window object, while gco can be used to get the handle of the current object. Extraction of other handles will be shown later.

The following examples can be used to demonstrate the design process and skills of graphical user interfaces.

■ **Example 2.23**   *Suppose we need to add two control items in a blank window: one a button and the other a text box. What we are expecting is that, when the button is clicked, the "Hello World!" string will appear in the text box. The GUI design can be achieved with the following procedures:*

1) Draw the prototype window. Open a blank prototype window and draw the two control items onto the window, by directly selecting and dragging from the control item palette, and the layout of the prototype windows will be as shown in Fig. 2.20(a).

*2) Control property modification. Double click the text box control item, and the property editor windows will be displayed. We can set the **String** property to an empty one, which means nothing initially will be displayed, before the button is clicked. Also, the **Tag** property of the control item should be set so that the handle of it can be found easily by other objects. Here the **Tag** property can be set to **txtHello**, as shown in Fig. 2.20(b). Please note that a unique string should be assigned to each object, so that other objects can find its handle easily and without any conflict. By convention*

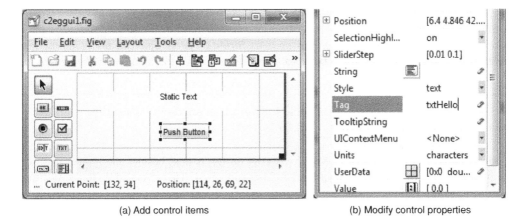

(a) Add control items   (b) Modify control properties

Figure 2.20   Prototype windows design and modifications.

here, the string of the **Tag** *property can be composed of two strings, the first one describing its type, the second one indicating its use. For instance, we have used* **txtHello** *here. Meanwhile we can set the tag of the button to* **btnOK***.*

3) Automatic generation of the framework of the program. *When the prototype window is designed, it can be saved into a .fig file, say, to file c2eggui1.fig. The framework of the MATLAB function, named c2eggui1.m can be generated, and the main part of it is show below.*

```
function varargout = c2eggui1(varargin)
gui_Singleton = 1;
gui_State = struct('gui_Name',      mfilename, ...
                'gui_Singleton',  gui_Singleton, ...
                'gui_OpeningFcn', @c2eggui1_OpeningFcn, ...
                'gui_OutputFcn',  @c2eggui1_OutputFcn, ...
                'gui_LayoutFcn',  [] , ...
                'gui_Callback',   []);
if nargin && ischar(varargin{1})
   gui_State.gui_Callback = str2func(varargin{1});
end
if nargout
   [varargout{1:nargout}]=gui_mainfcn(gui_State, varargin{:});
else
   gui_mainfcn(gui_State, varargin{:});
end
function c2eggui1_OpeningFcn(hObject, eventdata, handles, varargin)
handles.output = hObject; guidata(hObject, handles);
function varargout = c2eggui1_OutputFcn(hObject, eventdata, handles)
varargout{1} = handles.output;
% End of the main framework. No modification of the previous code is advised to
% be modified. The rest of the code are the frameworks of the callback functions
function btnOK_Callback(hObject, eventdata, handles)
```

4) Writing callback functions. *We can rewrite the requests given earlier: When the button is clicked, the string "Hello World!" is to be displayed in the text box. The request can be rephrased as* *follows, "when the button is clicked" means that we should write a callback function for the button. "Display the string in the text box" means that in the callback function, we should find the handle of the text box first, then we can assign the string to its **String** property. Since the tag of the text box is **txtHello**, the handle of it is **handles.txtHello**. The following callback function can be written for the button object*

```
function varargout = btnOK_Callback(hObject, eventdata, handles)
set(handles.txtHello,'String','Hello World!');
```

Another important thing to note in MATLAB GUI design is that, besides the typical callback functions, other types of callback functions are also supported. The so-called callback function actually means that, when an event happens to an object, a function is invoked automatically by MATLAB's internal mechanisms. Other commonly used callback functions are:

- **CloseRequestFcn**: The callback function when a window is closed.
- **KeyPressFcn**: The callback function when a key in keyboard is pressed.
- **WindowButtonDownFcn**: The callback function when a mouse button is clicked.
- **WindowButtonMotionFcn**: The callback function when the mouse is moved.
- **WindowButtonUpFcn**: The callback function when a mouse button is released.
- **CreateFcn** and **DeleteFcn**: The callback functions when an object is created or deleted.

Some of the callback functions are related to the window object, and some are for specific control items. If one masters the skills of callback function programming, the efficiency of MATLAB GUI programming can be improved.

Various properties are also provided for different types of control items. Some of the commonly used properties for the control items are listed below:

- **Units** and **Position** properties: The definitions are the same as the ones given in window properties. It should be noted that the lower left corner here refers to the corner of the window, not of the screen.
- **String** property: The property is used to assign a string in the control object. Normally it is used for labels or prompts.
- **CallBack** function: This is the most important property in GUI design. If an object is selected or an action is done to the object, for instance, a button is clicked, the callback function can be executed automatically.
- **Enable** property: This indicates whether the control item is enabled, with options **on** (default) and **off**. **Visible** property is also provided.
- **CData** property: A true color bitmap is to be drawn on the control item. It is a three-dimensional array.
- **TooltipString** property: This stores a string variable for displaying help information when the mouse is moved on top of the control object, when the mouse button is not clicked.
- **UserData** property: This can be used to store and exchange information with other controls.
- **Interruptible** property: This indicates whether interrupts are allowed in the execution of the callback functions, with available options **on** (default) and **off**.
- The properties related to the font, such as **FontAngle**, **FontName** and so on.

Functions gco and gcbo can be used to get the handles of the current object. All the property names and property values of the current object can be listed either with the command set(gco), or by the **View → Object Browser** menu item in the Guide window. In the latter case the users can modify the properties interactively.

2.8.3 Menu System Design

With the facilities provided in Guide, not only dialog boxes, but also windows with menu systems can be designed. The menu editor can be launched by choosing the **Tools → Menu Editor** menu item, as shown in Fig. 2.21(a), and the menu editor interface is shown in Fig. 2.21(b). Menu tools also provide facilities such as **Align Objects** and **Grid and Ruler**.

(a) Tools menu   (b) Menu editor

Figure 2.21   Menu tool and menu editor.

With the menu editor interface, the menu system shown in Fig. 2.22(a) can be easily designed. The execution results shown in Fig. 2.22(b) are produced. The program can be saved in the file c2eggui2.m.

(a) Menu editor results   (b) Menu system

Figure 2.22   Interface design and modifications.

2.8.4 Illustrative Examples in GUI Design

An illustrative example is given in the section to show the ideas and methodologies of object oriented programming and GUI design with MATLAB, together with the skills of programming.

■ **Example 2.24**   *GUI design in MATLAB uses object oriented programming techniques. Suppose that we want a three-dimensional plotting interface, and a sketch of the final layout of the interface is shown in Fig. 2.23. Other specific requirements are*

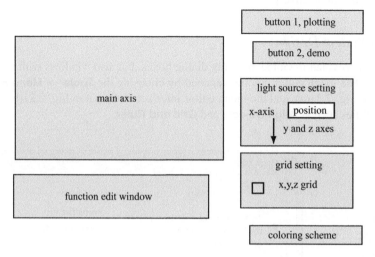

Figure 2.23   A sketch of the interface to be established.

- *A main axis object is needed in the window for use with three-dimensional graphics.*
- *An edit box is needed for accepting data for plots.*
- *Two push buttons are to be used, one for invoking the plot drawing process, and the other for invoking demonstration facilities.*
- *A group of three edit boxes, which can be used to accept the light source position, specified as coordinates.*
- *A group of three checkboxes to set whether or not the grids on the plot axis are displayed, one for each axis.*
- *A listbox is needed to allow the user to select different color shadings.*

Based on the above programming requests, it can be seen that the prototype windows can be established with Guide, as shown in Fig. 2.24(a), where the edit boxes and checkboxes can be created aligned with the ✛ icon, and the window for alignment tools will appear as shown in Fig. 2.24(b).

*From the above specifications, different tasks can be assigned to the control objects. This is the nature of object oriented programming techniques. The sketch of task assignment is shown in Fig. 2.25. It can be seen from the sketch that the objects in zones A and B do not have any tasks assigned. They only provide places for accepting data and drawing plots. Thus, their handles, identified through tags, are very useful in the program design. We can assign the **Tag** properties respectively to **axMain** and **edtCode**. To allow multiple lines to be input in the edit box **edtCode**, its **Max** property must be set to a number larger than 1, say, set it to 100.*

The tags of other control objects, useful in programming, are described as

- *The tags of the buttons in zones C and D are assigned to **btnDraw** and **btnDemo**.*
- *The tags of the three edit boxes in zone E are assigned to **edtX**, **edtY** and **edtZ**.*

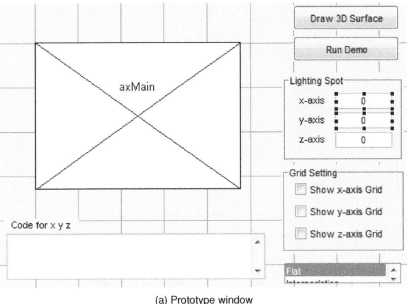

(a) Prototype window

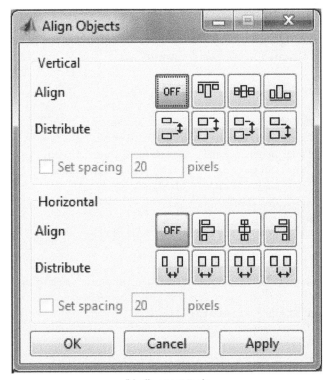

(b) alignment tools

Figure 2.24 Drawing the interface with Guide.

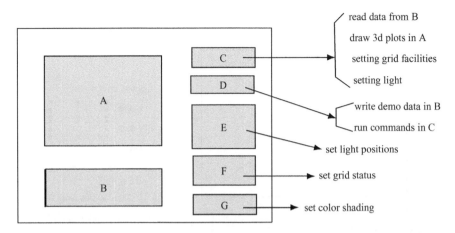

Figure 2.25 Illustration of task assignment for the control objects.

- *The tags of the checkboxes in zone F are assigned to* **chkX**, **chkY** *and* **chkZ**.
- *The tag of the listbox in zone G is assigned to* **lstFill**. *Also, with a click of the edit button* 🔳 *the* **String** *property in* **lstFill** *object can be set to* **Flat** → **Interpolation** → **Faceted**.

According to the task assignment sketch, the main interface windows can be established, and the filename is specified as c2fgui3(), *with the framework*

```
function varargout = c2eggui3(varargin)
gui_Singleton = 1;
gui_State = struct('gui_Name',       mfilename, ...
                   'gui_Singleton',  gui_Singleton, ...
                   'gui_OpeningFcn', @c2eggui3_OpeningFcn, ...
                   'gui_OutputFcn',  @c2eggui3_OutputFcn, ...
                   'gui_LayoutFcn',  [] , ...
                   'gui_Callback',   []);
if nargin && ischar(varargin{1})
   gui_State.gui_Callback = str2func(varargin{1});
end
if nargout
   [varargout{1:nargout}] = gui_mainfcn(gui_State, varargin{:});
else
   gui_mainfcn(gui_State, varargin{:});
end
```

It can be seen that the main framework generated here is exactly the same as that generated in the previous example. The differences between the programs of the two examples lie in the *.fig file, where the details of the GUIs are described. There is no difference in the main interface framework, while the callback functions for different control objects are different.

According to the requests assigned for the object in zone C, the callback function for the button can be organized as follows: the string in control, tagged **edtCode** *can be extracted first, then the* *data is used to draw the surface plot on the three-dimensional axis, which is tagged as* **axMain**). *The listing of the callback function is*

```
function btnDraw_Callback(hObject, eventdata, handles)
try
   str=get(handles.edtCode,'String'); str0=[];
   for i=1:size(str,1) % join the strings together to form a command
      str0=[str0, deblank(str(i,:))];
   end
   eval(str0); axes(handles.axMain); surf(x,y,z);
catch
   errordlg('Error in code');
end
```

Note that the try···catch *structure is used, so that it can ensure that correct data generating statements are filled in the edit box, tagged* **edtCode**. *If there are unwanted characters typed into the edit box, an error message will be displayed. This is a good application of the trial structure.*

Now let us consider the callback function programming in zone D. From the assignment, it can be seen that it should first set the demo commands to **edtCode**, *then call the callback function of* **btnDraw**. *The callback function can be written as*

```
function btnDemo_Callback(hObject, eventdata, handles)
str1='[x,y]=meshgrid(-3:0.1:3, -2:0.1:2);';
str2='z=(x.^2-2*x).*exp(-x.^2-y.^2-x.*y);';          % write the demo data
set(handles.edtCode,'String',str2mat(str1,str2)); % assign data
btnDraw_Callback(hObject, eventdata, handles)        % invoke btnDraw callback
```

The callback functions of the three edit boxes in zone E can be programmed together. The coordinates filled in the three control objects, tagged **edtX**, **edtY** *and* **edtZ**, *can be extracted and then the light source position can be set in the axis tagged* **axMain**. *The following callback function can be written*

```
function edtX_Callback(hObject, eventdata, handles)
try
   xx=str2num(get(handles.edtX,'String'));   % read in light source position
   yy=str2num(get(handles.edtY,'String'));
   zz=str2num(get(handles.edtZ,'String'));
   axes(handles.axMain); light('Position',[xx,yy,zz]); % set light source
catch
   errordlg('Wrong data in Lighting Spot Positions');
end
```

For simplicity, it is not necessary to write the same callback functions for all the three control objects. Write the above code for the **edtX** *control object and assign the other two control objects,* **edtY** *and* **edtZ**, *as follows*

```
function edtY_Callback(hObject, eventdata, handles)
edtX_Callback(hObject, eventdata, handles)
```

Similarly, the callback functions for the control objects in zone F can be written as

```
function chkX_Callback(hObject, eventdata, handles)
xx=get(handles.chkX,'Value'); yy=get(handles.chkY,'Value');
zz=get(handles.chkZ,'Value'); % extract the checkboxes
set(handles.axMain,'XGrid',onoff(xx),'YGrid',onoff(yy),...
    'ZGrid',onoff(zz)) % a subfunction is written to convert 0,1 to 'off' and 'on'
% a subfunction is written to convert 0,1 to 'off' and 'on'
function out=onoff(in)
out='off'; if in==1, out='on'; end
```

The function reads the status from the three checkboxes and sets the grid status of the three axes. Since the status is measured from the checkboxes, not the edit boxes, no error is possible. There is no need then to use the try $\cdots$ catch *structure. A subfunction* onoff() *is written to convert 0 and 1 into the strings* 'off' *and* 'on'. *Also the* chkY_Callback() *can be arranged to invoke the above written callback function* chkX_Callback.

Finally let us consider the callback function in zone G. The **Value** *option in* **lstFill** *is extracted, and according its value, different shading colors are specified. The callback function can be written as*

```
function lstFill_Callback(hObject, eventdata, handles)
v=get(handles.lstFill,'Value'); axes(handles.axMain);
switch v
    case 1, shading flat;      % set shading format to flat
    case 2, shading interp;    % interpolation to make surface smooth
    case 3, shading faceted;   % set the shading to default format
end
```

On running the program c2eggui3.m, the interface shown in Fig. 2.26 will be displayed.

2.8.5   Toolbar Design

Toolbars can also be designed for MATLAB interfaces. The menu item **Tools** → **Toolbar Editor** in the main window of Guide can be selected to launch the toolbar editor shown in Fig. 2.27. Standard icons in the toolbar can be used directly, and new icons can also be designed.

In the toolbar editor, the button **P** allows the user to design toolbar buttons, and button **T** allows the user to design tangle buttons. The new icons of the buttons can be described by the **CData** data type, or can use the existing icons provided.

🖳 **Example 2.25**   *We can design a new user interface where a sinusoidal curve can be generated automatically on the interface. A toolbar system can be designed for this new interface, with several standard buttons on it. Also, we can add more toolbar buttons to it, including the zooming buttons* 🔍, 🔍 *for the x and y axes.*

A blank prototype window can be designed first with Guide. An axis object can be added to it, with a tag of **axPlot**. *The interface framework can be saved to the file c2eggui4.fig. Open the toolbar editor, and copy directly some of the standard icons to the editor. This can be done by first*

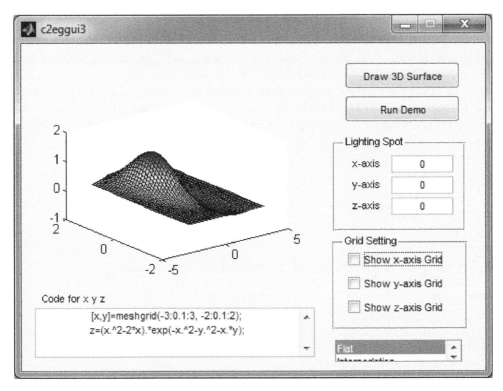

Figure 2.26   The effect when button **Run Demo** is clicked.

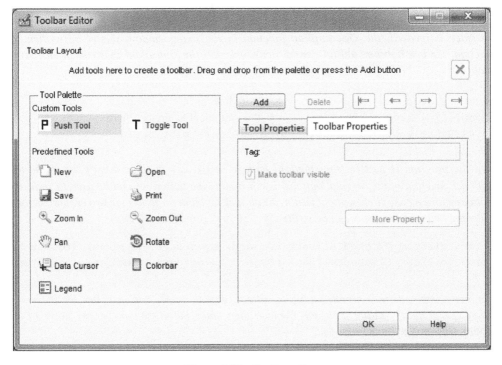

Figure 2.27   Toolbar editor.

*selecting the icons to be added, then clicking the **Add** button. A series of standard icons can be added to the **Toolbar Layout** in this way. If we want to add the custom icon ⊘ button in the new toolbar, we can click the **P** button, and then click the **Add** button. Then the **Tag** property of the new custom button can be assigned to **tolXZoom**. Click the **Edit** button, and then select an existing bitmap file or **CData** data, and the picture can be assigned directly to the icon of the object. The designed toolbar is shown in Fig. 2.28. In the same way, the custom buttons ⊘ , ⚲ and ⚲ can be added to the new toolbar, and their tags can be assigned to **tolYZoom**, **tolZoom** and **tolZOff**, respectively.*

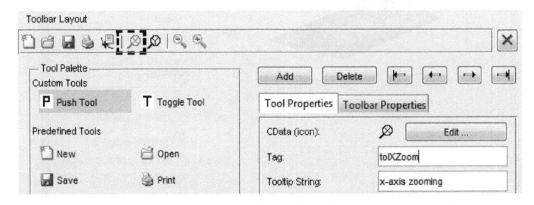

Figure 2.28   The designed toolbar.

*A file c2eggui4.m can be generated automatically by Guide, and the listings are exactly the same as the other framework files. If we want to automatically draw the sinusoidal curve when the interface is invoked, the **OpeningFcn** function in the main window should be written. In the function, the tag **handles.axPlot** can be activated, and the sinusoidal plot can be displayed automatically.*

```
function varargout = c2eggui4_OpeningFcn(hObject, eventdata, handles)
varargout{1} = handles.output;      % this statement is already there
t=0:0.01:2*pi; y=sin(t); axes(handles.axPlot); plot(t,y)
```

When the program is written, the standard buttons in the new toolbar inherit the properties and callback function of the original buttons. For instance, the button ⬚ can be used to show the coordinates with a mouse click, without writing a callback function for it. For the two custom buttons, the following callback functions can be written

```
function tolXZoom_ClickedCallback(hObject, eventdata, handles), zoom xon
function tolYZoom_ClickedCallback(hObject, eventdata, handles), zoom yon
```

Even more simply, there is no need to write callback functions. We can just click the ⊘ button to initiate the property editing interface. We can then enter the command `'zoom xon'` *to the **tolYZoom_ClickedCallback** property.*

2.8.6 Embedding ActiveX Components in GUIs

ActiveX usually refers to reusable components, such as the Windows Media Player components, database components, developed by software providers, such as Microsoft. The ActiveX components can be adopted and embedded in the MATLAB GUI to make the interfaces more powerful, without the need for low-level programming. These ActiveX components can be embedded in MATLAB GUIs by clicking the ☒ button, when a dialog box shown in Fig. 2.29 will be displayed, to allow the user to select a proper ActiveX components from the listbox. An example is given below to show how to use ActiveX in GUI design.

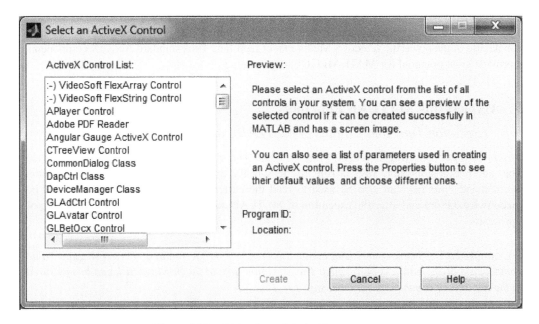

Figure 2.29   Dialog box of ActiveX components.

■ Example 2.26   *A blank prototype GUI window can be opened with the* guide *command, and an ActiveX control can be drawn on the window. The **Windows Media Player** item can be selected from the ActiveX listbox, as shown in Fig. 2.30 (a). Press the **Create** button, a Windows Media Player control will appear on the prototype window, and the **Tag** property can be set as **activex3**, as shown in Fig. 2.30 (b). A button can be drawn beside it, with the **Tag** set to **btnFile**, and the **String** property set to empty. We can also get the standard icon ⌂, and save it to the file btnFile.bmp. With the* W = imread('btnFile.bmp') *command, the image can be loaded into MATLAB workspace as variable W, and it can be set to the **CData** property, and the prototype window can be established as shown in Fig. 2.30 (b).*
   The callback function can also be established as (see file c2mmplay.m)

```
function btnLoad_Callback(hObject, eventdata, handles)
[f,p]=uigetfile('*.*','Select a media file');
if f ~ =0, set(handles.activex3,'URL',[p,f]); end
```

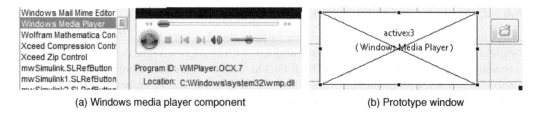

(a) Windows media player component   (b) Prototype window

Figure 2.30   Development of an interface with a Windows Media Player component.

It can be seen from the example that, with only a few simple MATLAB statements, the interface can make use of the powerful Windows Media Player facilities. Thus standard ActiveX components can provide great potential for MATLAB GUI programming.

2.9   Accelerating MATLAB Functions

2.9.1   Execution Time and Profiles of MATLAB Functions

Two sets of time-measuring commands can be used. The commands `tic` and `toc` can be used to start and read a stopwatch, respectively, to measure the execution time of the MATLAB code. The command `cputime` can also be used to read the current CPU time. The command can be invoked twice, before and after the execution of MATLAB code, to measure the execution time of the program.

Example 2.27   *The aim here is to create a 1000×1000 Hilbert matrix. Then a singular value decomposition of the matrix is performed. The execution time of the previous task can be measured with the two sets of commands; they measure 14~15s.*

```
>> tic, t=cputime;            % start watch and record current CPU time
   A=hilb(1000); [a b c]=svd(A);  % execute the MATLAB code
   toc, cputime-t             % measure the execution time and display them
```

It can be seen that the execution times measured by the two sets of commands are rather close. Since the `tic` *and* `toc` *commands do not generate additional variables, they are more practical in real programming.*

Profiling facilities are also provided in MATLAB functions, using the `profile` command. The profiles obtained by different versions of MATLAB may have some differences. The fundamental structure of the `profile` command are

```
profile on                   % start profiling facilities
the program to be measured   % execute the MATLAB program
profile report, profile off  % generate the profiling reports
```

After the profiling process, a report will be generated and displayed. For instance, the following commands can be used to measure the CtrlLAB program developed by the authors [4]

```
>> profile on;  ctrllab;  profile report
```

and the profiling results are generated as shown in Fig. 2.31. The time consumed on each statement is measured, from which the most time-consuming command can be found. The user can then modify the statements to increase the efficiency of the program.

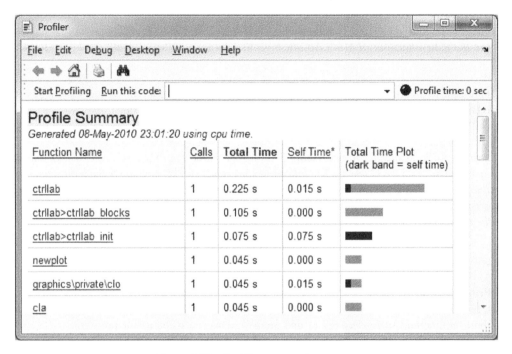

Figure 2.31   Profiling report of a program.

2.9.2   *Suggestions for Accelerating MATLAB Functions*

Since MATLAB language is an interactive and descriptive language, the efficiency is sometimes not very high. The following suggestions are made for speeding up MATLAB programs.

1) Avoid loop structures whenever possible. Loop structures are regarded as a bottleneck in MATLAB programming. To speed up MATLAB programs, the following two suggestions are made:

a) Use vectorized computation to replace loops whenever possible. An example will be given to show the difference between loop structures and vectorized computation.

Example 2.28   *Consider the evaluation of the sum*

$$I = \sum_{n=1}^{\infty} \left(\frac{1}{2^n} + \frac{1}{3^n} \right).$$

If the first 100 000 terms are examined (in practical applications, to accurately find the sum for this example, 20~30 terms are sufficient, but here the example is used for only demonstration purposes), the following loop structure can be used

```
>> tic, s=0; for i=1:100000, s=s+(1/2^i+1/3^i); end, s, toc
```

and the sum obtained is s = 1.5, and the time used is 0.063 s. We can create a vector first and, based on that, the problem can be solved in vectorized form, and the time required reduces to 0.038 s.

```
>> tic, i=1:100000; s=sum(1./2.^i+1./3.^i), toc
```

It can be seen that vectorized computation is more effective than the ordinary loop structure. However, in the new versions of MATLAB, the loop structure is significantly improved and speeded up.

b) If multiple loops are involved, and the execution time has a significant difference, it is suggested to have larger loops in the inner loop, and smaller loops can be used as the outer ones, so as to have higher efficiency.

Example 2.29   *Consider to generating a 10000×5 Hilbert matrix, and each element can be evaluated from $h_{i,j} = 1/(i + j - 1)$. The following statements can be used to try the loop sequence, when the $j=1:5$ is executed in inner loop and outer loop respectively. The time used in the two cases are respectively 0.45s and 0.09s. It can be seen that when the larger loop is placed as the inner loop, the effectiveness is much higher.*

```
>> tic, for i=1:10000, for j=1:5, H(i,j)=1/(i+j-1); end, end, toc
   tic, for j=1:5, for i=1:10000, L(i,j)=1/(i+j-1); end, end, toc
```

2) Pre-allocation of large arrays. Dynamic array allocation is a time-consuming job. It is suggested that when large matrices are involved, the size of it must first be pre-allocated, with the use of built-in functions such as `zeros()` and `ones()`. This can speed up the program.

Example 2.30   *Consider again the problem in Example 2.29. The following command measures the execution time when pre-allocation of a large matrix A is made. The time required is 0.0029s, which is faster than the method above.*

```
>> tic, H=zeros(10000,5);
   for j=1:5, for i=1:10000, H(i,j)=1/(i+j-1); end, end, toc
```

If the pre-allocation method is used together with the vectorized programming technique, the time required reduces to 0.0007s. It can be seen that with these methods, the time required has reduced from 0.45s to 0.0007s, an effectiveness increase of hundreds of times.

```
>> tic, H=zeros(10000,5); for i=1:5, H(:,i)=1./[i:i+9999]'; end, toc
```

3) Built-in functions should be considered first. Built-in functions in MATLAB are always regarded as high-quality functions. The efficiency of these functions is usually much higher than normal M-functions. They are in fact even more effective than the C code developed by ordinary users, since optimized low-level programming is implemented in the built-in functions.

4) High efficiency algorithms should be used. In practical applications, different algorithms can be used in scientific computations. For instance, two functions `quad()` and `quadl()` are provided for solving numerical integral problems, where the latter is more effective in both precision and speed than the former [5]. Thus in applications, more efficient algorithms should be adopted first.

5) Use MEX techniques if possible. If all the possible MATLAB measures are adopted, and the effectiveness of the program cannot be improved, for instance, if time-consuming loops are inevitable, then MEX techniques should be used, with C, Fortran or other languages. On the other hand, if there is already a professional program developed in another language, there is no need to rewrite it in MATLAB. A feasible solution is to write an interface to it, and to get data from MATLAB, then process it with the professional code, then write the results back to MATLAB. In this way, the efficiency of the program can be improved. MEX programming techniques will be presented later.

2.9.3   MEX Interface Design

In C, a MATLAB array data type, mxArray, is defined. It can be used to describe matrices, vectors and scalars. It can also be used to describe new data types such as structured arrays and cells. A variable A can be declared in C with the statement mxArray *A. String data type can be defined with mxChar. The related definition is provided in file mex.h, so the file should be included first. Other information regarding data types can be obtained with the following functions in C.

- **Find out the data type of a variable**, with the following C function

    mxClassID k = mxGetClassID(mxArray *ptr)

  and this function measures the actual class identifier of the mxArray variable pointed to by ptr, and the returned variable k describes the data type of the variable. In MEX C programming, the variable data types and class identifiers are given in Table 2.3.

Table 2.3   Class identifiers supported in MATLAB.

identifier	data type	class	identifier	data type	class
mxDOUBLE_CLASS	double precision	'double'	mxSINGLE_CLASS	single precision	'single'
mxINT8_CLASS	8 bit integer	'int8'	mxUINT8_CLASS	8 bit unsigned	'uint8'
mxINT16_CLASS	16 bit integer	'int16'	mxUINT16_CLASS	16 bit unsigned	'uint16'
mxINT32_CLASS	32 bit integer	'int32'	mxUINT32_CLASS	32 bit unsigned	'uint32'
mxCHAR_CLASS	string	'char'	mxSTRUCT_CLASS	structure	'struct'
mxCELL_CLASS	cell	'cell'	mxUNKNOWN_CLASS	unused	

- **Get the total number of elements in an input variable**. The number of elements of the variable pointed to by ptr can be measured with the following function

    int n = mxGetNumberOfElements(mxArray *ptr);

  where n returns the total number of elements in the variable. The value of it corresponds to the MATLAB command prod(size(A)), or length(A(:)).

- **Measure the size of an input variable**. The size of the variable pointed to by ptr can be measured with the function

    int m = mxGetNumberOfDimensions(mxArray *ptr);

  where m is the size of the multi-dimensional array. The sizes in each dimension can be measured with the function

    int *ndims = mxGetDimensions(mxArray *ptr)

  and the returned variable ndims is an integer array, where ndims[i] is the size of the $(i+1)$th dimension. The multi-dimensional array is in fact an ndims[0] $\times$ ndims[1] $\times \cdots \times$ ndims[$m-1$] array.

- **Check whether a variable is a certain class or not**. For instance, the following function checks whether the variable pointed to by `ptr` is a string or not. If it is, then the returned logical variable k is 1, otherwise it is 0.

    bool k = mxIsChar(mxArray *ptr)

  Similar functions in this category are `mxIsCell()`, `mxIsClass()` and `mxIsNaN()`, and the meaning of the function is very straightforward.

The header file `mex.h` should be included since all the related definitions are specified in this file. With those definitions, the MATLAB data types can be referenced and executed in the C language.

C compilers supporting 32 bit programming, including Microsoft Visual C++, Watcom C++ and the LCC-win32 provided in MATLAB, can be used to compile and link MEX files. Before the compiling process is started for the first time, the command `mex -setup` should be executed first. Then MATLAB command

    mex options source C program name

can be used in either MATLAB or in a DOS command box. The suffix .c should be used in the command, otherwise an error may occur. All the possible options can be listed with the `mex -h` command. For instance, the option `-c` means compile only: no executable file is to be generated. If no option is given, an executable file with the same name will be generated. To save the executable into a different file, the option `-output new_file` should be used, the new executable file is then generated in `new_file.mexw32`, and copied to the current MATLAB directory. These MEX files can be executed directly with a MATLAB command, just the same as an ordinary MATLAB function.

The MEX C file structure is shown in Fig. 2.32. Under such a structure, a MEX C interface function, named `mexFunction()`, should be prepared. Then the input and output arguments and pointers can be obtained and loaded in C. The main body of the C program can then be executed.

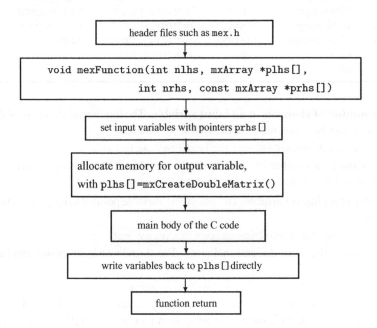

Figure 2.32   Fundamental structures of MEX files.

The results will be written back to the MATLAB environment. The MEX interface function is led by the name `mexFunction()`, and the structure of the function is fixed.

```
void mexFunction( int nlhs, mxArray *plhs[],
                  int nrhs, const mxArray *prhs[])
```

In Fig. 2.32 `nlhs` and `nrhs` are the numbers of input and output arguments, respectively. They are similar to MATLAB commands `nargout` and `nargin`. The variables `*plhs[]` and `*prhs[]` are respectively the pointers to the output and input arguments. Please note that `*prhs[0]` is the pointer of the first input argument. Note too that the index to a vector in C is different from that in MATLAB. The indexes in C start from 0, while in MATLAB the indexes start from 1. The following C functions are commonly used in MEX programming.

- **Get the numbers of rows and columns in a matrix**. The two numbers can be extracted from the functions `mxGetM()` and `mxGetN()` respectively. For instance, the dimension of the kth input variable can be obtained with `mxGetM(prhs[k−1])` and `mxGetN(prhs[k−1])`. These functions are a simplified form of the function `mxGetDimensions()`, in matrix description.

- **Get the pointers of a matrix variable**. The pointers of variables can be obtained with the `mxGetPr()` function. For instance, the pointer of the kth input variable can be obtained from `mxGetPr(prhs[k−1])`. If output variable allocation is completed, the function `mxGetPr(plhs[k−1])` can be used to get the pointer to the kth output variable.

  If the kth variable is a scalar, the functions `mxGetM()` and `mxGetN()` can be bypassed, and `mxGetPr()` can be simplified to `mxGetScalar(prhs[k−1])`.

  It is worth pointing out that, although the input variables are matrices, their C representation still has to be specified as vectors, which are rearranged from the matrix in a column-wise format.

- **Check whether a matrix is complex or not**. The `mxIsComplex(prhs[k−1])` function can be used to check if the kth input argument has complex imaginary parts. If the matrix is complex, the returned variable is 1, otherwise it is 0. The pointer of the imaginary part of the matrix can be written as `mxGetPi(prhs[k−1])`.

- **Dynamical allocation of the pointers of output arguments**. The following function can be used to allocate pointers for matrix output arguments

  ```
  plhs[k − 1] = mxCreateDoubleMatrix(mrows,ncols, mxREAL);
  ```
  where `mrows` and `ncols` are respectively the numbers of rows and columns of the matrix to be returned. The constant `mxREAL` means that the returned matrix is real. If a complex matrix is to be returned, the constant `mxCOMPLEX` should be used instead. After the function call, memory is allocated for the kth returned variable. The actual pointer of the variable can be assigned by the `mxGetPr()` function.

In the above function call, the pointers `plhs` and `prhs` are used to indicate variable pointers. In fact, the pointers can be processed by the functions and are not restricted only to them; they can be any pointer of the `mxArray` MATLAB arrays.

Example 2.31 *Assume that we want to write a matrix multiplication MEX function in C, such that the calling syntax of the function is $C = $ `c2exmex2(A, B)`. The following code can be written.*

```
#include "mex.h"
void mat_multiply(double *A, double *B, double *C,
    int mA, int nA, int mB, int nB)   /* matrix multiplication */
```

```
{
   int i,j,k,m=0;
   for (i=0; i<mA; i++){ for (j=0; j<nB; j++){ C[j*mA+i]=0;
   for (k=0; k<mB; k++) C[j*mA+i]+=A[k*mA+i]*B[j*mB+k];}}}
/* The main interface function with MEX fixed structure */
void mexFunction(int nlhs,mxArray *plhs[],
                 int nrhs,const mxArray *prhs[])
{
   double *Ap, *Bp, *Cp; int mA,nA,mB,nB,mC,nC;
   Ap=mxGetPr(prhs[0]); Bp=mxGetPr(prhs[1]);
   mA=mxGetM(prhs[0]); nA=mxGetN(prhs[0]);
   mB=mxGetM(prhs[1]); nB=mxGetN(prhs[1]);
   plhs[0]=mxCreateDoubleMatrix(mA,nB,mxREAL);
   Cp=mxGetPr(plhs[0]); mat_multiply(Ap,Bp,Cp,mA,nA,mB,nB);
}
```

where the interface function is still mexFunction(). In this function, we need to get the pointers and sizes for the input arguments *A* and *B*. Then the size of matrix *C* can be determined, and a pointer and allocation space for such a matrix can be established. The subfunction mat_multiply can then be used to perform matrix multiplication, and the result is stored to the pointer just created, thus to be returned automatically to MATLAB environment. This concludes the calling process.

We need to issue command mex c2exmex2.c under the MATLAB prompt to compile and link the program. An executable file c2exmex2.mexw32 will be created. The following statements can be used to evaluate the multiplication of different matrices.

```
>> A=[1 2 3; 4 5 6]; B=[1 2; 3 4];
   C=c2exmex2(A',B), D=A'*B, E=c2exmex2(A,B)
```

It can be seen that the two matrices, $A^T B$ and AB can be obtained as

$$C = \begin{bmatrix} 13 & 18 \\ 17 & 24 \\ 21 & 30 \end{bmatrix}, \quad D = \begin{bmatrix} 13 & 18 \\ 17 & 24 \\ 21 & 30 \end{bmatrix}, \quad E = \begin{bmatrix} 7 & 10 \\ 19 & 28 \end{bmatrix}$$

where *C* and *D* are exactly the same, although these results are obtained by different methods. It is strange to note, however, that we managed to find the product *E*, even though it is known that the matrices *A* and *B* are mathematically not compatible. There must be some peculiar things happened in the above C MEX code. Thus to avoid the problem, the compatibility of the matrices should be examined first. If they are not compatible, a warning or error messages should be given.

Error messages can be generated with the mexErrMsgTxt() function, which is very similar to the function error() in MATLAB. The program may also be terminated while calling such a function. The following program, which includes a compatibility check, can be written

```
#include "mex.h"
void mat_multiply(double *A, double *B, double *C,
                  int mA, int nA, int mB, int nB)
```

```
{ int i,j,k,m=0;
   for (i=0; i<mA; i++){ for (j=0; j<nB; j++){ C[j*mA+i]=0;
      for (k=0; k<mB; k++) C[j*mA+i]+=A[k*mA+i]*B[j*mB+k];}}}
void mexFunction(int nlhs,mxArray *plhs[],int nrhs,
                 const mxArray *prhs[])
{ double *Ap, *Bp, *Cp; int mA,nA,mB,nB,mC,nC;
  Ap=mxGetPr(prhs[0]); Bp=mxGetPr(prhs[1]);
  mA=mxGetM(prhs[0]);  nA=mxGetN(prhs[0]);
  mB=mxGetM(prhs[1]);  nB=mxGetN(prhs[1]);
  if (nA!=mB) mexErrMsgTxt("Matrix dimensions not compatible!");
  plhs[0]=mxCreateDoubleMatrix(mA,nB,mxREAL);
  Cp=mxGetPr(plhs[0]); mat_multiply(Ap, Bp, Cp, mA, nA, mB, nB);
}
```

It seems that when the compatibility check is introduced, the problems indicated earlier can be solved successfully. If the two matrices are not compatible, an error message will be displayed and the program will be terminated.

In fact, when studying carefully the modified program, it can be seen that the new program still has bugs. For instance, if either of the two matrices is a scalar, the compatibility check will gave an error message, while the two matrices are in fact compatible. Thus the program should be updated again. There are further limitations, since the matrix multiplication function thus created applies only to handling real matrices, so if either or both of the two matrices are complex, the whole program would have to be rewritten.

It can be concluded from the above example that, to write a program in C, you have to consider many tedious things. A slight carelessness may result in the program being unusable. However, when MATLAB is used, the programmers can concentrate on solving more important problems, rather than spending valuable time on trivial things like this.

In formal programming, apart from the executable file, a .m file with the same name should also be established to provide help information. For instance, a help file for this example can be established

```
function c2exmex2a()
%C2EXMEX2A is a help file for the executable function C2EXMEX2A.mexw32
```

It can be seen that there are two files with the same name under MATLAB's search path. They do not conflict at all due to MATLAB's internal mechanism. If an on-line help command is issued, the .m file is found first, but if a function call is performed, the executable file has the priority.

From the above presentation, the application procedure of the C MEX technique is summarized below:

- The function mexFunction() is the interface of the whole MEX function.
- The header functions should be included first. The function mxGetPr() can be used to get the pointers of the input arguments, and the sizes of the input arguments can be obtained with mxGetM() and mxGetN() functions. The input variables specified in MATLAB can then be transferred into the C function.
- Write the main functional part in the C MEX program.

- The function `mxCreateDoubleMatrix()` can be used to allocate memory for the returned variables. The function `mxGetPr()` can be used to assign pointers to them, so that the returned variables can be found from MATLAB.

- After the C MEX program is written, the command `mex` can be executed such that an executable file can be created.

- A .m file with the same name can be written for on-line help purposes. Such developed MEX executables can be used in the same way as other MATLAB functions.

Exercises

2.1 Start the MATLAB environment, and type in the following commands
`tic, A=rand(500); B=inv(A); norm(A*B-eye(500)), toc`
   Run the commands and observe the results. The on-line help system can be used to find more information on the commands that you are not familiar with, and interpret the above code line by line.

2.2 Enter the following two matrices into MATLAB workspace

$$A = \begin{bmatrix} 1 & 2 & 3 & 3 \\ 2 & 3 & 5 & 7 \\ 1 & 3 & 5 & 7 \\ 3 & 2 & 3 & 9 \\ 1 & 8 & 9 & 4 \end{bmatrix}, \quad B = \begin{bmatrix} 1+4i & 4 & 3 & 6 & 7 & 8 \\ 2 & 3 & 3 & 5 & 5 & 4+2i \\ 2 & 6+7i & 5 & 3 & 4 & 2 \\ 1 & 8 & 9 & 5 & 4 & 3 \end{bmatrix}.$$

Evaluate the multiplication matrix C, and assign the 2×3 submatrix at the bottom-right corner of it to matrix D. Once all these tasks are completed, observe the status of the MATLAB workspace.

2.3 Find all the integers that have a reminder of 2 for modulo 13 operation, in the range of $0 \sim 1000$.

2.4 Solve the following linear equation

$$\begin{bmatrix} 5 & 7 & 6 & 5 & 1 \\ 7 & 10 & 8 & 7 & 2 \\ 6 & 8 & 10 & 9 & 3 \\ 5 & 7 & 9 & 10 & 4 \\ 1 & 2 & 3 & 4 & 5 \end{bmatrix} X = \begin{bmatrix} 24 & 96 \\ 34 & 136 \\ 36 & 144 \\ 35 & 140 \\ 15 & 60 \end{bmatrix}.$$

2.5 Consider the matrix A in Example 2.2. If a further command `A(5,6) = 3` is given, what kind of result can be obtained? Observe the actual result and explain it.

2.6 Execute the command `A=[1 2 NaN Inf -Inf 5 NaN]`, then observe and explain what kind of result can be obtained with the following function calls: `isnan(A)`, `isfinite(A)`, `isinf(A)`, `any(A)` and `all(A)`.

2.7 Enter the following Jordanian matrix in simple ways

$$A = \begin{bmatrix} A_{11} & 0 & 0 \\ 0 & A_{22} & 0 \\ 0 & 0 & A_{33} \end{bmatrix},$$

where

$$A_{11} = \begin{bmatrix} -2 & 1 & 0 \\ 0 & -2 & 1 \\ 0 & 0 & -2 \end{bmatrix}, \quad A_{22} = \begin{bmatrix} -3 & 1 & 0 \\ 0 & -3 & 1 \\ 0 & 0 & -3 \end{bmatrix}, \quad A_{33} = \begin{bmatrix} -4 & 1 & 0 \\ 0 & -4 & 1 \\ 0 & 0 & -4 \end{bmatrix}.$$

2.8 Write an M-function to describe the following piecewise function

$$y = f(x) = \begin{cases} h, & x > D \\ h/Dx, & |x| \le D \\ -h, & x < -D. \end{cases}$$

2.9 An iterative sequence is given by

$$x_{n+1} = \frac{x_n}{2} + \frac{3}{2x_n},$$

with $x_1 = 1$; if n is large enough, the sequence may converge to a certain value X. Select a suitable n and find the steady-state value X, with the error tolerance of 10^{-14}.

2.10 Find all the prime numbers less than 1000 using a loop structure. What other methods could be adopted to solve the same problem without loops?

2.11 Find the value of S

$$S = \prod_{n=1}^{\infty} \left(1 + \frac{2}{n^2} \right),$$

with an error tolerance of $\epsilon = 10^{-12}$.

2.12 For the formula

$$\arctan(x) = x - \frac{x^3}{3} + \frac{x^5}{5} - \frac{x^7}{7} + \cdots,$$

if one selects $x = 1$, the value of π can be approximated as

$$\pi \approx 4 \left(1 - \frac{1}{3} + \frac{1}{5} - \frac{1}{7} + \frac{1}{9} - \frac{1}{11} + \cdots \right).$$

Find the approximate value with MATLAB statements, such that an error tolerance of 10^{-6} can be reached.

2.13 Evaluate the sum $K = \sum\limits_{i=0}^{63} 2^i = 1 + 2 + 2^2 + 2^3 + \cdots + 2^{62} + 2^{63}$ respectively with `for` and `while` loops, and compare the results. Find a vectorized way to solve the same problem which can avoid the use of loop structures. If the exact results cannot be obtained with the above method, what kind of measure can be taken to find accurate results?

2.14 Assume that a function $f(x)$ can be approximated by a series

$$f(x) = \lim_{N \to \infty} \sum_{n=1}^{N} (-1)^n \frac{x^{2n}}{(2n)!}.$$

If N is large enough, the function $f(x)$ converges to a certain $\hat{f}(x)$. Write a MATLAB function to draw the curve of $\hat{f}(x)$ in $x \in (0, \pi)$, and observe the curve to judge what kind of function $f(x)$ is. Validate your observation graphically.

2.15 Calculate the first 20 terms of the Fibonacci sequence defined as $a_{k+2} = a_k + a_{k+1}$, $k = 1, 2, \cdots$, with initial terms $a_1 = 1$, $a_2 = 1$, using a loop structure. Solve the same problem with a recursive function call and observe the results.

2.16 Write an M-function to implement the bisection method for solving nonlinear equation with a single variable. The bisection method is used to find the solution of a nonlinear equation $f(x) = 0$ in the interval $[a, b]$, if $f(a)f(b) < 0$.

2.17 Expand the polynomial $P(s) = (s^2 + 1)^3(s + 5)^2(s^4 + 4s^2 + 7)$ by different methods and compare the final results.

2.18 Select suitable step size to draw the following curves, and validate the results.
(a) $\sin(1/t)$, where $t \in (-1, 1)$, (b) $\sin(\tan t) - \tan(\sin t)$, where $t \in (-\pi, \pi)$.

2.19 Draw a sinusoidal curve with MATLAB, and then, with the use of the graphics editing tool, to change the line color to green, and set the width of the line to 7. Also reverse the direction of y axis downward.

2.20 Select a suitable range of θ to draw the following polar plots.
(a) $\rho = \cos(7\theta/2)$, (b) $\rho = 1 - \cos^3(7\theta)$.

2.21 Solve equations $\begin{cases} x^2 + y^2 = 3xy^2 \\ x^3 - x^2 = y^2 - y \end{cases}$ by a graphical method.

2.22 For the given iterative model

$$\begin{cases} x_{k+1} = 1 + y_k - 1.4x_k^2 \\ y_{k+1} = 0.3x_k, \end{cases}$$

with initial values $x_0 = 0$, $y_0 = 0$, write an M-function to implement the iteration for 30,000 points to form two vectors x and y, then draw all the coordinates x_k and y_k with mark `'o'`.

2.23 Draw the three-dimensional surface of the function

$$z = f(x, y) = \frac{1}{\sqrt{(1-x)^2 + y^2}} + \frac{1}{\sqrt{(1+x)^2 + y^2}}.$$

2.24 Draw the surface and contour plot of the functions xy and $\sin xy$.

2.25 Three-dimensional plots can also be decorated with pseudo color techniques, with the function `pcolor()`. Use different schemes of pseudo colors to redraw the surface in Problem 2.23.

2.26 In MATLAB graphics, if there exists `NaN` in the data, the point is omitted from the plot. In practice, `NaN` may also be introduced deliberately to omit certain parts of a plot. Consider again the surface plot in Problem 2.23, and cut off the central part described by $x^2 + y^2 < 0.5$.

2.27 Design a graphical window with its own original menu system, and set its background to green. When the left mouse is clicked, display, on the window, "Left Mouse Button".

2.28 An Excel-like control object is available in MATLAB's GUI design tool. Build a matrix processor with such a control, to accept a matrix in a visual way. Also add push buttons to find the matrix analysis results of A^{-1}, e^A and $\sin A$.

2.29 Design a graphical user interface, with an ActiveX control to show a digital clock in the window, which can display time automatically in real-time.

2.30 Use C MEX format to write a program to generate a $10,000 \times 5$ Hilbert matrix. Compare computation speed with the MATLAB implementation presented in the examples.

References

[1] CAD Center. GINO-F Users' manual, 1976

[2] The MathWorks Inc. Creating graphical user interfaces, 2001

[3] L Lamport. LaTeX: a document preparation system – user's guide and reference manual. Reading MA: Addison-Wesley Publishing Company, 2nd Edition , 1994

[4] D Xue, Y Q Chen, D P Atherton. Linear feedback control: Analysis and design with MATLAB®. Philadelphia: SIAM Press, 2007

[5] D Xue, Y Q Chen. Solving applied mathematical problems with MATLAB®. Boca Raton: CRC Press, 2008

3

MATLAB Applications in Scientific Computations

System simulation problems involve a large amount of scientific computation, and computers have the capability to solve some complicated and tedious mathematical problems. The scientific computation capacities depend on algorithms and their implementation. MATLAB is an effective and high precision scientific computation language. It can be used to solve seemingly complicated mathematical problems in an easy manner.

In this chapter, numerical computation and analytical problem solution with MATLAB are covered. Section 3.1 mainly discusses the analytical solutions and numerical solutions to mathematical problems, and explores the necessity of using numerical methods. Section 3.2 concentrates on the presentation of linear algebra problems, including the input of special matrices, matrix analysis, similarity transformation, decomposition, eigenvalue problems, algebraic equation solutions and matrix function evaluations. The use of MATLAB in the solution of linear algebra problems is very straightforward and reliable. In Section 3.3, calculus problem solution is presented, including numerical solutions on difference, differentiation, integration and multiple integral problems, as well as analytical calculus problem solutions. In Section 3.4, the fundamentals of dynamical system simulation techniques, – numerical solutions to ordinary differential equations in MATLAB, are presented. Examples will be given to show the solutions of stiff equations, implicit differential equations and differential algebraic equations. Transformation methods and analytical solutions of differential equations are also dealt with. In particular, the numerical Laplace transform technique is introduced for solving complicated differential equations. Section 3.5 presents the numerical solution methods in optimization problems. A universal nonlinear equation solver is presented, for finding with ease the possible multiple solutions to nonlinear equations, together with other approaches. Unconstrained optimization problems are explored, followed by linear programming problems and quadratic programming problems as well as ordinary nonlinear programming problems. Dynamic programming techniques and their use in path planning problem applications are discussed in Section 3.6. Section 3.7 introduces data processing methods, including one- and two-dimensional interpolation problems and least squares curve fitting problems. Data sorting, pseudo random number generator problems are also presented. Fast Fourier series transformation and spectrum analysis problem solutions are also presented. More on mathematical problem solutions with MATLAB can be found in [1].

System Simulation Techniques with MATLAB® and Simulink®, First Edition. Dingyü Xue and YangQuan Chen.
© 2014 John Wiley & Sons, Ltd. Published 2014 by John Wiley & Sons, Ltd.

3.1 Analytical and Numerical Solutions

The development of modern science and engineering is heavily dependent on mathematics. The main interests of pure mathematicians are quite different from those of scientists and engineers. Mathematicians are interested in analytical solutions, or closed-form solutions, as well as the theoretical proof of uniqueness and existence, and do not usually care what the solutions are. Engineers, on the other hand, are more interested in how to find a solution to a specific problem at hand. If the solution cannot be found easily, the engineers are interested in an approximate solution, and may not care much as to how the solution is found, as long as the solution is usable and reliable. In real applications, at least there are two cases where numerical techniques must be employed.

1) An analytical solution does not exist

It is common to see that the analytical solution to a mathematical problem may not exist. For instance, the definite integral $\int_a^b e^{-x^2/2} dx$ does not have an analytical solution. Although mathematicians have introduced a special function $\mathrm{erf}(\cdot)$ to denote the "analytical" solution to the original problem, it is useless for engineers, since they want to have an idea about what $\mathrm{erf}(0.5)$ is, rather than a mere notation. Although the exact solution does not exist, a good approximation is sufficient for engineers. In this case, numerical methods should be adopted.

Another example is that the circumference ratio π does not have an analytical solution. Chinese mathematician Zu Chongzhi (AD429–500), also known as *Tsu Ch'ung-chih*, pointed out that the value of π lies between 3.1415926 and 3.1415927. Mathematicians are still trying to get more digits of π and by October 1995, 6442450938 decimal digits had been worked out, but the solution obtained is not an analytical solution. In engineering, the value of π obtained by Zu Chongzhi is sufficient. More digits may only increase the burden for computers. In fact, for the purpose of estimation, the value 3.14 obtained by Archimedes in 250BC is also sufficient. There is no need to pursue analytical solutions here.

2) Analytical solutions exist but are not practical

For instance, the determinant of an $n \times n$ matrix can be obtained mathematically by the algebraic complement method. In this method, the determinant of an $n \times n$ matrix can be converted to the evaluation of n determinants of $(n-1) \times (n-1)$ matrices, which can further be converted to the determinants of $(n-2) \times (n-2)$ matrices, and so on. Theoretically speaking, the determinant of $n \times n$ matrix can be evaluated analytically, whatever the value of n is.

Unfortunately, the above conclusion neglects the computational load problem. The computational load for such an evaluation task could be vast, requiring $(n-1)(n+1)! + n$ arithmetic operations. For instance, when $n = 25$, the number of arithmetic operations reaches 3.7227×10^{26}. This means that the fastest mainframe computer in the world, doing 200 million billion floating-point operations per second (flops), would take 59 years. In practical engineering computations, for instance in civil engineering, we usually have to solve problems with 500×500 matrices or even much larger matrices. The analytical solution method cannot be used at all. So, numerical methods must be used instead.

Numerical solutions to scientific computational problems are used in all engineering disciplines. For instance, in mechanics, finite element methods are used to solve partial differential equations. In aerospace and control engineering, numerical linear algebra and numerical solutions to ordinary differential equations are extensively used. In simulation of engineering and non-engineering fields, numerical solutions to difference and differential equations are widely used. In hi-tech areas, the fast Fourier transform (FFT) is an essential tool for dealing with signals.

Although MATLAB can be used for solving computation problems of almost all disciplines, in this chapter, only a few essential topics, such as linear algebra, calculus, differential equations, optimization and data processing related to system simulation are covered. More on scientific computation can be found in [1].

3.2   Solutions to Linear Algebra Problems

3.2.1   Inputting Special Matrices

Before starting matrix computation, it is necessary to learn how to present various forms of matrices in MATLAB.

1) Zero, one and identity matrices

The matrix of zeros, the matrix of ones and the identity matrix can be entered with the following statements.

A = zeros(m,n), A = ones(m,n), A = eye(m,n), % $n \times m$ matrices
A = zeros(n), A = ones(n), A = eye(n),   % $n \times n$ square matrices

where m and n are respectively the rows and columns of matrix A. If B has already been specified in MATLAB, the MATLAB command A = zeros(size(B)) can be used to enter a zero matrix whose size is exactly the same as the matrix B.

The functions zeros() and ones() can also be used to enter multi-dimensional arrays in MATLAB. For instance, zeros$(3, 4, 5)$ can be used to generate a $3 \times 4 \times 5$ array whose elements are all zeros.

2) Random matrices

Random matrices are matrices whose elements are generated randomly. Commonly used random matrices can be entered by

A = rand(n,m),   % uniformly distributed in $[0, 1]$
A = $a + (b - a)$*rand(n,m),   % uniformly distributed in $[a, b]$
A = randn(n,m),   % normally distributed $N(0, 1)$
A = $\mu + \sigma$*randn(n,m),   % normally distributed $N(\mu, \sigma)$

Note that such generated random numbers are in fact pseudo random numbers, since they are generated with particular mathematical formulae. The statistical behavior of the pseudo random numbers is satisfied, but the random numbers in theory repeat.

3) Diagonal matrices

Diagonal matrices are special matrices whose diagonal terms are specified as $\alpha_1, \alpha_2, \cdots, \alpha_n$. The mathematical form of the diagonal matrix is diag$(\alpha_1, \alpha_2, \cdots, \alpha_n)$.

In MATLAB, a vector $v = [\alpha_1, \alpha_2, \cdots, \alpha_n]$ can be specified first, then the diagonal matrix can be entered into MATLAB with the following syntaxes

V = diag(v),   % generate diagonal matrix V for given vector v
v = diag(V),   % extract diagonal vector v from given matrix V
V = diag(v,k),   % matrix whose kth diagonal elements are in vector v

4) The Hilbert matrix and its inverse

The Hilbert matrix is a special kind of matrix whose (i, j)th element is specified by $h_{i,j} = 1/(i + j - 1)$. An $n \times n$ square Hilbert matrix can be written as

$$H = \begin{bmatrix} 1 & 1/2 & 1/3 & \cdots & 1/n \\ 1/2 & 1/3 & 1/4 & \cdots & 1/(n+1) \\ \vdots & \vdots & \vdots & \ddots & \vdots \\ 1/n & 1/(n+1) & 1/(n+2) & \cdots & 1/(2n-1) \end{bmatrix}. \tag{3.1}$$

The Hilbert matrix can be specified in MATLAB with the function A = hilb(n). When n is large, the Hilbert matrix is known to be an ill-conditioned matrix. The inverse Hilbert matrix can be obtained with $\boldsymbol{B} = \texttt{invhilb}(n)$, but the function is not ideal for large n's. In this case, symbolic computations can be performed, if necessary.

5) Companion matrix

Suppose that a monic polynomial is specified as

$$P(s) = s^n + a_1 s^{n-1} + a_2 s^{n-2} + \cdots + a_{n-1} s + a_n. \tag{3.2}$$

Its corresponding companion matrix can be written as

$$\boldsymbol{A}_{\mathrm{c}} = \begin{bmatrix} -a_1 & -a_2 & \cdots & -a_{n-1} & -a_n \\ 1 & 0 & \cdots & 0 & 0 \\ 0 & 1 & \cdots & 0 & 0 \\ \vdots & \vdots & \ddots & \vdots & \vdots \\ 0 & 0 & \cdots & 1 & 0 \end{bmatrix}. \tag{3.3}$$

Companion matrices can be entered into MATLAB with the command $\boldsymbol{B} = \texttt{compan}(\boldsymbol{p})$, where $\boldsymbol{p}$ is a vector containing the coefficients of a monic polynomial in descending order of s.

6) Hankel matrices

The Hankel matrix is a symmetric matrix, and its elements in each back diagonal are the same. The Hilbert matrix is a special Hankel matrix. The standard form of a Hankel matrix is

$$\boldsymbol{H} = \begin{bmatrix} c_1 & c_2 & \cdots & c_n \\ c_2 & c_3 & \cdots & c_{n+1} \\ \vdots & \vdots & \ddots & \vdots \\ c_n & c_{n+1} & \cdots & c_{2n-1} \end{bmatrix}. \tag{3.4}$$

In MATLAB, given a vector $\boldsymbol{c}$, a Hankel matrix can be generated with $\texttt{hankel}(\boldsymbol{c})$. The generated matrix is a triangular matrix. An alternative MATLAB function call can be issued with $\boldsymbol{H} = \texttt{hankel}(\boldsymbol{c}, \boldsymbol{r})$ to specify the vectors $\boldsymbol{c}$ and $\boldsymbol{r}$. It is required that the first element in $\boldsymbol{r}$ is the same as the last element in vector $\boldsymbol{c}$.

Example 3.1   *Hankel matrices can be generated with the commands*

```
>> C=[1,2,3]; H1=hankel(C),
   C=[1,2,3]; R=[3,9,10,11,12,13]; H2=hankel(C,R)
```

and two Hankel matrices can be generated as

$$\boldsymbol{H}_1 = \begin{bmatrix} 1 & 2 & 3 \\ 2 & 3 & 0 \\ 3 & 0 & 0 \end{bmatrix}, \quad \boldsymbol{H}_2 = \begin{bmatrix} 1 & 2 & 3 & 9 & 10 & 11 \\ 2 & 3 & 9 & 10 & 11 & 12 \\ 3 & 9 & 10 & 11 & 12 & 13 \end{bmatrix}.$$

7) Vandermonde matrices

For a given vector $c = [c_1, c_2, \cdots, c_n]^T$, the corresponding Vandermonde matrix is defined such that the (i, j)th element is $v_{i,j} = c_i^{n-j}, i, j = 1, 2, \cdots, n$

$$V = \begin{bmatrix} c_1^{n-1} & c_1^{n-2} & \cdots & c_1 & 1 \\ c_2^{n-1} & c_2^{n-2} & \cdots & c_2 & 1 \\ \vdots & \vdots & \ddots & \vdots & \vdots \\ c_n^{n-1} & c_n^{n-2} & \cdots & c_n & 1 \end{bmatrix}. \tag{3.5}$$

For a given vector c, the Vandermonde matrix can be entered into MATLAB with the command $V = $ vander(c).

8) Matrix data type conversion

Symbolic matrices can be used to perform matrix analysis in an analytical way, while the double matrix in MATLAB can be used in numerical matrix analysis. The two commonly used data types can be converted with the statements $B = $ sym(A) and $A = $ double(B), where A is a numerical matrix while B is a symbolic matrix.

3.2.2 Matrix Analysis and Computation

MATLAB provides a large number of matrix analysis and computation functions. Here, some basic matrix analysis functions are presented. Unless specifically indicated, the functions illustrated in the following can be used for finding both numerical and analytical solutions.

1) Determinant

The determinant of matrix $A = \{a_{ij}\}$ is defined as

$$D = |A| = \det(A) = \sum_{k_1 k_2 \cdots k_n} (-1)^k a_{1k_1} a_{2k_2} \cdots a_{nk_n}, \tag{3.6}$$

where $k_1, k_2, \cdots, k_n$ is the permutation of the entities $1, 2, \cdots, n$. There are many ways of finding the determinants of a matrix. The LU factorization method is implemented and used in MATLAB. A MATLAB built-in function is provided $d = $ det(A) to find the determinant of matrix A.

Example 3.2 *Assuming that a matrix is given by*

$$A = \begin{bmatrix} 1 & 2 & 3 \\ 4 & 5 & 6 \\ 7 & 8 & 0 \end{bmatrix},$$

the following statements can be used to find that the determinant is 27.

```
>> A=[1,2,3; 4 5 6; 7 8 0]; det(A)
```

2) Trace

For a given square matrix $A = \{a_{ij}\}, i, j = 1, 2, \cdots, n$, its trace is defined as the sum of its diagonal elements

$$\text{tr}(A) = \sum_{i=1}^{n} a_{ii}. \tag{3.7}$$

The trace of the matrix is the same as the sum of its eigenvalues. In MATLAB, the trace can be evaluated with `trace(A)`.

3) Rank

If a matrix has a maximum number r_c of linearly independent columns, r_c is referred to as the column rank of A. Similarly the row rank r_r can also be defined. It can be shown that when these two ranks are the same, $\text{rank}(A) = r_c = r_r$ is referred to as the rank of the matrix.

It is extremely difficult to find the rank of large-scale rectangular matrix A. In MATLAB, the rank of a matrix A is evaluated using the singular value decomposition algorithm [2]. A built-in function is provided in MATLAB to find the rank of matrix A, with `rank(A)`. Sometimes, numerical methods can be used to find the rank of the matrix with `trace(A,ε)`, where ε is the error tolerance. For the matrix used in Example 3.2, the rank of the matrix can be found with the `rank(A)` command, and the result is 3.

4) The norms of matrices

Before introducing the norms of a matrix, we need to look at the norms of a vector.

For a vector x, if there exists a scalar $\rho(x)$, such that the following three conditions are satisfied:

a) $\rho(x) \geq 0$, and $\rho(x) = 0$ if and only is $x = 0$.

b) $\rho(ax) = |a|\rho(x)$, where a is any scalar.

c) For vectors x and y, $\rho(x + y) \leq \rho(x) + \rho(y)$.

the value $\rho(x)$ is referred to as the norm of the vector x. There are various forms of norms for a given vector. It can be seen that the following set of quantities satisfy the above mentioned three conditions:

$$||x||_p = \left(\sum_{i=1}^{n} |x_i|^p\right)^{1/p}, p = 1, 2, \cdots, \text{ and } ||x||_\infty = \max_{1 \leq i \leq n} |x_i|, \tag{3.8}$$

and the notation $||x||_p$ is used to indicate p norms.

The definition of a matrix norm is little bit involved. For an arbitrarily given nonzero vector x, the norms of the matrix A are defined as

$$||A|| = \sup_{x \neq 0} \frac{||Ax||}{||x||}. \tag{3.9}$$

The commonly used matrix norms are defined as

$$||A||_1 = \max_{1 \leq j \leq n} \sum_{i=1}^{n} |a_{ij}|, ||A||_2 = \sqrt{s_{\max}(A^T A)}, ||A||_\infty = \max_{1 \leq i \leq n} \sum_{j=1}^{n} |a_{ij}|, \tag{3.10}$$

where $s(X)$ is the eigenvalues of matrix X, and $s_{\max}(A^T A)$ denotes the maximum eigenvalue of matrix $A^T A$. In fact, $||A||_2$ is equal to the maximum singular value of A.

MATLAB provides a function `norm()` which can be used to evaluate different norms of the matrix, with $N = \text{norm}(A, \text{options})$, and the options can be 1, 2, `inf` and `'fro'`, corresponding respectively to $||A||_1, ||A||_2, ||A||_\infty$ and the Frobinius norm: $||A||_F = \sqrt{\text{tr}(A^T A)}$. The norms of the matrix given in Example 3.2 can be obtained with the following commands

```
>> [norm(A,2), norm(A,1), norm(A,Inf), norm(A,'fro')]
```

with $||A||_2 = 13.2015$, $||A||_1 = 15$, $||A||_\infty = 15$, $||A||_F = 14.2829$.

It should be noted that the function `norm()` can only be used for evaluating the norms of a matrix containing numbers only. If the matrix is specified as a symbolic matrix, the `double()` function should be called first (in earlier versions of MATLAB) to convert it to a double-precision matrix.

5) Characteristic polynomial and eigenvalues

Constructing a matrix $s\boldsymbol{I} - \boldsymbol{A}$, its determinant is a polynomial $C(s)$, expressed by

$$C(s) = \det(s\boldsymbol{I} - \boldsymbol{A}) = s^n + c_1 s^{n-1} + \cdots + c_{n-1} s + c_n, \tag{3.11}$$

and the polynomial $C(s)$ is referred to as the characteristic polynomial of matrix $\boldsymbol{A}$, and the coefficients $c_i, i = 1, 2, \cdots, n$ are referred to as the coefficients of the characteristic polynomial.

MATLAB provides a function `poly()` to find the characteristic polynomial of a matrix, with the syntax $\boldsymbol{c} = $ `poly(`$\boldsymbol{A}$`)`, and the returned variable $\boldsymbol{c}$ is a row vector, containing the coefficients of the polynomial. An alternative way of calling the function `poly()` is that when $\boldsymbol{A}$ is a vector, a matrix can be established.

■ **Example 3.3** *Considering the matrix $\boldsymbol{A}$ given in Example 3.2, the function* `poly(`$\boldsymbol{A}$`)` *can be used to find its characteristic polynomial, and the vector $\boldsymbol{B} = [1, -6, -72, -27]^{\mathrm{T}}$ can be obtained*

```
>> A=[1,2,3; 4 5 6; 7 8 0]; B=poly(A)
```

From the results, it can be seen that the characteristic polynomial is $P(s) = s^3 - 60s^2 - 72s - 27$. In fact, the directly obtained characteristic polynomials may have an error, since numerical methods were used. The relative error of the results obtained is 5.6538×10^{-15}.

```
>> P=[1, -6 -72, -27]; norm((P-B)./P)
```

From the above result, it can be seen that, for the function `poly()` *provided in MATLAB, slight numerical error may exist in the function. The help information for* `poly.m` *points out that the iterative function in* `eig()` *is used, hence the small errors.*

In real applications, there are other better algorithms for finding the characteristic polynomial coefficients. For instance, the Leverrier–Faddeev recursive algorithm can be used

$$c_{k+1} = -\frac{1}{k}\mathrm{tr}(\boldsymbol{A}\boldsymbol{R}_k), \ \boldsymbol{R}_{k+1} = \boldsymbol{A}\boldsymbol{R}_k + c_{k+1}\boldsymbol{I}, \quad k = 1, \cdots, n, \tag{3.12}$$

where the initial values are assigned as $\boldsymbol{R}_1 = \boldsymbol{I}$, and $c_1 = 1$. In this algorithm, an identity matrix $\boldsymbol{I}$ is created and assigned to $\boldsymbol{R}_1$ first. Then the polynomial coefficients $c_1, c_2, \cdots, c_n$ can be found recursively. The MATLAB implementation of the algorithm can be written as

```
function c=poly1(A)
[nr,nc]=size(A);
if nc==nr, I=eye(nc); R=I; c=[1 zeros(1,nc)];
    for k=1:nc, c(k+1)=-1/k*trace(A*R); R=A*R+c(k+1)*I; end
else, error('Argument must be a square matrix.'); end
```

With the use of the new `poly1(`$\boldsymbol{A}$`)` function, accurate characteristic polynomial coefficients can be obtained, numerically.

The eigenvalues of a matrix A can also be obtained by solving the polynomial equation $C(s) = 0$, with MATLAB function $V = \texttt{roots}(p)$. For instance, the eigenvalues of the matrix given in Example 3.2 can be obtained with the following MATLAB statements, and the three eigenvalues are 12.1229, -5.7345, -0.3884.

```
>> A=[1,2,3; 4 5 6; 7 8 0]; c=poly1(A); roots(c)
```

6) Polynomials and polynomial matrix evaluation

Polynomials can be evaluated with $C = \texttt{polyval}(a,x)$, where a is the coefficient vector of the polynomial: $a = [a_1, a_2, \cdots, a_n, a_{n+1}]$.

A polynomial matrix is mathematically defined as

$$B = a_1 A^n + a_2 A^{n-1} + \cdots + a_n A + a_{n+1} I, \tag{3.13}$$

where I is an identity matrix of the same size as matrix A. It can be evaluated with the function $B = \texttt{polyvalm}(a, A)$.

Example 3.4 *The Cayley–Hamilton theorem is very important in matrix theory. It states that, if the characteristic polynomial of the matrix A can be written as*

$$\lambda(s) = \det(sI - A) = a_1 s^n + a_2 s^{n-1} + \cdots + a_n s + a_{n+1}, \tag{3.14}$$

then, $\lambda(A) = 0$, that is

$$a_1 A^n + a_2 A^{n-1} + \cdots + a_n A + a_{n+1} I = 0. \tag{3.15}$$

Assume that matrix A is given in Example 3.2. The following statements can be used to validate the Cayley–Hamilton theorem.

```
>> A=[1,2,3; 4,5,6; 7,8,0]; aa=poly(A); B=polyvalm(aa,A); norm(B)
```

It can be seen that the norm of the error matrix is 2.9932×10^{-13}, which is very small. The small error comes from function `poly()`. *Note that if the function* `poly1()` *is used instead, the error matrix is a zero matrix. This validates the Cayley–Hamilton theorem.*

```
>> aa1=poly1(A); B1=polyvalm(aa1,A); norm(B1)
```

3.2.3 *Inverse and Pseudo Inverse of Matrices*

For a square $n \times n$ nonsingular matrix A, if there exists a matrix C of the same size, satisfying

$$AC = CA = I, \tag{3.16}$$

where I is an identity matrix, matrix C is referred to as the inverse matrix of A, denoted by $C = A^{-1}$. MATLAB provides a matrix inverse function inv(), with the syntax $B = \texttt{inv}(A)$.

■ **Example 3.5** *Consider again the matrix in Example 3.2. Its inverse matrix can be obtained by either a numerical method or an analytical method, and the results are*

$$B = \begin{bmatrix} -1.7778 & 0.88889 & -0.11111 \\ 1.5556 & -0.77778 & 0.22222 \\ -0.11111 & 0.22222 & -0.11111 \end{bmatrix}, \quad C = \begin{bmatrix} -16/9 & 8/9 & -1/9 \\ 14/9 & -7/9 & 2/9 \\ -1/9 & 2/9 & -1/9 \end{bmatrix}.$$

```
>> A=[1,2,3; 4,5,6; 7,8,0]; B=inv(A), C=inv(sym(A)), norm(A*B-eye(3))
```

The inverse matrix obtained by the numerical method has an error of 1.9984×10^{-15}.

If the original matrix A is singular and/or rectangular, the concept of a generalized inverse matrix should be used. If there exists a matrix N, such that

$$ANA = A, \tag{3.17}$$

then N is referred to as the generalized inverse matrix of A, denoted by $N = A^-$. However, it can be shown that there are an infinite number of such N matrices. To make it unique, the following minimum norm criterion is introduced

$$\min_{M} \|AMA - A\|, \tag{3.18}$$

and it can be shown that there is a unique matrix M, satisfying simultaneously the following three conditions:
1) $AMA = A$.
2) $MAM = M$.
3) AM and MA are both Hermitian symmetric matrices.
The matrix M is referred to as the Moore–Penrose generalized inverse, or pseudo inverse, of the matrix A, denoted by $M = A^+$. If A is an $n \times m$ rectangular matrix, then M is an $m \times n$ matrix.
The Moore–Penrose generalized inverse matrix of A matrix can be obtained with the MATLAB function $B = $ `pinv(A)`. If the A matrix is nonsingular, the pseudo inverse becomes its inverse matrix.

■ **Example 3.6** *For a rectangular matrix*

$$A = \begin{bmatrix} 6 & 1 & 4 & 2 & 1 \\ 3 & 0 & 1 & 4 & 2 \\ -3 & -2 & -5 & 8 & 4 \end{bmatrix},$$

the Moore–Penrose generalized inverse matrix can be obtained with the following statements, and the three conditions are validated, since the norms of error matrices are all around 10^{-14}.

$$A^+ = \begin{bmatrix} 0.073025 & 0.041301 & -0.022147 \\ 0.010774 & 0.0019952 & -0.015563 \\ 0.04589 & 0.017757 & -0.038508 \\ 0.032721 & 0.043097 & 0.063847 \\ 0.016361 & 0.021548 & 0.031923 \end{bmatrix}.$$

```
>> A=[6,1,4,2,1; 3,0,1,4,2; -3,-2,-5,8,4]; rank(A)
   iA=pinv(A) % pseudo inverse and it is validated
   norm(iA-iA*A*iA), norm(A*iA*A-A), norm(iA*A-A'*iA'),
   norm(A*iA-iA'*A')
```

Taking the Moore–Penrose generalized inverse for the above obtained matrix, it can be seen that the original matrix A is restored, from which it can be concluded that $(A^+)^+ = A$, with the norm of error matrix 9.3256×10^{-15}.

```
>> iiA=pinv(iA), norm(iiA-A)
```

3.2.4 Similarity Transform and Decomposition of Matrices

3.2.4.1 Similarity Transform and Orthogonal Matrices

For an $n \times n$ square matrix A, if there exists a nonsingular matrix T of the same size, the original matrix A can be transformed

$$\widehat{A} = T^{-1}AT. \tag{3.19}$$

This kind of transformation is referred to as the similarity transformation of A. It can be seen that the eigenvalues of the transformed matrix $\widehat{A}$ are identical to those of matrix A, that is, the similarity transformation does not change the eigen structure of the original matrix.

For a class of special transformation matrix T, if it satisfies $T^{-1} = T^*$, where T^* is the Hermitian transpose of T, then T is referred to as an orthogonal matrix, denoted by $Q = T$. The orthogonal matrix Q satisfies the following conditions

$$Q^*Q = I, \text{ and } Q Q^* = I, \tag{3.20}$$

where I is an $n \times n$ identity matrix.

There is a special kind of orthogonal matrix. If matrix A is not a full-rank matrix, and Z is an orthogonal matrix such that $AZ = 0$, then matrix Z is referred to as a null space, which can be used to find the basic set of solutions of linear homogeneous equations.

MATLAB provides the functions orth() and null() to find the orthogonal matrix and the null space matrix, respectively. The syntaxes of the two functions are $Q = $ orth(A) and $Z = $ null(A), where the former finds the orthogonal basis, while the latter finds the null space.

■ **Example 3.7** *Consider again the matrix in Example 3.2. The following results can be obtained with the following statements, and the result is validated where $\|Q^T Q - I\| = 5.6023 \times 10^{-16}$, $\|Q Q^T - I\| = 5.1660 \times 10^{-16}$.*

$$Q = \begin{bmatrix} -0.23036 & -0.39607 & -0.88886 \\ -0.60728 & -0.65521 & 0.44934 \\ -0.76036 & 0.6433 & -0.089596 \end{bmatrix}.$$

```
>> A=[1,2,3; 4 5 6; 7 8 0]; Q=orth(A), I=eye(3);
   norm(Q*Q'-I), norm(Q'*Q-I)
```

3.2.4.2   Triangular Factorization

Triangular factorization of matrices is also called LU factorization, since the original matrix can be expressed as the product of a lower triangular matrix L and an upper triangular matrix U, that is $A = LU$, where L and U matrices can be written as

$$L = \begin{bmatrix} 1 & & & \\ l_{21} & 1 & & \\ \vdots & \vdots & \ddots & \\ l_{n1} & l_{n2} & \cdots & 1 \end{bmatrix}, \quad U = \begin{bmatrix} u_{11} & u_{12} & \cdots & u_{1n} \\ & u_{22} & \cdots & u_{2n} \\ & & \ddots & \vdots \\ & & & u_{nn} \end{bmatrix}. \tag{3.21}$$

In MATLAB, LU factorization can be performed with `lu()` function

```
[L,U]=lu(A),      % simple syntax
[L,U,P]=lu(A),    % syntax with permutation matrix P
```

where L and U are respectively the lower and upper triangular matrices. Since pivot selections are considered in the `lu()` function, sometimes the matrix L obtained is not really a lower triangular matrix, and the permutation matrix P is also returned such that $A = P^{-1}LU$.

The `lu()` function is a built-in file, and cannot be used to handle symbolic matrices. Based on the LU factorization algorithm summarized in [1], a function `lusym()` for symbolic variables is written below. The listing of it is written as

```
function [L,U]=lusym(A)
n=length(A); U=sym(zeros(size(A))); L=sym(eye(size(A)));
U(1,:)=A(1,:); L(:,1)=A(:,1)/U(1,1);
for i=2:n,
    for j=2:i-1, L(i,j)=(A(i,j)-L(i,1:j-1)*U(1:j-1,j))/U(j,j); end
    for j=i:n, U(i,j)=A(i,j)-L(i,1:i-1)*U(1:i-1,j); end
end
```

Note that, since no pivot selection is involved, the function may sometimes fail due to the fact that 0 may be used as a denominator. One can improve this function by considering pivot when necessary. The syntax of the new function is $[L, U] =$ `lusym`(A).

■ **Example 3.8**   *Considering again the LU factorization problem in Example 3.2, two methods can be used to solve the problem, with the following statements*

$$L_1 = \begin{bmatrix} 0.14286 & 1 & 0 \\ 0.57143 & 0.5 & 1 \\ 1 & 0 & 0 \end{bmatrix}, \quad U_1 = \begin{bmatrix} 7 & 8 & 0 \\ 0 & 0.85714 & 3 \\ 0 & 0 & 4.5 \end{bmatrix}$$

$$L = \begin{bmatrix} 1 & 0 & 0 \\ 0.14286 & 1 & 0 \\ 0.57143 & 0.5 & 1 \end{bmatrix}, \quad U = \begin{bmatrix} 7 & 8 & 0 \\ 0 & 0.85714 & 3 \\ 0 & 0 & 4.5 \end{bmatrix}, \quad P = \begin{bmatrix} 0 & 0 & 1 \\ 1 & 0 & 0 \\ 0 & 1 & 0 \end{bmatrix}.$$

```
>> A=[1,2,3; 4,5,6; 7,8,0]; [L1,U1]=lu(A), [L,U,P]=lu(A)
```

Note that since **P** *is not an identity matrix, the pivot selection method is arranged for permutation of the original matrix. Thus the* **L** *matrix is not a lower triangular matrix. In the latter case,* **LU** $\neq$ **A**.

With the new symbolic LU factorization function, the following statements can be used and the result shown below is obtained

$$L = \begin{bmatrix} 1 & 0 & 0 \\ 4 & 1 & 0 \\ 7 & 2 & 1 \end{bmatrix}, \quad U = \begin{bmatrix} 1 & 2 & 3 \\ 0 & -3 & -6 \\ 0 & 0 & -9 \end{bmatrix}.$$

```
>> A=[1,2,3; 4,5,6; 7,8,0]; [L,U]=lusym(A)
```

3.2.4.3   Cholesky Factorization of Symmetric Matrices

If matrix A is symmetric, LU factorization can also be performed. In this case, $L = U^T$. Let $D = L$ be a lower triangular matrix, matrix A can be expressed by

$$A = DD^T = \begin{bmatrix} d_{11} & & & \\ d_{21} & d_{22} & & \\ \vdots & \vdots & \ddots & \\ d_{n1} & d_{n2} & \cdots & d_{nn} \end{bmatrix} \begin{bmatrix} d_{11} & d_{21} & \cdots & d_{n1} \\ & d_{22} & \cdots & d_{n2} \\ & & \ddots & \vdots \\ & & & d_{nn} \end{bmatrix} \tag{3.22}$$

where D can be understood as the square root of the original matrix A. This factorization approach is also known as Cholesky factorization.

In MATLAB, function `chol()` is provided to perform the Cholesky factorization to find the matrix D of the original matrix, with $[D, P] = $ `chol`(A), where the returned variable D is the Cholesky factorization matrix, such that $A = DD^T$. $P - 1$ returns the size of positive definite submatrix in matrix A. If A is a positive definite matrix, $P = 0$. The syntax $D = $ `chol`(A) can be used to check whether a matrix is positive definite or not. If it is not, an error message is given. In Reference [1], a function `cholsym()` was written for processing symbolic symmetric matrices.

```
function L=cholsym(A)
n=length(A); L(1,1)=sqrt(A(1,1)); L(2:n,1)=A(2:n,1)/L(1,1);
for i=2:n, k=1:i-1; L(i,i)=sqrt(A(i,i)-sum(L(i,k).^2));
   for j=i+1:n, L(j,i)=(A(j,i)-sum(L(j,k).*L(i,k)))/L(i,i); end
end
```

Example 3.9   *Consider a* 3×3 *Hilbert matrix* **A**. *The following MATLAB commands can be issued, and the Cholesky factorization matrix can be found with* $P = 0$. *This means that matrix* **A** *is a positive definite matrix.*

$$D = \begin{bmatrix} 1 & 0.5 & 0.33333 \\ 0 & 0.28868 & 0.28868 \\ 0 & 0 & 0.074536 \end{bmatrix}, \quad D_1 = \begin{bmatrix} 1 & 0 & 0 \\ 1/2 & \sqrt{3}/6 & 0 \\ 1/3 & \sqrt{3}/6 & \sqrt{5}/30 \end{bmatrix}.$$

```
>> A=hilb(3); [D,P]=chol(A), D1=cholsym(sym(A))
```

3.2.4.4 Singular Value Decomposition

Singular values of a matrix can be regarded as a measure of the matrix. For an $n \times m$ matrix A

$$A^T A \geq 0, A A^T \geq 0, \tag{3.23}$$

and $\text{rank}(A^T A) = \text{rank}(A A^T) = \text{rank}(A)$. It can further be shown that the $A^T A$ and $A A^T$ matrices have the same nonnegative eigenvalues λ_i. In mathematics, the square roots of these eigenvalues are referred to as the singular values of matrix A, denoted as $\sigma_i(A) = \sqrt{\lambda_i(A^T A)}$.

Assume that A is an $n \times m$ matrix, and $\text{rank}(A) = r$. A can be decomposed as

$$A = L \begin{bmatrix} \Delta & 0 \\ 0 & 0 \end{bmatrix} M^T, \tag{3.24}$$

where L and M are orthogonal matrices, and $\Delta = \text{diag}(\sigma_1, \cdots, \sigma_r)$ is a diagonal matrix whose diagonal elements $\sigma_1, \sigma_2, \cdots, \sigma_r$ satisfy the inequality

$$\sigma_1 \geq \sigma_2 \geq \cdots \geq \sigma_r > 0. \tag{3.25}$$

In MATLAB, the function svd() is provided with the syntax $[L, A_1, M]$=svd(A), where A is the original matrix, and in the returned variables, A_1 is a diagonal matrix, while L and M are orthogonal matrices, satisfying $A = L A_1 M^T$.

The value of singular values normally determines the status of the matrix. If the difference in the singular values varies too much, a slight change in one element of the matrix may lead to a significant change in the properties of the matrix. This type of matrix is often referred to as a ill-conditioned matrix. If there exist zero singular values, the matrix is then a singular matrix. The ratio of maximum singular value $\sigma_{\max}$ to the minimum singular value $\sigma_{\min}$ is referred as the condition number, denoted by $\text{cond}(A)$, that is, $\text{cond}(A) = \sigma_{\max}/\sigma_{\min}$. The maximum and minimum singular values are often denoted by $\bar{\sigma}(A)$ and $\underline{\sigma}(A)$. In MATLAB, function cond(A) can be used to evaluate the condition number of matrix A.

Example 3.10 *Consider again the matrix A studied in Example 3.2. The following MATLAB statements can be used to perform singular value decomposition of the matrix. The matrices L, A_1 and M can be obtained easily, and the condition number can be found.*

```
>> A=[1, 2, 3; 4, 5, 6; 7, 8, 0]; [L, A1, M]=svd(A)
   B=A'*A; C=sqrt(eig(B)); [cond(A), A1(1)/A1(end) C(end)/C(1)]
```

The decomposed matrices are obtained as follows, and the condition number is 35.1059.

$$L = \begin{bmatrix} -0.23036 & -0.39607 & -0.88886 \\ -0.60728 & -0.65521 & 0.44934 \\ -0.76036 & 0.6433 & -0.089596 \end{bmatrix}, \quad A_1 = \begin{bmatrix} 13.201 & 0 & 0 \\ 0 & 5.4388 & 0 \\ 0 & 0 & 0.37605 \end{bmatrix},$$

$$M = \begin{bmatrix} -0.60463 & 0.27326 & 0.74817 \\ -0.72568 & 0.19824 & -0.65886 \\ -0.32836 & -0.94129 & 0.078432 \end{bmatrix}.$$

Example 3.11   *For a rectangular matrix A*

$$A = \begin{bmatrix} 1 & 3 & 5 & 7 \\ 2 & 4 & 6 & 8 \end{bmatrix},$$

singular value decomposition can be performed with the svd() *function, with the following MATLAB statements*

$$L = \begin{bmatrix} -0.6414 & -0.7672 \\ -0.7672 & 0.6414 \end{bmatrix}, \quad A_1 = \begin{bmatrix} 14.269 & 0 & 0 & 0 \\ 0 & 0.6268 & 0 & 0 \end{bmatrix},$$

$$M = \begin{bmatrix} -0.1525 & 0.8227 & -0.3945 & -0.3800 \\ -0.3499 & 0.4214 & 0.2428 & 0.8007 \\ -0.5474 & 0.0201 & 0.6979 & -0.4614 \\ -0.7448 & -0.3812 & -0.5462 & 0.0407 \end{bmatrix}.$$

```
>> A=[1, 3, 5, 7; 2, 4, 6, 8]; [L,A1,M]=svd(A), norm(L*A*M-A1)
```

It can be seen that LAM^T *can basically transform the original matrix A to a diagonal matrix* A_1, *with the relatively small error of* 7.0358×10^{-15}.

3.2.5   Eigenvalues and Eigenvectors of Matrices

For a given matrix A, if there exists a nonzero vector x and a scalar λ satisfying

$$Ax = \lambda x, \tag{3.26}$$

then λ is referred to as an eigenvalue of matrix A, while the vector x is referred to as the eigenvector of A corresponding to the eigenvalue λ. A MATLAB function eig() can be used to find the eigenvalues and eigenvectors of the matrix, with the syntaxes

```
d=eig(A),        % the eigenvalues are returned in the column vector d
[V,D] =eig(A),   % returns eigenvalues D and eigenvector matrix V.
```

Example 3.12   *Consider the matrix A in Example 3.2. The two syntaxes of the* eig() *function can be used to find the eigenvalues and eigenvectors*

$$v = \begin{bmatrix} -0.2998 & -0.7471 & -0.2763 \\ -0.7075 & 0.6582 & -0.3884 \\ -0.6400 & -0.0931 & 0.8791 \end{bmatrix}, \quad d = \begin{bmatrix} 12.123 & 0 & 0 \\ 0 & -0.3884 & 0 \\ 0 & 0 & -5.7345 \end{bmatrix},$$

$$d_1 = \begin{bmatrix} 12.123 \\ -0.3884 \\ -5.7345 \end{bmatrix}.$$

```
>> A=[1,2,3; 4,5,6; 7,8,0]; [v,d]=eig(A), d1=eig(A)
```

3.2.6 Solution of Matrix Equations

In this section, solutions to linear algebraic equations, Lyapunov equations, Sylvester equations and Riccati quadratic equations are presented.

3.2.6.1 Solution of Linear Algebraic Equations

Consider the linear algebraic equation described as

$$Ax = B, \tag{3.27}$$

where A and B are compatible matrices

$$A = \begin{bmatrix} a_{11} & a_{12} & \cdots & a_{1n} \\ a_{21} & a_{22} & \cdots & a_{2n} \\ \vdots & \vdots & \ddots & \vdots \\ a_{m1} & a_{m2} & \cdots & a_{mn} \end{bmatrix}, \quad B = \begin{bmatrix} b_{11} & b_{12} & \cdots & b_{1p} \\ b_{21} & b_{22} & \cdots & b_{2p} \\ \vdots & \vdots & \ddots & \vdots \\ b_{m1} & b_{m2} & \cdots & b_{mp} \end{bmatrix}. \tag{3.28}$$

From matrix theory, it is known that three possibilities in the solutions should be considered: unique solution, infinite number of solutions and no solution.

3.2.6.2 The equation has a unique solution

If A is a nonsingular square matrix, the unique solution to the original equation can be obtained

$$x = A^{-1}B. \tag{3.29}$$

The unique solution to the linear algebraic equation can be obtained with the MATLAB command $x = \text{inv}(A) * B$, and the analytical solution to the equation can be obtained from $x = \text{inv}(\text{sym}(A)) * B$.

If A is not a nonsingular square matrix, a solution judgement matrix C can be constructed from the matrices A B such that

$$C = \begin{bmatrix} a_{11} & a_{12} & \cdots & a_{1n} & b_{11} & b_{12} & \cdots & b_{1p} \\ a_{21} & a_{22} & \cdots & a_{2n} & b_{21} & b_{22} & \cdots & b_{2p} \\ \vdots & \vdots & \ddots & \vdots & \vdots & \vdots & \ddots & \vdots \\ a_{m1} & a_{m2} & \cdots & a_{mn} & b_{m1} & b_{m2} & \cdots & b_{mp} \end{bmatrix}. \tag{3.30}$$

3.2.6.3 The equation has an infinite number of solutions

If $\text{rank}(A) = \text{rank}(C) = r < n$, the equation (3.27) has an infinite number of solutions. With the null space facilities, the basic set of solutions x_i, $i = 1, 2, \cdots, n - r$ can be found. A general solution to the corresponding homogeneous equation $AX = 0$ can be constructed as

$$\hat{x} = \alpha_1 x_1 + \alpha_2 x_2 + \cdots + \alpha_{n-r} x_{n-r}, \tag{3.31}$$

where the coefficients α_i, $i = 1, 2, \cdots, n - r$ are arbitrary constants. The null space of the matrix can be found with $Z = \text{null}(A)$. Function $\text{null}()$ can also be used in finding analytical solutions.

We can also find a particular solution x_0 to (3.27), with the $x_0 = \text{pinv}(A) * B$ command. The general solution to the original equation can be expressed as $x = \hat{x} + x_0$.

Example 3.13 *Solve the following linear algebraic equation*

$$\begin{bmatrix} 1 & 2 & 3 & 4 \\ 2 & 2 & 1 & 1 \\ 2 & 4 & 6 & 8 \\ 4 & 4 & 2 & 2 \end{bmatrix} X = \begin{bmatrix} 1 \\ 3 \\ 2 \\ 6 \end{bmatrix}.$$

We can first enter matrices A and B into the MATLAB workspace. It can be found that matrix A is singular. Thus C can be constructed, and we can find that the ranks of the two matrices are the same, both equal to 2, which means that the original equation has an infinite number of solutions.

```
>> A=[1 2 3 4; 2 2 1 1; 2 4 6 8; 4 4 2 2]; B=[1;3;2;6];
   C=[A B]; [rank(A), rank(C)]
```

The symbolic method can be used to construct the general form of the infinite number of solutions

```
>> Z=null(sym(A)), x0=sym(pinv(A))*B, syms a1 a2; x=Z*[a1; a2]+x0
```

From the basic set of solutions, the general solution can be found with x

$$Z = \begin{bmatrix} 0 & 1 \\ 1 & 0 \\ -6 & -7 \\ 4 & 5 \end{bmatrix}, \quad x_0 = \frac{1}{131}\begin{bmatrix} 125 \\ 96 \\ -10 \\ -39 \end{bmatrix},$$

$$x = a_1 \begin{bmatrix} 0 \\ 1 \\ -6 \\ 4 \end{bmatrix} + a_2 \begin{bmatrix} 1 \\ 0 \\ -7 \\ 5 \end{bmatrix} + \frac{1}{131}\begin{bmatrix} 125 \\ 96 \\ -10 \\ -39 \end{bmatrix} = \begin{bmatrix} a_2 + 125/131 \\ a_1 + 96/131 \\ -6a_1 - 7a_2 - 10/131 \\ 4a_1 + 5a_2 - 39/131 \end{bmatrix}$$

where a_1 and a_2 are arbitrary constants.

The reduced row echelon method to matrix C can be performed with the MATLAB function `rref()`, *so that the analytical solution to the linear equation can be found as*

$$C_1 = \begin{bmatrix} 1 & 0 & -2 & -3 & 2 \\ 0 & 1 & 5/2 & 7/2 & -1/2 \\ 0 & 0 & 0 & 0 & 0 \\ 0 & 0 & 0 & 0 & 0 \end{bmatrix}.$$

```
>> C1=rref(sym(C))
```

From the above result, if we let $x_3 = b_1$, $x_4 = b_2$, where b_1, b_2 are arbitrary constants, the analytical solution of the equation can be written as

$$x_1 = 2b_1 + 3b_2 + 2, \quad x_2 = -5b_1/2 - 7b_2/2 - 1/2.$$

3.2.6.4 No solution to the equation

If rank(A) < rank(C), the equation in (3.27) is an inconsistent equation. There is no solution to such an equation. The Moore–Penrose generalized inverse can be used to solve the original equation $x = \text{pinv}(A) * B$. The solution x does not satisfy the original equation, while it can minimize the norm of the error matrix $||Ax - B||$.

Example 3.14 *If matrix B in the previous example is changed to $B = [1, 2, 3, 4]^{\text{T}}$, the following commands can be used*

```
>> B=[1:4]'; C=[A B]; [rank(A), rank(C)]
```

It can be seen that rank(A) = 2, *while* rank(C) = 3. *This means that the original equation is an inconsistent equation. There is no solution to it. With the Moore–Penrose generalized inverse, the least squares solution to the original equation can be obtained with*

$$x = \begin{bmatrix} 0.5465648855 \\ 0.4549618321 \\ 0.04427480916 \\ -0.04732824427 \end{bmatrix}, \text{ with error } \begin{bmatrix} 0.4 \\ 8.8818 \times 10^{-16} \\ -0.2 \\ 1.7764 \times 10^{-15} \end{bmatrix}.$$

```
>> x=pinv(A)*B, e=A*x-B
```

Obviously the solution obtained does not satisfy the original equation, while it can minimize the overall error, that is the norm of the error matrix.

3.2.6.5 Kronecker Product and Solutions of Matrix Equations

Consider again the following linear algebra equation

$$AX = C, \tag{3.32}$$

where A is an $n \times n$ matrix, while C is an $n \times m$ matrix. Of course, the original equation can still be solved with $X = A^{-1}C$, and the MATLAB solution can be found from $X = \text{inv}(A) * C$. Here an alternative method is introduced.

In order to introduce the Kronecker product in the solution process, the matrices can be represented as a one-dimensional array, such that

$$X = \begin{bmatrix} x_1 & x_2 & \cdots & x_m \\ x_{m+1} & x_{m+2} & \cdots & x_{2m} \\ \vdots & \vdots & \ddots & \vdots \\ x_{(n-1)m+1} & x_{(n-1)m+2} & \cdots & x_{nm} \end{bmatrix}, \quad C = \begin{bmatrix} c_1 & c_2 & \cdots & c_m \\ c_{m+1} & c_{m+2} & \cdots & c_{2m} \\ \vdots & \vdots & \ddots & \vdots \\ c_{(n-1)m+1} & c_{(n-1)m+2} & \cdots & c_{nm} \end{bmatrix}, \tag{3.33}$$

With such defined matrices, the original equation can be transformed into the following form, with the unknown matrix X transformed to a vector x

$$(A \otimes I_m)x = c, \tag{3.34}$$

where $\otimes$ is the Kronecker product operator, and x and c are column vectors

$$x^{\mathrm{T}} = [x_1 x_2 \cdots x_{nm}], \quad c^{\mathrm{T}} = [c_1 c_2 \cdots c_{nm}]. \tag{3.35}$$

With knowledge of the Kronecker product and its application in the transformation of equations, more complicated equations can also be solved in a similar way. Consider the generalized Lyapunov equation, also known as the Sylvester equation

$$AX + XB = -C, \tag{3.36}$$

where A is an $n \times n$ matrix, and B is an $m \times m$ matrix, and X is an $n \times m$ matrix. With the Kronecker product, the above equation can be transformed into the following form

$$(A \otimes I_m + I_n \otimes B^{\mathrm{T}})x = -c. \tag{3.37}$$

Based on the above algorithm, the following MATLAB function can be written, so that the analytical solutions to the Sylvester equations can be found.

```
function X=lyapsym(A,B,C)
if nargin==2, C=B; B=A'; end
[n,m]=size(C); A0=kron(A,eye(m))+kron(eye(n),B');
try, C1=C'; x0=-inv(A0)*C1(:); X=reshape(x0,m,n)';
catch, error('singular matrix found.'), end
```

Various of Lyapunov equations can be solved analytically by the above code, with the syntaxes

$X = \mathrm{lyapsym}(A, C)$ % Lyapunov $AX + XA^{\mathrm{T}} = -C$
$X = \mathrm{lyapsym}(A), -\mathrm{inv}(A'), Q^*\mathrm{inv}(A')$ % discrete $AXA^{\mathrm{T}} - X + Q = 0$
$X = \mathrm{lyapsym}(A, B, C)$ % Sylvester $AX + XB = -C$

A numerical solution function `lyap()` to those equations is provided in the MATLAB Control System Toolbox. The syntaxes for the three equations are the same.

Example 3.15 *Assume that the matrices A, B and C in (3.36) are given as*

$$A = \begin{bmatrix} 1 & 2 & 3 \\ 4 & 5 & 6 \\ 7 & 8 & 0 \end{bmatrix}, \quad B = A^{\mathrm{T}}, \quad C = \begin{bmatrix} 1 & 5 & 4 \\ 5 & 6 & 7 \\ 4 & 7 & 9 \end{bmatrix}.$$

The following statements can be used to solve the Lyapunov equation

```
>> A=[1 2 3;4 5 6; 7 8 0];  B=A'; C=[1, 5, 4; 5, 6, 7; 4, 7, 9];
   X1=lyap(A,C), X2=lyap(sym(A),C), norm(A*X1+X1*A'-C)
```

and the numerical and analytical solutions are obtained as

$$X_1 = \begin{bmatrix} 1.5556 & -1.1111 & 0.38889 \\ -1.1111 & 1.2222 & 0.22222 \\ 0.38889 & 0.22222 & 0.38889 \end{bmatrix}, \quad X_2 = \begin{bmatrix} 14/9 & -10/9 & 7/18 \\ -10/9 & 11/9 & 2/9 \\ 7/18 & 2/9 & 7/18 \end{bmatrix},$$

and the error in the numerical solution is 1.1597×10^{-14}.

3.2.6.6   Solutions of Riccati Equations

The following quadratic equation is referred to as an algebraic Riccati equation (ARE)

$$A^{\mathrm{T}}X + XA - XBX + C = 0, \tag{3.38}$$

where A, B and C are known matrices, and B is nonnegative definite symmetric, and C is symmetric. Algebraic Riccati equations can be solved directly with the `are()` function in the MATLAB Control System Toolbox, with the syntax $X = $ `are`(A, B, C), and the solution X is also a symmetric matrix.

Example 3.16   *Consider the following Riccati equation*

$$\begin{bmatrix} -2 & -1 & 0 \\ 1 & 0 & -1 \\ -3 & -2 & -2 \end{bmatrix} X + X \begin{bmatrix} -2 & 1 & -3 \\ -1 & 0 & -2 \\ 0 & -1 & -2 \end{bmatrix} - X \begin{bmatrix} 2 & 2 & -2 \\ -1 & 5 & -2 \\ -1 & 1 & 2 \end{bmatrix} X + \begin{bmatrix} 5 & -4 & 4 \\ 1 & 0 & 4 \\ 1 & -1 & 5 \end{bmatrix} = 0.$$

Comparing the above equation with the standard form in (3.38), it can be seen that

$$A = \begin{bmatrix} -2 & 1 & -3 \\ -1 & 0 & -2 \\ 0 & -1 & -2 \end{bmatrix}, \quad B = \begin{bmatrix} 2 & 2 & -2 \\ -1 & 5 & -2 \\ -1 & 1 & 2 \end{bmatrix}, \quad C = \begin{bmatrix} 5 & -4 & 4 \\ 1 & 0 & 4 \\ 1 & -1 & 5 \end{bmatrix}.$$

The following statements can be used to solve the ARE. The result is also validated, with an error of 1.6008×10^{-14}.

$$X = \begin{bmatrix} 0.98739 & -0.79833 & 0.41887 \\ 0.57741 & -0.13079 & 0.57755 \\ -0.28405 & -0.073037 & 0.69241 \end{bmatrix}.$$

```
>> A=[-2,1,-3; -1,0,-2; 0,-1,-2]; B=[2,2,-2; -1 5 -2; -1 1 2];
   C=[5 -4 4; 1 0 4; 1 -1 5]; X=are(A,B,C); norm(A'*X+X*A-X*B*X+C)
```

The above solution leads to other questions: How many solutions are there for the quadratic equation and how can we find all of them? If the typical form of the equation is changed, are we able to solve them? The questions will be answered in the Section 3.5.

3.2.7   Nonlinear Matrix Functions

Two types of nonlinear operations on matrices are supported in MATLAB. One is the element-wise nonlinear operation, which is similar to the dot operation in MATLAB. In this book, we shall use the term nonlinear function for this type of nonlinear operation. The other type is referred to as a matrix function, and it performs nonlinear operations on the whole matrix. In this section, these two types of nonlinear operation are presented.

3.2.7.1   Nonlinear Function Evaluations

MATLAB provides many functions for evaluating nonlinear function of matrices. Some of the commonly used ones are given in Table 3.1. The syntax of these functions is quite straightforward:

```
B = fun_name(A),    % e.g., B = sin(A);
```

Table 3.1   Element-wise nonlinear functions of matrices.

Function name	Explanation	Function name	Explanation
abs()	absolute value	asin(), acos()	arc sine and cosine
sqrt()	square root	log(), log10()	logarithmic
exp()	exponential	real(), imag(), conj()	for complex
sin(), cos()	sine, cosine	round(), floor(), ceil(), fix()	extract integers

3.2.7.2   Nonlinear Matrix Functions

Apart from the element-wise nonlinear function evaluation, we can also evaluate matrix functions. For instance, the exponential function can be evaluated with different algorithms. For instance, Reference [3] presents 19 different algorithms. In MATLAB, function expm() is provided, and it can also be used for the symbolic A matrix exponential.

Example 3.17   *Consider a Jordanian block matrix*

$$A = \begin{bmatrix} -2 & 1 & 0 & & \\ 0 & -2 & 1 & & \\ 0 & 0 & -2 & & \\ & & & -5 & 1 \\ & & & 0 & -5 \end{bmatrix}.$$

To get the exponential function, the following statements can be issued

$$e^A = \begin{bmatrix} 0.1353 & 0.1353 & 0.06767 & 0 & 0 \\ 0 & 0.1353 & 0.1353 & 0 & 0 \\ 0 & 0 & 0.1353 & 0 & 0 \\ 0 & 0 & 0 & 0.006738 & 0.006738 \\ 0 & 0 & 0 & 0 & 0.006738 \end{bmatrix}.$$

```
>> A=[-2 1 0; 0 -2 1; 0 0 -2]; A(4:5,4:5)=[-5 1; 0 -5]; expm(A)
```

The exponential functions e^A and e^{At} can also be obtained directly with

$$e^A = \begin{bmatrix} e^{-2} & e^{-2} & e^{-2}/2 & 0 & 0 \\ 0 & e^{-2} & e^{-2} & 0 & 0 \\ 0 & 0 & e^{-2} & 0 & 0 \\ 0 & 0 & 0 & e^{-5} & e^{-5} \\ 0 & 0 & 0 & 0 & e^{-5} \end{bmatrix}, \quad e^{At} = \begin{bmatrix} e^{-2t} & te^{-2t} & t^2e^{-2t}/2 & 0 & 0 \\ 0 & e^{-2t} & te^{-2t} & 0 & 0 \\ 0 & 0 & e^{-2t} & 0 & 0 \\ 0 & 0 & 0 & e^{-5t} & te^{-5t} \\ 0 & 0 & 0 & 0 & e^{-5t} \end{bmatrix}.$$

```
>> expm(sym(A)), syms t; expm(A*t)
```

With the powerful exponential function expm(), sinusoidal and cosine matrix functions can also be evaluated. For instance, the Euler formula can be used to realize the following operations, which are also applicable to matrices

$$\sin A = \frac{1}{j2}\left(e^{jA} - e^{-jA}\right), \quad \cos A = \frac{1}{2}\left(e^{jA} + e^{-jA}\right). \tag{3.39}$$

Example 3.18   *For the matrix*

$$A = \begin{bmatrix} -5 & 1 & 1 \\ 1 & -4 & 1 \\ -1 & -1 & -6 \end{bmatrix},$$

the following statements can be used to evaluate $\cos At$

```
>> A=[-5,1,1; 1,-4,1; -1,-1,-6]; syms t; j=sqrt(-1);
   cA=(expm(j*A*t)+expm(-j*A*t))/2; simple(cA)
```

which yields

$$\cos At = \begin{bmatrix} \cos 5t & t\sin 5t & t\sin 5t \\ t\sin 5t & \cos 5t + t\sin 5t - t^2\cos 5t/2 & t\sin 5t - t^2\cos 5t/2 \\ -t\sin 5t & -t\sin 5t + t^2\cos 5t/2 & \cos 5t + t^2\cos 5t/2 - t\sin 5t \end{bmatrix}.$$

Apart from exponential and other related functions, other functions can also be evaluated. There are some numerical solution functions such as `logm()`, `sqrtm()`, and a general-purpose function `funm()` that can be directly used. However, for matrices with repeated eigenvalues, these functions may have errors. An overload function `funm()` for symbolic computation is provided in Reference [1] and it can be used to deal with even more complicated matrix functions, such as $e^{A\cos At}$.

3.3   Solutions of Calculus Problems

3.3.1   *Analytical Solutions to Calculus Problems*

Many functions for solving calculus problems are provided in the Symbolic Math Toolbox, and they can be used for easily solving limit, differentiation, integration, Taylor series expansion and series problems. The syntaxes for the related function are

```
L=limit(f,x,x₀)        % evaluate limits
d=diff(f,x,n)          % high-order derivative dⁿf/dxⁿ
I=int(f,x)             % indefinite integration
I=int(f,x,a,b)         % evaluate ∫ᵇₐ f(x)dx, and a, b can be set to inf
f₁=taylor(f,x,n,a)     % Taylor series expansion
s=symsum(fₙ,n,a₁,a_f)  % evaluate series ∑ᵃ_f_ₙ₌ₐ₁ fₙ
```

The following examples are used to illustrate the use of these functions to solve calculus problems.

Example 3.19   *The limit problem*

$$\lim_{x\to\infty}\left(\frac{3x^2 - x + 1}{2x^2 + x + 1}\right)^{x^3/(1-x)}$$

can be solved with the following statements, and the result is 0.

```
>> limit(((3*x^2-x+1)/(2*x^2+x+1))^(x^3/(1-x)),x,inf)
```

Example 3.20 *For the function* $y(x) = \dfrac{\sin x}{x^2 + 4x + 3}$, *the following statements can be used to evaluate its second-order derivative*

$$2\frac{(2x+4)^2 \sin x}{\left(x^2+4x+3\right)^3} - 2\frac{(2x+4)\cos x}{\left(x^2+4x+3\right)^2} - 2\frac{\sin x}{\left(x^2+4x+3\right)^2} - \frac{\sin x}{x^2+4x+3}.$$

```
>> syms x; y=sin(x)/(x^2+4*x+3); y1=diff(y,x,2)
```

If the result is integrated twice, the original function can be restored.

```
>> y2=int(int(y1,x),x), y2=simple(y2)
```

The following MATLAB function can be used to find the first 10 terms of Taylor series expansion of the original function $f(x)$

$$\frac{1}{3}x - \frac{4}{9}x^2 + \frac{23}{54}x^3 - \frac{34}{81}x^4 + \frac{4087}{9720}x^5 - \frac{3067}{7290}x^6 + \frac{515273}{1224720}x^7 - \frac{386459}{918540}x^8 + \frac{37100281}{88179840}x^9.$$

```
>> y4=taylor(y,x,10)
```

Example 3.21 *The infinite series*

$$I = 2\sum_{n=1}^{\infty}\left(\frac{1}{2^n} + \frac{1}{3^n}\right)$$

can be obtained with the following MATLAB statements, with the sum 3.

```
>> syms n; 2*symsum(1/2^n+1/3^n,n,1,inf)
```

A more complicated series problem

$$I = 2\sum_{n=0}^{\infty}\frac{1}{(2n+1)(2x+1)^{2n+1}}$$

can be obtained with the following statements, and the simplified result is $\ln[(x+1)/x]$.

```
>> syms n x; s1=2*symsum(1/((2*n+1)*(2*x+1)^(2*n+1)),n,0,inf), simple(s1)
```

Example 3.22 *The double integral problem*

$$J = \int_{-1}^{1}\int_{-\sqrt{1-y^2}}^{\sqrt{1-y^2}} e^{-x^2/2}\sin(x^2+y)\mathrm{d}x\mathrm{d}y$$

can be solved with the following statements

```
>> syms x y; f1=exp(-x^2/2)*sin(x^2+y);    % define the integrand
   f2=int(f1,x,-sqrt(1-y^2),sqrt(1-y^2))   % inner integral
   f3=int(f2,-1,1); vpa(f3)                % double integral
```

and "Warning: Explicit integral could not be found" is displayed. This means that the analytical solution of the original problem cannot be found. The command vpa() *can be used to find the high precision numerical solution 0.53686038269880787557759384929130.*

3.3.2 Numerical Difference and Differentiation

3.3.2.1 Numerical Difference

For the vector y given by $\{y_i\}, i = 1, 2, \cdots, n$, difference evaluation can be performed with the MATLAB function diff(), and the syntax $y_1 = \text{diff}(y)$ is straightforward. The function diff() can be used to get the new vector $\{y_{i+1} - y_i\}, i = 1, 2, \cdots, n - 1$. It can be seen that the length of the new vector y_1 is one less than the original vector y.

3.3.2.2 Numerical Differentiation

Commonly used numerical differentiation evaluation methods include the forward difference method, the backward difference method and the central-point difference method.

- The first-order forward difference method is defined as

$$y_i' = \frac{\Delta y_i}{\Delta t} = \frac{y_{i+1} - y_i}{\Delta t}. \tag{3.40}$$

- The first-order backward difference method is defined as

$$y_i' = \frac{\Delta y_i}{\Delta t} = \frac{y_i - y_{i-1}}{\Delta t}. \tag{3.41}$$

The accuracy of the two methods is of $o(\Delta t)$. A high-precision central-point difference method with an accuracy of $o(\Delta t^4)$ is given by

$$
\begin{aligned}
y_i' &= \frac{-y_{i+2} + 8y_{i+1} - 8y_{i-1} + y_{i-2}}{12\Delta t} \\[2mm]
y_i'' &= \frac{-y_{i+2} + 16y_{i+1} - 30y_i + 16y_{i-1} - y_{i-2}}{12\Delta t^2} \\[2mm]
y_i''' &= \frac{-y_{i+3} + 8y_{i+2} - 13y_{i+1} + 13y_{i-1} - 8y_{i-2} + y_{i-3}}{8\Delta t^3} \\[2mm]
y_i^{(4)} &= \frac{-y_{i+3} + 12y_{i+2} - 39y_{i+1} + 56y_i - 39y_{i-1} + 12y_{i-2} - y_{i-3}}{6\Delta t^4}.
\end{aligned}
\tag{3.42}
$$

Based on the central-point differentiation algorithm, a MATLAB function can be written

```
function [dy,dx]=diff_centre(y,h,n,key)
yx1=[y 0 0 0 0 0]; yx2=[0 y 0 0 0 0]; yx3=[0 0 y 0 0 0];
yx4=[0 0 0 y 0 0]; yx5=[0 0 0 0 y 0]; yx6=[0 0 0 0 0 y];
switch n
case 1
    dy=(-diff(yx1)+7*diff(yx2)+7*diff(yx3)-diff(yx4))/(12*h); L1=4;
case 2
    dy=(-diff(yx1)+15*diff(yx2)-15*diff(yx3)+diff(yx4))/(12*h^2); L1=4;
case 3
    dy=(-diff(yx1)+7*diff(yx2)-6*diff(yx3)-6*diff(yx4)+...
        7*diff(yx5)-diff(yx6))/(8*h^3); L1=6;
case 4
    dy=(-diff(yx1)+11*diff(yx2)-28*diff(yx3)+28*diff(yx4)-...
        11*diff(yx5)+diff(yx6))/(6*h^4); L1=6;
end
dy=dy(L1:end-L1-1); dx=(([1:length(dy)]+L1-3-(n>2))*h;
```

The syntax of the function is $[d_y, d_x] = \texttt{diff\_centre}(y, h, n)$, where y is a given set of data points, h is the step size and n is the expected order between 1~4. With these parameters, the nth order numerical differentiation $d_y\, d_x$ can be found.

▉ Example 3.23   *For the given function*

$$f(x) = \frac{\sin x}{x + \cos 2x},$$

selecting a step size of $h = 0.05$, a set of sample points can be generated. With these points, numerical differentiation can be obtained with the following statements

```
>> h=0.05; x0=0:h:pi; y=sin(x0)./(x0+cos(2*x0));
   [y1,x1]=diff_centre(y,h,1); [y2,x2]=diff_centre(y,h,2);
   [y3,x3]=diff_centre(y,h,3); [y4,x4]=diff_centre(y,h,4);
```

The following statements can be used to find the analytical solutions of the original function.

```
>> syms x; f=sin(x)/(x+cos(2*x));
   ya=diff(f); y10=subs(ya,x,x0); ya=diff(f,x,2); y20=subs(ya,x,x0);
   ya=diff(f,x,3); y30=subs(ya,x,x0); ya=diff(f,x,4); y40=subs(ya,x,x0);
   subplot(221),plot(x0,y10,x1,y1,':')
   subplot(222),plot(x0,y20,x2,y2,':')
   subplot(223),plot(x0,y30,x3,y3,':')
   subplot(224),plot(x0,y40,x4,y4,':')
```

The analytical solution and numerical solutions to the differentiation problem can be obtained as shown in Fig. 3.1. It can be seen that the numerical differentiation results are vary close to the exact ones. In fact, they cannot be distinguished from the plots.

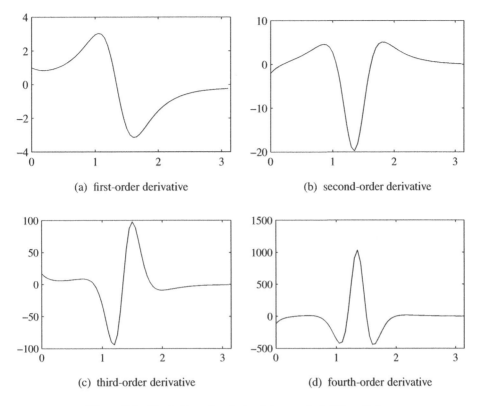

(a) first-order derivative

(b) second-order derivative

(c) third-order derivative

(d) fourth-order derivative

Figure 3.1   Illustration of central-point numerical differentiation.

3.3.3   Numerical Integration

Consider the definite integral

$$I = \int_a^b f(x)\mathrm{d}x. \tag{3.43}$$

If the integrand $f(x)$ is very complicated, even with the symbolic function `int()`, the analytical solutions cannot be found. Numerical methods should be used. There are many numerical integration methods such as the trapezoidal method, the Simpson method and the Romberg method. The fundamental idea of the integration algorithm is to divide the interval $[a, b]$ into several subintervals $[x_i, x_{i+1}], i = 1, 2, \cdots, N$, where $x_1 = a, x_{N+1} = b$. The integral problem can be approximated by

$$\int_a^b f(x)\mathrm{d}x \approx \sum_{i=1}^{N} \int_{x_i}^{x_{i+1}} f(x)\mathrm{d}x. \tag{3.44}$$

There are several effective numerical integration functions such as `quad()`, `quadl()`, `quadgk()` and `quadv()`. In MATLAB 8.0, a new function `integral()` is provided and recommended, and the syntax is $y = $ `integral(Fun, a, b)`, where `Fun` is the MATLAB representation of the integrand function. There are three ways of representing the integrand function: the M-function method, the anonymous function method and the inline function method. The variables a and b are the lower and upper bound of the definite integral. The vectorized numerical integral will be demonstrated later.

■ **Example 3.24** *For the definite integral problem* $\dfrac{1}{\sqrt{2\pi}} \displaystyle\int_{-\infty}^{\infty} \mathrm{e}^{-x^2/2}\mathrm{d}x$, *the following method can be used in solving the problem.*

From calculus it is known that there is no analytical solution for the definite integral. Three methods can be used to describe the integrand function:

1) Anonymous function *This can be written* $f = $ `@(x)1/sqrt(2*pi)*exp(-x.^;2/2)`, *where x is the independent variable, and the integrand can be expressed in the function, with dot operation notations used. The variable f is used in the numerical solver.*

2) M-function *For the integrand, this is written as*
`function y=myerrf(x), y=1/sqrt(2*pi)*exp(-x.^;2/2);`

3) Inline function *This is written as* $f = $ `inline('1/sqrt(2*pi)*exp(-x.^;2/2)','x')`. *It can be seen that the method is not as good as the anonymous function method and it is not recommended in computation.*

The `integral()` *function can be used to deal directly with infinite integrals as well. The following statements can be used to evaluate the integral, and the result is 1.000000000000038.*

```
>> format long; f=@(x)1/sqrt(2*pi)*exp(-x.^2/2); y=integral(f,-inf,inf)
```

3.3.4 Numerical Multiple Integration

Consider the double integral problem

$$I = \int_{y_\mathrm{m}}^{y_\mathrm{M}} \int_{x_\mathrm{m}}^{x_\mathrm{M}} f(x,y)\mathrm{d}x\mathrm{d}y. \tag{3.45}$$

A new function `integral2()` is provided in MATLAB 8.0 to evaluate the double integral problem, with the syntax $I = $ `integral2(Fun,`x_m`,`x_M`,`y_m`,`y_M`)`. In earlier versions, `dblquad()` with the same syntax can be used.

■ **Example 3.25** *For the double integral problem*

$$J = \int_{-1}^{1} \int_{-2}^{2} \mathrm{e}^{-x^2/2} \sin(x^2 + y)\mathrm{d}x\mathrm{d}y,$$

an anonymous function can be written to describe the integrand function, and the integral can be numerically evaluated with the following statements, and the result is 1.574498159253890.

```
>> f=@(x,y)exp(-x.^2/2).*sin(x.^2+y); I=integral2(f,-2,2,-1,1)
```

Unfortunately, the functions provided in MATLAB cannot be used to solve the following double integral problem when the area is not rectangular

$$I = \int_{y_\mathrm{m}}^{y_\mathrm{M}} \int_{x_\mathrm{m}(y)}^{x_\mathrm{M}(y)} f(x,y)\mathrm{d}x\mathrm{d}y. \tag{3.46}$$

Alternatively, the freely downloadable Numerical Integration Toolbox (NIT), developed by Howard Wilson and Bryce Gardner, can be used to solve this kind of problem. The function gquad2dggen() can be used to solve the problem in (3.46), with the syntax

$I =$ quad2dggen(Fun, Fun$_m$, Fun$_M$, y_m, y_M, ϵ)

Three user functions should be written to describe the integrand and the two boundary functions.

■ Example 3.26   *Consider the double integral problem studied in Example 3.22*

$$J = \int_{-1}^{1} \int_{-\sqrt{1-y^2}}^{\sqrt{1-y^2}} e^{-x^2/2} \sin(x^2 + y) \mathrm{d}x\mathrm{d}y.$$

The following statements can be used to describe the three user functions for the particular problem, and the integral can be found directly. The result is 0.5369. Of course, it is not possible to make the accuracy as high as with symbolic computation, but the advantage is that it can be used to solve any double integration problems.

```
>> Fun=@(x,y)exp(-x.^2/2).*sin(x.^2+y); fm=@(y)-sqrt(1-y.^2);
   fM=@(y)sqrt(1-y.^2); y=quad2dggen(Fun,fm,fM,-1,1)
```

The NIT toolbox also provides a function quadndg(), which can be used to solve an n-dimensional integral with hyper-rectangular regions. The efficiency of the simple integral solver quadg is much higher than that of the MATLAB functions such as quadl().

3.4   Solutions of Ordinary Differential Equations

Numerical solutions to ordinary differential equations (ODEs) are, in fact, the fundamentals of dynamical system simulation techniques. Assume that the vector first-order differential equation is described by

$$\dot{x}_i = f_i(t, \boldsymbol{x}), \quad i = 1, 2, \cdots, n, \tag{3.47}$$

where $\boldsymbol{x}$ is the state variable vector, $\boldsymbol{x} = [x_1, x_2, \cdots, x_n]^{\mathrm{T}}$, t is the time variable, $\boldsymbol{x}(0)$ is the initial state vector and n is the order of the system. The functions $f_i(\cdot)$ are arbitrary nonlinear functions. Numerical algorithms can be used to solve the ODEs.

There are various of algorithms for numerical ODE solutions. The commonly used ones are the Euler algorithm, the Runge–Kutta algorithm, the Adams multi-step method and the Gear algorithm. There are also many algorithms for stiff equations and other types of ODEs. In this section, MATLAB-based solutions for different kinds of ODEs are presented.

3.4.1   Numerical Methods of Ordinary Differential Equations

For a better understanding of the numerical solution to ODE problems, the simplest Euler algorithm is presented. For simplicity, assume that the vector form of the ODE is written as

$$\dot{\boldsymbol{x}} = \boldsymbol{f}(t, \boldsymbol{x}), \tag{3.48}$$

where $f(\cdot) = [f_1(\cdot), f_2(\cdot), \cdots, f_n(\cdot)]^T$. Assume that, at initial time t_0, the initial state vector is given by x_0. Selecting a step size h, the state vector at time $t_0 + h$ can be approximated by

$$x(t_0 + h) = \hat{x}(t_0 + h) + R_0 = x_0 + h f(t, x_0) + R_0. \tag{3.49}$$

Denote simply $x_1 = x(t_0 + h)$, $\hat{x}_1 = \hat{x}(t_0 + h)$ is the approximate vector x_1 at time $t_0 + h$. The approximation is known as a numerical solution of the ODEs. R_0 is the numerical error in the solution.

Assume that, at time t_k, the state vector is written as x_k, then the numerical solution at time $t_k + h$ can be written as

$$x_{k+1} = x_k + h f(t_k, x_k). \tag{3.50}$$

Thus an iterative method can be used to find the numerical solutions to the ODE in the interval $t \in [0, T]$, at time instances $t_0 + h, t_0 + 2h, \cdots$.

It can be seen that one feasible way to increase accuracy is to reduce the step size h. However, this approach is not always feasible, since we cannot indefinitely reduce the step size for two main reasons:

- **Computation load**. For a fixed interval $[0, T]$, reducing the step size means an increase in the number of points in computation. If the step size is selected too small, the computation load required may be too heavy to be practical.
- **Accumulative error**. No matter how small a step size is selected, there are still errors in approximation in each step. Increasing the number of computation points means an increased cumulative error.

Thus, in dynamical system simulation, the following precautions should be taken:

- **Select a suitable step size**. As in the case of the Euler algorithm, the step size cannot be selected too large or too small. However, it is very difficult to select an appropriate size. This method is not very realistic.
- **Use improved algorithms**. Since the Euler algorithm only has accuracy of $o(h)$, other high-precision algorithms should be used, such as the Runge–Kutta algorithm, which has an accuracy of $o(h^4)$, or others such as the Adams algorithm.
- **Use a variable step size approach**. The phrase "suitable step size" has been mentioned previously. It is very difficult to find one that is suitable. In fact, many ODE solvers support a variable size methodology. When the error is large, smaller step size should be chosen, while when the error is small, larger step size should be selected.

3.4.2 MATLAB Solutions to ODE Problems

In MATLAB, several ODE solvers are provided, such as `ode23()`, `ode45()`, `ode15s()` and `ode113()`, where Runge–Kutta–Felhberg, Adams–Bashforth–Moulton, are implemented. These methods all support variable step size schemes. The syntaxes of the functions are the same

$[t,x]$ = ode23(Fun,tspan,x_0,options,additional parameters)
$[t,x]$ = ode45(Fun,tspan,x_0,options,additional parameters)
$[t,x]$ = ode15s(Fun,tspan,x_0,options,additional parameters)
$[t,x]$ = ode113(Fun,tspan,x_0,options,additional parameters

where `options` can be accessed with the `odeget()` and `odeset()` commands. Other useful options are given in Table 3.2. All the related options can be listed with the `odeset` command.

Table 3.2 Commonly used control parameters of ODE solvers.

Options	Parameter description
RelTol	relative error tolerance, with a default value of 0.001, i.e., the relative error is 0.1%.; in some applications, smaller values should be used
AbsTol	absolute error tolerance, with a default value of 10^{-6}
MaxStep	maximum allowed step size
Mass	mass matrix in differential algebraic equations
Jacobian	Jacobian matrix $\partial f/\partial x$ function name; if the Jacobian matrix is known, the speed of computation is increased

Normally speaking, there is no need to modify the default values in the property template, unless necessary. Three methods can be used to describe the ODEs. `Fun` can be expressed by M-functions, anonymous functions and inline functions. It will be referred to as the "ODE descriptive function".

The variable `tspan` can be used to describe the range of simulation, that is `tspan`$= [t_0, t_f]$, where t_0 and t_f are respectively the starting and terminating time instances. It should be noted that in MATLAB, t_0 is allowed to be larger than t_f. In this way, final value problems can be solved.

With these parameters, the ODE solvers can be used to solve the systems directly. After the function call, two arguments t and x are returned, where t is the vector of time instances. Since variable step size is allowed, it may not be evenly spaced. Another variable x returns the states of the simulation results, where the ith column represents the values of the state $x_i(t)$ at all time instances. The state versus time plot can be drawn with the command `plot(t,x)`, and the phase space trajectory can be drawn with `plot(x(:,i),x(:,j))`.

The syntax of the ODE descriptive functions is quite standard. The leading statement in the ODE descriptive function is

`function` x_1 = Fun(t,x,additional parameters)

where t is the time variable, x is the state vector and x_1 is the derivative of the state vector. Note that, even though the ODE is time invariant, the variable t should still be used to hold its place.

If there are additional parameters to be transferred, they should be given in both the solvers and the ODE descriptive function. The number of variables and the format in the two functions should be the same.

Several examples are used here to show the ODE solving methods, and to indicate the possible issues in problem solving schemes.

Example 3.27 *Assume that the Lorenz equation is described by*

$$\begin{cases} \dot{x}_1(t) = -8x_1(t)/3 + x_2(t)x_3(t) \\ \dot{x}_2(t) = -10x_2(t) + 10x_3(t) \\ \dot{x}_3(t) = -x_1(t)x_2(t) + 28x_2(t) - x_3(t), \end{cases}$$

and the initial states $x_1(0) = x_2(0) = 0$, $x_3(0) = \epsilon$ are known, where $\epsilon = 10^{-10}$. The original differential equation can be expressed by the following anonymous function and the original ODE can be solved with the ode45() *function*

```
>> f=@(t,x)[-8/3*x(1)+x(2)*x(3); -10*x(2)+10*x(3);
            -x(1)*x(2)+28*x(2)-x(3)];
   t_final=100; x0=[0;0;1e-10];
   [t,x]=ode45(f,[0,t_final],x0); plot(t,x)
   figure; plot3(x(:,1),x(:,2),x(:,3)); axis([10 40 -20 20 -20 20]);
```

where t_final *is the terminating time for the simulation. The vector $\boldsymbol{x}_0$ is the initial states. The state versus time plot and the phase space trajectory can be obtained, as shown in Figs. 3.2(a) and (b) respectively. It can be seen that the seemingly complicated ODE problem can be solved easily with just a few MATLAB statements.*

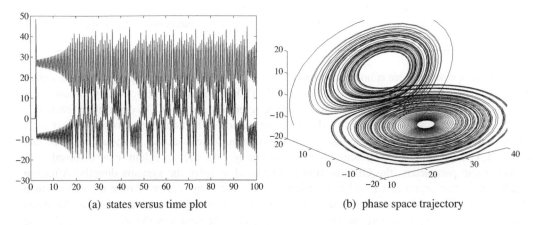

(a) states versus time plot   (b) phase space trajectory

Figure 3.2   Simulation results of the Lorenz equation.

Example 3.28   *Consider the well-known Van der Pol equation $\ddot{y} + \mu(y^2 - 1)\dot{y} + y = 0$. Selecting state variables $x_1 = y$, and $x_2 = \dot{y}$, the original equation can be transformed into the form*

$$\begin{bmatrix} \dot{x}_1 \\ \dot{x}_2 \end{bmatrix} = \begin{bmatrix} x_2 \\ -\mu(x_1^2 - 1)x_2 - x_1 \end{bmatrix}.$$

Since μ can be assigned as an additional variable, it might be easier to change the value of μ without rewriting the ODE descriptive function. We can use the anonymous function to describe the original ODE. Then the solver ode45() *can be used to solve numerically the ODE described by f. The value of μ can be assigned in the MATLAB workspace, before the solution process. Assuming that the initial states are given by $\boldsymbol{x}_0 = [-0.2; -0.7]$, the solutions of the ODE can be completed with the following MATLAB statements*

```
>> f=@(t,x,mu)[x(2); -mu*(x(1)^2-1)*x(2)-x(1)];
   h_opt=odeset; x0=[-0.2; -0.7]; t_final=20;
   mu=1; [t1,y1]=ode45(f,[0,t_final],x0,h_opt,mu);
   mu=2; [t2,y2]=ode45(f,[0,t_final],x0,h_opt,mu);
```

```
plot(t1,y1,t2,y2,'--'),
figure; plot(y1(:,1),y1(:,2),y2(:,1),y2(:,2),'--')
```

The time responses and phase space trajectories for $\mu = 1, 2$ are obtained as shown in Figs. 3.3(a) and (b).

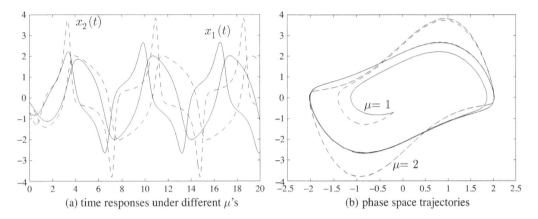

(a) time responses under different μ's   (b) phase space trajectories

Figure 3.3   Van der Pol equation solutions for different values of μ's.

If the μ value is changed, e.g., $\mu = 1000$, and the terminate time is set to 3000, the solutions of the Van der Pol equation can be tried with the following commands

```
>> h_opt=odeset; x0=[2;0]; t_final=3000;
   mu=1000; [t,y]=ode45(f,[0,t_final],x0,h_opt,mu);
```

*The waiting time will be long (and it is worth using **Ctrl-C** to terminate the solution), the following error messages "??? Error using ==> vertcat" will be displayed.*

In fact, since the variable step size scheme is used, to satisfy the required error tolerance, the step size may be selected extremely small, so that the memory may not be enough. This ODE is a stiff equation and should be solved with other algorithms.

Example 3.29   *Assume that an implicit differential equation is given by*

$$\begin{cases} \sin x_1 \dot{x}_1 + \cos x_2 \dot{x}_2 + x_1 = 1 \\ -\cos x_2 \dot{x}_1 + \sin x_1 \dot{x}_2 + x_2 = 0. \end{cases}$$

Selecting $x = [x_1, x_2]^{\mathrm{T}}$, the original equations can be rewritten in a matrix form $A(x)\dot{x} = B(x)$, where

$$A(x) = \begin{bmatrix} \sin x_1 & \cos x_2 \\ -\cos x_2 & \sin x_1 \end{bmatrix}, \quad B(x) = \begin{bmatrix} 1 - x_1 \\ -x_2 \end{bmatrix}.$$

If the matrix $A(x)$ can be proved to be a nonsingular matrix, the equation can easily be solved, since the original ODE can be converted to the standard form $\dot{x} = A^{-1}(x)B(x)$. In fact, since *trying to find the inverse of a singular matrix may lead to warnings, we can try to solve the above transformed ODE directly. If a warning message appears during the solution process, the ODE solution may not be successful. If there is no warning message, the solution process has been successful. The following statements can be used to describe the ODE and solve it, and the time responses obtained are shown in Fig. 3.4. Since there is no warning message, the results obtained are acceptable.*

```
>> f=@(t,x)inv([sin(x(1)) cos(x(2));-cos(x(2)) sin(x(1))])*[1-x(1);-x(2)];
   [t,x]=ode45(f,[0,10],[0; 0]); plot(t,x)
```

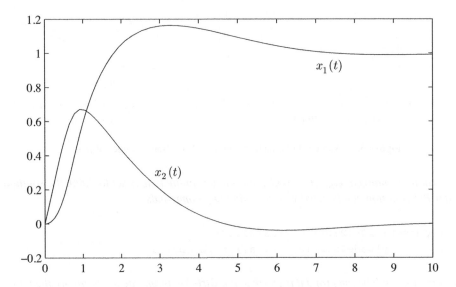

Figure 3.4   Time responses of the implicit differential equation.

In many fields, there exists a class of special ODEs, where some of the solutions vary very fast, and others vary very slowly, and the differences are very significant. This type of ODE is referred to as stiff ordinary differential equations. It has been shown that stiff ODEs are not suitable to be solved with solvers such as `ode45()`. Stiff equation solvers such as `ode15s()` and `ode113()` should be used instead. The syntaxes of these functions are exactly the same as that of `ode45()`.

Example 3.30   *Consider again the Van der Pol equation, with $\mu = 1000$. It has been seen from Example 3.28 that with `ode45()` the equation cannot be solved. However, with the stiff equation solver, the solution can be found within 0.6 s.*

```
>> h_opt=odeset; x0=[2;0]; t_final=3000;
   f=@(t,x,mu)[x(2); -mu*(x(1)^2-1)*x(2)-x(1)];
   tic, mu=1000; [t,y]=ode15s(f,[0,t_final],x0,h_opt,mu); toc
   plot(t,y(:,1)); figure; plot(t,y(:,2))
```

The two states are shown separately in Figs. 3.5(a) and (b). It can be seen that the curve $x_1(t)$ is smooth while $x_2(t)$ is too steep at some time instances. Thus when $\mu = 1000$, the Van der Pol equation is a typical stiff equation. Stiff equation solvers must be used instead in this case.

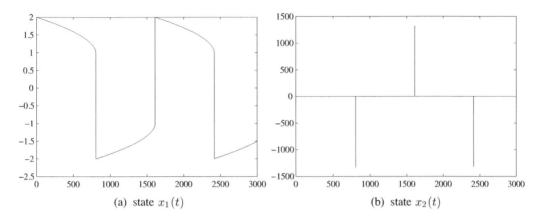

(a) state $x_1(t)$ (b) state $x_2(t)$

Figure 3.5 Solutions of Van der Pol equation with $\mu = 1000$.

Example 3.31 *In ODE textbooks [4], the following ODE is regarded as a stiff equation*

$$\dot{y} = \begin{bmatrix} -21 & 19 & -20 \\ 19 & -21 & 20 \\ 40 & -40 & -40 \end{bmatrix} y, \quad y_0 = \begin{bmatrix} 1 \\ 0 \\ -1 \end{bmatrix}.$$

Note that the eigenvalues of the system matrix are $-2, -40 \pm j40$, which span a wide range. Since the analytical solution can be expressed as $y(t) = e^{At}y(0)$, with the following commands the analytical solution of the ODE can be written as

$$y(t) = \begin{bmatrix} 0.5e^{-2t} + 0.5e^{-40t}(\cos 40t + \sin 40t) \\ 0.5e^{-2t} - 0.5e^{-40t}(\cos 40t + \sin 40t) \\ -e^{-40t}(\cos 40t - \sin 40t). \end{bmatrix}.$$

```
>> A=[-21,19,-20; 19,-21,20; 40,-40,-40]; syms t; y=expm(A*t)*[1;0;-1]
```

For the original problem, the numerical solutions to the ODE can be obtained with the following statements, and the step size of 0.1 s is used. The norm of the error vector is 2.97×10^{-6}.

```
>> f=@(t,x)A*x; ff=odeset; ff.RelTol=1e-6;
   tic,[t0,y1]=ode45(f,[0,1],[1;0;-1],ff); toc,
   y2=subs(y',t,t0); plot(t0,y2,t0,y1,'--'), norm(y2-y1)
```

The analytical solutions and the numerical ones are shown in Fig. 3.6. It can be seen that the accuracy is acceptable with a non-stiff equation solver.

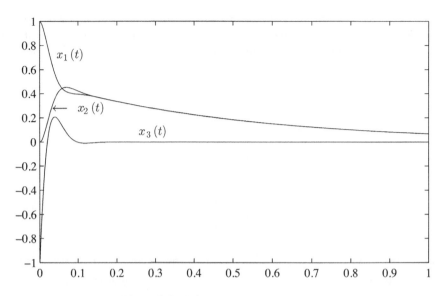

Figure 3.6   Solution of a linear stiff ODE.

■ Example 3.32   *Consider the following stiff ODE*

$$\begin{cases} \dot{y}_1 = 0.04(1 - y_1) - (1 - y_2)y_1 + 0.0001(1 - y_2)^2 \\ \dot{y}_2 = -10^4 y_1 + 3000(1 - y_2)^2, \end{cases}$$

where the initial values are $y_1(0) = 0$, $y_2(0) = 1$. *Select a computation interval* $t \in (0, 100)$. *An anonymous function is used to describe the original equation. The following commands can be used to solve the original problem. The solution process for this example takes 47.7 s with total 356941 computation points. The time responses of the states obtained are shown in Fig. 3.7(a).*

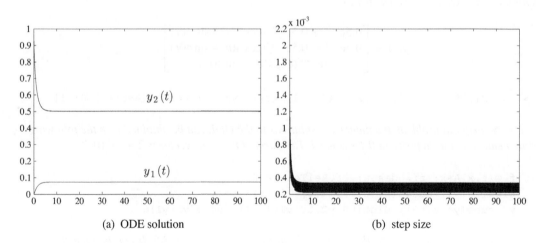

(a) ODE solution   (b) step size

Figure 3.7   Solutions with Runge–Kutta method.

```
>> f=@(t,y)[0.04*(1-y(1))-(1-y(2))*y(1)+0.0001*(1-y(2))^2;
            -10^4*y(1)+3000*(1-y(2))^2];
   tic,[t2,y2]=ode45(f,[0,100],[0;1]); toc, length(t2), plot(t2,y2)
```

It can be seen that with the `ode45()` *function, the time elapsed is too long and the points calculated are too many. The step size plot can also be obtained with the commands, as shown in Fig. 3.7(b).*

```
>> [min(diff(t2)), max(diff(t2))], plot(t2(1:end-1),diff(t2))
```

If `ode15s()` *is used to replace the* `ode45()` *function in the above statements, the solution plots are almost identical to the one shown in Fig. 3.7(a), while the time required is reduced to 0.24 s, and the number of computation points is reduced to 56. The stiff equation solver has significantly improved the efficiency of this problem.*

```
>> tic,[t1,y1]=ode15s(f,[0,100],[0;1]); toc, length(t1), plot(t1,y1)
```

3.4.3 Conversion of ODE Sets

From the above description, it can be seen that ODE solvers such as `ode45()` can only be used to solve numerically ODEs in the standard form shown in (3.47). If an ODE is given in other forms, such as high-order ODEs, they should be converted first to the standard form. Two types pf ODE conversions are considered here.

1) Converting a single high-order ODE

Assume that a high-order ODE is given by

$$y^{(n)} = f(t, y, \dot{y}, \cdots, y^{(n-1)}), \tag{3.51}$$

and the initial values of $y(t)$ and its derivatives are given as $y(0), \dot{y}(0), \cdots, y^{(n-1)}(0)$. A set of state variables can be selected as $x_1 = y, x_2 = \dot{y}, \cdots, x_n = y^{(n-1)}$. The original high-order ODE can be converted to the following standard form

$$\begin{cases} \dot{x}_1 = x_2 \\ \dot{x}_2 = x_3 \\ \vdots \\ \dot{x}_n = f(t, x_1, x_2, \cdots, x_n), \end{cases} \tag{3.52}$$

where $x_1(0) = y(0), x_2(0) = \dot{y}(0), \cdots, x_n(0) = y^{(n-1)}(0)$. The converted ODE can be solved directly with MATLAB solvers, such as `ode45()`.

2) Converting high-order ODE set

Suppose that the ODE set consists of two high-order ODEs in explicit form as

$$\begin{cases} x^{(m)} = f(t, x, \dot{x}, \cdots, x^{(m-1)}, y, \cdots, y^{(n-1)}) \\ y^{(n)} = g(t, x, \dot{x}, \cdots, x^{(m-1)}, y, \cdots, y^{(n-1)}). \end{cases} \tag{3.53}$$

Still one can select the state variables

$$x_1 = x, x_2 = \dot{x}, \cdots, x_m = x^{(m-1)}, x_{m+1} = y, x_{m+2} = \dot{y}, \cdots, x_{m+n} = y^{(n-1)}. \quad (3.54)$$

The original ODE set can be converted to the following standard form, such that the converted ODE can be solved directly with the relevant MATLAB ODE solvers.

$$\begin{cases} \dot{x}_1 = x_2 \\ \vdots \\ \dot{x}_m = f(t, x_1, x_2, \cdots, x_{m+n}) \\ \dot{x}_{m+1} = x_{m+2} \\ \vdots \\ \dot{x}_{m+n} = g(t, x_1, x_2, \cdots, x_{m+n}). \end{cases} \quad (3.55)$$

■ Example 3.33   *The trajectory of the Apollo satellite (x, y) satisfies the following ODE [5]*

$$\begin{cases} \ddot{x} = 2\dot{y} + x - \dfrac{\mu^*(x + \mu)}{r_1^3} - \dfrac{\mu(x - \mu^*)}{r_2^3} \\ \ddot{y} = -2\dot{x} + y - \dfrac{\mu^* y}{r_1^3} - \dfrac{\mu y}{r_2^3}, \end{cases}$$

where $\mu = 1/82.45, \mu^* = 1 - \mu, r_1 = \sqrt{(x + \mu)^2 + y^2}, r_2 = \sqrt{(x - \mu^*)^2 + y^2}$. The initial conditions are given by $x(0) = 1.2, \dot{x}(0) = 0, y(0) = 0, \dot{y}(0) = -1.04935751$.

We can select a set of state variables $x_1 = x, x_2 = \dot{x}, x_3 = y, x_4 = \dot{y}$, and the original equation can be converted to the following standard form

$$\begin{cases} \dot{x}_1 = x_2 \\ \dot{x}_2 = 2x_4 + x_1 - \mu^*(x_1 + \mu)/r_1^3 - \mu(x_1 - \mu^*)/r_2^3 \\ \dot{x}_3 = x_4 \\ \dot{x}_4 = -2x_2 + x_3 - \mu^* x_3/r_1^3 - \mu x_3/r_2^3, \end{cases}$$

where $r_1 = \sqrt{(x_1 + \mu)^2 + x_3^2}, r_2 = \sqrt{(x_1 - \mu^*)^2 + x_3^2}$, and $\mu = 1/82.45, \mu^* = 1 - \mu$.

The following M-function can be established to describe the converted ODE

```
function dx=apolloeq(t,x)
mu=1/82.45; mu1=1-mu;
r1=sqrt((x(1)+mu)^2+x(3)^2); r2=sqrt((x(1)-mu1)^2+x(3)^2);
dx=[x(2);
    2*x(4)+x(1)-mu1*(x(1)+mu)/r1^3-mu*(x(1)-mu1)/r2^3;
    x(4);
    -2*x(2)+x(3)-mu1*x(3)/r1^3-mu*x(3)/r2^3];
```

With the `ode45()` function, the converted ODE can be solved numerically and the trajectory of the satellite can be obtained as shown in Fig. 3.8(a). The total number of points computed is 689, and the time required is 0.16 s.

```
>> x0=[1.2; 0; 0; -1.04935751];
   tic, [t,y]=ode45(@apolloeq,[0,20],x0); toc,
   length(t), plot(y(:,1),y(:,3))
```

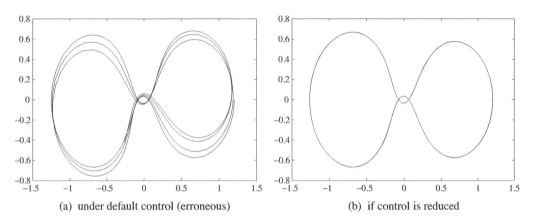

(a) under default control (erroneous)   (b) if control is reduced

Figure 3.8   Apollo trajectories under different control options.

In fact, the obtained solution is not correct, due to the fact that the default control option RelTol *is too large. If this option is set to* 10^{-6}, *the following commands can be used and the new trajectory is obtained as shown in Fig. 3.8(b). The time used is 0.36 s, and the number of points calculated is 1873. It can be seen that the option* RelTol *can further be reduced in order to validate the results.*

```
>> options=odeset; options.RelTol=1e-6;
   tic, [t1,y1]=ode45(@apolloeq,[0,20],x0,options); toc
   length(t1), plot(y1(:,1),y1(:,3)),
   figure, plot(t1(1:end-1),diff(t1))
```

The step size plot can be obtained as shown in Fig. 3.9. It can be seen that at certain time instances, the step size is extremely small, while at other times, the step size is rather large. This ensures the efficiency of the simulation algorithm.

3.4.4   Validation of Numerical ODE Solutions

It has been shown in the previous examples that, if the error control options are not properly chosen, the ODE solutions obtained with MATLAB may not be correct. Thus a validation of simulation results must be made. One possible way is to set options to other values and see whether consistent results are obtained. An effective error control is the relative tolerance, RelTol: if the simulation results under different RelTol values are almost the same, the simulation results are considered as acceptable. If not, smaller RelTol should be tried, until consistent results are obtained. Since a variable step size scheme is used, the computation load may not increase much, even with much smaller RelTol controls. Also, different solvers may be used to validate the simulation results.

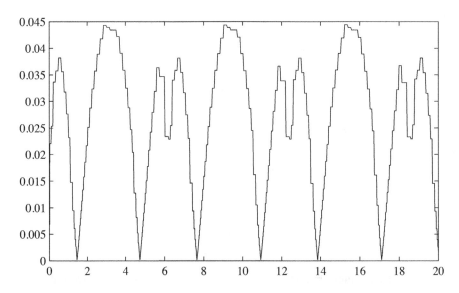

Figure 3.9   Step size plot in simulation.

3.4.5   Solutions to Differential Algebraic Equations

In the previous presentation, the ODEs of the standard form, and those that can be converted to the standard forms, can be solved easily with MATLAB. If some of the differential equations are reduced into algebraic equations, special algorithms should be used instead.

In certain differential equations, if some of the state variables satisfy certain algebraic constraints, the ODE is referred to as using differential algebraic equations (DAE). The standard form of the DAE can be expressed as

$$M(t, x)\dot{x} = f(t, x), \tag{3.56}$$

where the $f(t, x)$ vector is the same as in the conventional ODEs, and the matrix $M(t, x)$ is a singular matrix. The Mass property can be used to express the matrix $M(t, x)$. The differential algebraic equation can then be easily solved.

Example 3.34   *Consider the following differential algebraic equation*

$$\begin{cases} \dot{x}_1 = -0.2x_1 + x_2x_3 + 0.3x_1x_2 \\ \dot{x}_2 = 2x_1x_2 - 5x_2x_3 - 2x_2^2 \\ x_1 + x_2 + x_3 - 1 = 0, \end{cases}$$

where the initial states are given by $x_1(0) = 0.8$, $x_2(0) = x_3(0) = 0.1$. It can be seen that the last equation is an algebraic equation, indicating the relationship of the three state variables. The original equation can be expressed in the standard form of DAE as

$$\begin{bmatrix} 1 & 0 & 0 \\ 0 & 1 & 0 \\ 0 & 0 & 0 \end{bmatrix} \begin{bmatrix} \dot{x}_1 \\ \dot{x}_2 \\ \dot{x}_3 \end{bmatrix} = \begin{bmatrix} -0.2x_1 + x_2x_3 + 0.3x_1x_2 \\ 2x_1x_2 - 5x_2x_3 - 2x_2^2 \\ x_1 + x_2 + x_3 - 1 \end{bmatrix}.$$

*The right hand side of the equation can be described with an anonymous function, and the singular matrix **M** can be assigned to the* Mass *option. The following commands can be used in solving the DAE, and the results can be obtained as shown in Fig. 3.10.*

```
>> f=@(t,x) [-0.2*x(1)+x(2)*x(3)+0.3*x(1)*x(2);
             2*x(1)*x(2)-5*x(2)*x(3)-2*x(2)*x(2); x(1)+x(2)+x(3)-1];
   M=[1,0,0; 0,1,0; 0,0,0]; options=odeset; options.Mass=M;
   x0=[0.8; 0.1; 0.1]; [t,x]=ode15s(f,[0,20],x0,options); plot(t,x)
```

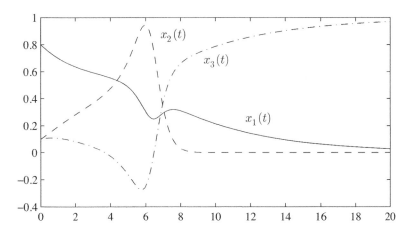

Figure 3.10   Solution of the DAE.

In fact, some DAEs can be converted to ODEs. For instance, from the algebraic constraint, it can be found that $x_3(t) = 1 - x_1(t) - x_2(t)$, and substituting it into the other two equations, the new ODE can be found

$$\begin{cases} \dot{x}_1 = -0.2x_1 + x_2(1 - x_1 - x_2) + 0.3x_1x_2 \\ \dot{x}_2 = 2x_1x_2 - 5x_2(1 - x_1 - x_2) - 2x_2^2. \end{cases}$$

Again the anonymous function can be used to describe the right hand side of the equation. The following statements can be used to solve the above equation, and it can be seen that the results obtained with the commands are the same as the one obtained with the DAE solver.

```
>> f=@(t,x) [-0.2*x(1)+x(2)*(1-x(1)-x(2))+0.3*x(1)*x(2);
             2*x(1)*x(2)-5*x(2)*(1-x(1)-x(2))-2*x(2)*x(2)];
   x0=[0.8; 0.1]; [t1,x1]=ode45(f,[0,20],x0);
   plot(t1,x1,t1,1-sum(x1'))
```

■ Example 3.35   *Consider the ODE given in Example 3.29, where $A(t, x)\dot{x}(t) = B(t, x)$, can be regarded as a DAE. If is not necessary to check whether **A** is singular or not. The following commands can be issued and the results are exactly the same as those in Example 3.29.*

```
>> B=@(t,x) [1-x(1); -x(2)];
   A=@(t,x) [sin(x(1)),cos(x(2)); -cos(x(2)),sin(x(1))];
   options=odeset; options.Mass=A; options.RelTol=1e-6;
   [t,x]=ode45(B,[0,10],[0;0],options); plot(t,x)
```

3.4.6   Solutions to Linear Stochastic Differential Equations

Consider a linear differential equation described by

$$\dot{y}(t) + ay(t) = \gamma(t), \tag{3.57}$$

where a is a given constant. Assume that $\gamma(t)$ is Gaussian white noise with zero mean and a variance of σ^2. It is known [6] that the output signal $y(t)$ is also Gaussian, with zero mean, and a variance of $\sigma_y^2 = \sigma^2/(2a)$. Assume that the input signal is kept a constant e_k within a computation step size, the original system can be discretized as

$$y_{k+1} = e^{-\Delta t/a} y_k + (1 - e^{-\Delta t/a})\sigma e_k, \tag{3.58}$$

where Δt is a computation step size, and e_k is a pseudo random number which satisfies standard normal distribution $N(0, 1)$. It can be seen that

$$E[y_{k+1}^2] = e^{-2\Delta t/a} E[y_k^2] + 2\sigma e^{-\Delta t/a} E[e_k y_k] + \sigma^2 (1 - e^{-\Delta t/a})^2 E[e_k^2]. \tag{3.59}$$

If the input and output signals are stationary processes, then $E[y_{k+1}^2] = E[y_k^2] = \sigma_y^2$, and since y_k and e_k are independent, then $E[y_k e_k] = 0$. Also $E[e_k^2] = 1$. It can be shown that

$$\sigma_y^2 = \frac{\sigma^2 (1 - e^{-\Delta t/a})^2}{(1 - e^{-2\Delta t/a})} = \frac{\sigma^2 (1 - e^{-\Delta t/a})}{1 + e^{-\Delta t/a}}. \tag{3.60}$$

If $\Delta t/a \to 0$, and the numerator and denominator in (3.60) are approximated by power series, it can be seen that

$$\sigma_y^2 = \lim_{\Delta t/a \to 0} \frac{\Delta t/a + o[(\Delta t/a)^2]}{2 + o(\Delta t/a)} \sigma^2 = \frac{\Delta t}{2a} \sigma^2. \tag{3.61}$$

It can be seen that the variance of the output signal depends on the computation step size Δt. Of course, the results are not correct. This means that, when the input signal is random, conventional methods cannot be used in simulation.

Because of this, in some simulation software, other kinds of processes are used to replace Gaussian white noise. For instance, in the ACSL [7] language, the Ornstein–Uhlenbeck process was used to approximate Gaussian white noise so that the input signal could maintain a constant within a certain frequency range. However, the simulation result thus obtained is not satisfactory either [8].

Assume that the linear state space representation

$$\dot{x}(t) = Ax(t) + B[d(t) + \gamma(t)], \quad y(t) = Cx(t), \tag{3.62}$$

where A is an $n \times n$ matrix, B is an $n \times m$ matrix, C is an $r \times n$ matrix, and $d(t)$ is an $m \times 1$ deterministic input signal, $\gamma(t)$ is an $m \times 1$ Gaussian white noise vector, satisfying

$$E[\gamma(t)] = 0, \quad E[\gamma(t)\gamma^{\mathrm{T}}(\tau)] = V_\sigma \delta(t - \tau). \tag{3.63}$$

Introducing a vector $\boldsymbol{y}_c(t) = \boldsymbol{B}\boldsymbol{y}(t)$, it can be shown that $\boldsymbol{y}_c(t)$ is also a Gaussian white noise, satisfying

$$E[\boldsymbol{y}_c(t)] = 0, \quad E[\boldsymbol{y}_c(t)\boldsymbol{y}_c^{\mathrm{T}}(\tau)] = \boldsymbol{V}_c\delta(t - \tau), \tag{3.64}$$

where $\boldsymbol{v}_c = \sigma \boldsymbol{B}\boldsymbol{V}\boldsymbol{B}^{\mathrm{T}}$ is an $m \times m$ covariance matrix, then (3.62) can be rewritten as

$$\dot{\boldsymbol{x}}(t) = \boldsymbol{A}\boldsymbol{x}(t) + \boldsymbol{B}\boldsymbol{d}(t) + \boldsymbol{y}_c(t), \, y(t) = \boldsymbol{C}\boldsymbol{x}(t). \tag{3.65}$$

The analytical solutions of the states can be written as

$$\boldsymbol{x}(t) = \mathrm{e}^{-\boldsymbol{A}t}\boldsymbol{x}(t_0) + \int_{t_0}^{t} \mathrm{e}^{\boldsymbol{A}(t-\tau)}\boldsymbol{d}(\tau)\boldsymbol{B}\mathrm{d}\tau + \int_{t_0}^{t} \boldsymbol{y}_c(t)\mathrm{d}\tau. \tag{3.66}$$

Assume that $t_0 = k\Delta t$, $t = (k + 1)\Delta t$, where Δt is the computation step size, and assume that within a computation step size, the deterministic input $d(t)$ is constant, that is, if $\Delta t \leq t \leq (k + 1)\Delta t$, $d(t) = d(k\Delta t)$. The discrete form of (3.66) can be written as

$$\boldsymbol{x}[(k + 1)\Delta t] = \boldsymbol{F}\boldsymbol{x}(k\Delta t) + \boldsymbol{G}\boldsymbol{d}(k\Delta t) + \boldsymbol{y}_d(k\Delta t), \quad y(k\Delta t) = \boldsymbol{C}\boldsymbol{x}(k\Delta t), \tag{3.67}$$

where $\boldsymbol{F} = \mathrm{e}^{\boldsymbol{A}\Delta t}$, $\boldsymbol{G} = \int_0^{\Delta t} \mathrm{e}^{\boldsymbol{A}(\Delta t-\tau)}\boldsymbol{B}\mathrm{d}\tau$, and

$$\boldsymbol{y}_d(k\Delta t) = \int_{k\Delta t}^{(k+1)\Delta t} \mathrm{e}^{\boldsymbol{A}[(k+1)\Delta t-\tau]}\boldsymbol{y}_c(t)\mathrm{d}\tau = \int_0^{\Delta t} \mathrm{e}^{\boldsymbol{A}t}\boldsymbol{y}_c[(k + 1)\Delta t - \tau]\mathrm{d}\tau. \tag{3.68}$$

It can be seen that $\boldsymbol{F}$ and $\boldsymbol{G}$ matrices are exactly the same as those in deterministic systems. When there exists a stochastic input, discretization of the system is slightly different. It can be shown that $\boldsymbol{y}_d(t)$ is also a Gaussian white noise vector, satisfying

$$E[\boldsymbol{y}_d(k\Delta t)] = 0, \, E[\boldsymbol{y}_d(k\Delta t)\boldsymbol{y}_d^{\mathrm{T}}(j\Delta t)] = \boldsymbol{V}\delta_{kj}, \tag{3.69}$$

where $\boldsymbol{V} = \int_0^{\Delta t} \mathrm{e}^{\boldsymbol{A}t}\boldsymbol{V}_c\mathrm{e}^{\boldsymbol{A}^{\mathrm{T}}t}\mathrm{d}t$. With Taylor series expansion,

$$\boldsymbol{V} = \int_0^{\Delta t} \sum_{k=0}^{\infty} \frac{\boldsymbol{R}^{(k)}(0)}{k!}t^k\mathrm{d}t = \sum_{k=0}^{\infty} \boldsymbol{V}_k, \tag{3.70}$$

where $\boldsymbol{R}^{(k)}(0)$ and $\boldsymbol{V}_k$ can recursively obtained as

$$\begin{cases} \boldsymbol{R}^{(k)}(0) = \boldsymbol{A}\boldsymbol{R}^{(k-1)}(0) + \boldsymbol{R}^{(k-1)}(0)\boldsymbol{A}^{\mathrm{T}} \\ \boldsymbol{V}_k = \dfrac{\Delta t}{k + 1}(\boldsymbol{A}\boldsymbol{V}_{k-1} + \boldsymbol{V}_{k-1}\boldsymbol{A}^{\mathrm{T}}), \end{cases} \tag{3.71}$$

with initial values $\boldsymbol{R}^{(0)}(0) = \boldsymbol{R}(0) = \boldsymbol{V}_c$, $\boldsymbol{V}_0 = \boldsymbol{V}_c\Delta t$. With the singular value decomposition technique, matrix $\boldsymbol{V}$ can be written as $\boldsymbol{V} = \boldsymbol{U}\boldsymbol{\Gamma}\boldsymbol{U}^{\mathrm{T}}$, where $\boldsymbol{U}$ is an orthogonal matrix, and $\boldsymbol{\Gamma}$ is a diagonal matrix containing nonzero elements. Cholesky factorization can be performed such that $\boldsymbol{V} = \boldsymbol{D}\boldsymbol{D}^{\mathrm{T}}$, and $\boldsymbol{y}_d(k\Delta t) = \boldsymbol{D}e(k\Delta t)$, where $e(k\Delta t)$ is an $n \times 1$ vector, and $e(k\Delta t) = [e_k, e_{k+1}, \cdots, e_{k+n-1}]^T$, such that the components e_k satisfy a standard normal distribution, that is, $e_k \sim N(0, 1)$. A recursive solution can be obtained

$$\boldsymbol{x}[(k+1)\Delta t] = \boldsymbol{F}\boldsymbol{x}(k\Delta t) + \boldsymbol{G}d(k\Delta t) + \boldsymbol{D}e(k\Delta t), \quad y(k\Delta t) = \boldsymbol{C}\boldsymbol{x}(k\Delta t). \tag{3.72}$$

Based on the above algorithm, the discretization algorithm for continuous linear stochastic systems can be written as

```
function [F,G,D,C]=sc2d(G,V,T)
G=ss(G); G=balreal(G); A=G.a; B=G.b; C=G.c; [F,G]=c2d(A,B,T);
V0=B*V*B'*T; Vd=V0; vmax=sum(sum(abs(Vd))); vv=vmax; i=1;
while (1)
    V1 = T/(i+1)*(A*V0+V0*A'); v0 = sum(sum(abs(V1)));
    Vd = Vd+V1; V0 = V1; vv = [vv v0]; i=i+1;
    if v0 < 1e-10*vmax, break; end
end
[U,S,V0]=svd(Vd); V0=sqrt(diag(S)); Vd=diag(V0); D=U*Vd;
```

In simulation, a set of pseudo random numbers can be generated, and the vector $e(k\Delta t)$ can be constructed. Then the state vector $\boldsymbol{x}[(k+1)\Delta t]$ at the next time instance can be evaluated, and the current output signal $y(k\Delta t)$ can be obtained

$$\dot{y}(t) = -\frac{1}{a}y(t) + \frac{1}{a}y_0(t). \tag{3.73}$$

The discrete form of the output signal can be written as

$$y_{k+1} = -e^{\Delta t/a}y_k + \sigma\sqrt{\frac{1}{2a}\left(1 - e^{-2\Delta t/a}\right)}e_k. \tag{3.74}$$

Example 3.36 *Consider a transfer function model defined as*

$$G(s) = \frac{s^3 + 7s^2 + 24s + 24}{s^4 + 10s^3 + 35s^2 + 50s + 24}.$$

If a white noise signal is used to excite the system, and we select a sampling interval $T = 0.001$, the discretized model can be obtained with the following MATLAB statements

```
>> G=tf([1,7,24,24],[1,10,35,50,24]); T=0.02; [F,G0,D,C]=sc2d(G,1,T)
```

The matrices in the discretized model can be found as

$$F = \begin{bmatrix} 0.9838 & -0.0067 & 0.0132 & 0.0013 \\ 0.0067 & 0.9883 & 0.0702 & 0.0036 \\ 0.0132 & -0.0702 & 0.8653 & -0.0257 \\ 0.0013 & -0.0036 & -0.0257 & 0.9684 \end{bmatrix}, \quad G_0 = \begin{bmatrix} 0.0182 \\ -0.0036 \\ -0.0076 \\ -0.0007 \end{bmatrix}, \quad D = \begin{bmatrix} -0.1303 & 0 & 0 & 0 \\ 0.0235 & 0 & 0 & 0 \\ 0.0594 & 0 & 0 & 0 \\ 0.0061 & 0 & 0 & 0 \end{bmatrix},$$

and $C = [0.9216, 0.1663, -0.4201, -0.0431]$. From the discrete model, 30 000 simulation points can be calculated with the following statements, and the time response of the system is shown in Fig. 3.11(a).

```
>> n_point=30000; r=randn(n_point+4,1); r=r-mean(r);
   y=zeros(n_point,1); x=zeros(4,1); d0=0;
   for i=1:n_point, x=F*x+G0*d0+D*r(i:i+3); y(i)=C*x; end
   t=0:.02:(n_point-1)*0.02; plot(t,y)
   figure; v=covar(G,1); xx=linspace(-2.5,2.5,30); yy=hist(y,xx);
   yy=yy/(30000*(xx(2)-xx(1))); yp=exp(-xx.^2/(2*v))/sqrt(2*pi*v);
   bar(xx,yy), hold on; plot(xx,yp)
```

It can be seen that the output signal behaves in a disorderly way. For systems with random signal, statistical analysis may be more informative. Histograms can be used to approximate probability density functions, as shown in Fig. 3.11(b). The result obtained from simulation data agrees well with the theoretical results, and this means that the simulation results are valid for stochastic systems.

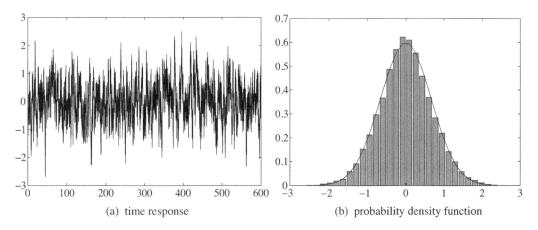

(a) time response   (b) probability density function

Figure 3.11   Response to stochastic inputs.

3.4.7   Analytical Solutions to ODEs

An analytical solution function `dsolve()` to differential equations is provided in the Symbolic Math Toolbox. Examples are given, demonstrating the use of this function.

■ **Example 3.37**   *Consider the linear ODE $y^{(4)} - 4y^{(3)} + 8\ddot{y} - 8\dot{y} + 4y = 2$. In the Symbolic Math Toolbox, the notation* `'D4y'` *can be used to denote the fourth-order derivative of signal $y(t)$. Thus the original ODE can be expressed by a string. The following statements can be used to find the analytical solutions to the given ODE*

$$y(t) = \frac{1}{2} + e^{t} \left(C_1 \sin t + C_2 \cos t + C_3 t \cos t + C_4 t \sin t \right).$$

```
>> syms y; X=dsolve('D4y-4*D3y+8*D2y-8*Dy+4*y=2')
```

Consider again the Van der Pol equation $\ddot{y} + \mu(y^2 - 1)\dot{y} + y = 0$ *in Example 3.28. The following statement can be given*

```
>> syms mu y; dsolve('D2y+mu*(y^2-1)*Dy+y=0')
```

and it will respond that there is no analytical solution. It can be seen that, for nonlinear ODEs, if there are no analytical solutions, they can only be studied with numerical simulation methods.

3.4.8   Numerical Laplace Transforms in ODE Solutions

Laplace and inverse Laplace transforms are mathematically defined as

$$F(s) = \mathscr{L}[f(t)] = \int_0^\infty f(t)e^{-st}dt \tag{3.75}$$

$$f(t) = \mathscr{L}^{-1}[F(s)] = \frac{1}{2\pi j} \int_{\sigma-j\infty}^{\sigma+j\infty} F(s)e^{st}ds. \tag{3.76}$$

These transforms can convert the function between t and s domains. For certain functions, the transforms can be performed symbolically with the functions `laplace()` and `ilaplace()`.

Laplace and inverse Laplace transforms are useful tools for solving certain differential equations. If the input–output relationship of the ordinary differential equation can be written as $Y(s) = G(s)U(s)$, where $G(s)$ is regarded as the complex gain of the ordinary differential equation, also known as the transfer function of the system, $Y(s)$ and $U(s)$ are respectively the Laplace transforms of the output $y(t)$ and the input $u(t)$. Once $Y(s)$ is formulated, the analytical solution can be found by performing an inverse Laplace transform to yield $y(t)$.

Example 3.38   *Consider the ordinary differential equation in Example 3.37, with zero initial conditions. Using the property of Laplace transform of* $\mathscr{L}[d^n y(t)/dt^n] = s^n \mathscr{L}[y(t)]$, *the original differential equation can be mapped to an algebraic equation* $(s^4 - 4s^3 + 8s^2 - 8s + 4)Y(s) = U(s)$, *from which the analytical solution of the equations can be obtained*

```
>> syms s; U=laplace(sym(2)); Y=U/(s^4-4*s^3+8*s^2-8*s+4); y=ilaplace(Y)
```

and the analytical solutions can be written as

$$y(t) = e^t \sin t - \frac{e^t \cos t}{2} - \frac{te^t \cos t}{2} - \frac{te^t \sin t}{2} + \frac{1}{2}.$$

Unfortunately, symbolic Laplace and inverse Laplace transforms are not always solvable for complicated systems, so numerical techniques have to be used instead. The `INVLAP()` function is a powerful tool for performing numerical inverse Laplace transforms [9, 10], with the syntax

$[t, y] = \text{INVLAP}(\text{fun}, t_0, t_f, n, \cdots)$

The essential input arguments are: `fun`, which is a string describing the Laplace transform form $F(s)$, and (t_0, t_f), which is the time interval, with n points.

There is a bug in the code when $t_0 = 0$. This will be fixed later in the `num_laplace()` function. The remaining input arguments are internal parameters and the use of default values is suggested.

🔲 **Example 3.39**   *Considering a fractional-order transfer function*

$$G(s) = \frac{(s^{0.4} + 0.4s^{0.2} + 0.5)}{\sqrt{s}(s^{0.2} + 0.02s^{0.1} + 0.6)^{0.4}},$$

where the input is $u(t) = e^{-0.2t} \sin t$, *the Laplace transform of it can be obtained and converted to a string. The output signal can be obtained as shown in Fig. 3.12. The execution speed of* INVLAP() *is extremely fast, the whole process needing only 0.3 s.*

```
>> G='(s^0.4+0.4*s^0.2+0.5)/(s^0.2+0.02*s^0.1+0.6)^0.4';
   syms t; u=exp(-0.2*t)*sin(t); Y=['(' G ')*' char(laplace(u))];
   [t,y]=INVLAP(Y,0.01,25,1000); plot(t,y)
```

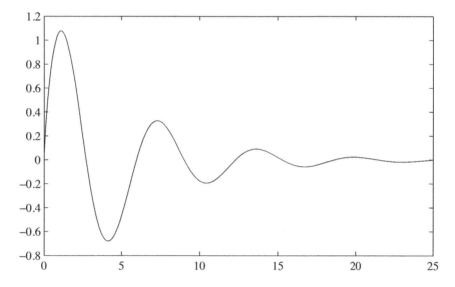

Figure 3.12   Response to stochastic inputs.

To further illustrate the topic, suppose that the input signal is given by a set of sample points. Of course, using the symbolic Laplace transform is not possible for functions given by sample points. Interpolation, and then numerical integration, should be performed to find the numerical Laplace transform. Based on the source code of INVLAP(), numerical Laplace transform functions can be embedded to solve complicated problems. A new MATLAB function is written, with the syntax

$[t, y] =$ num_laplace $(\text{fun}, t_0, t_f, n, x_0, u_0)$

where fun is a string describing the transfer function of the system. The extra arguments x_0 and u_0, specify the sample points of the input signal. Note that the function is extremely slow. The listing of the M-function is

```
function [t,y]=num_laplace(G,t0,tf,nnt,x0,u0)
FF=strrep(strrep(strrep(G,'*','.*'),'/','./'),'^','.^');
a=6; ns=20; nd=19; t=linspace(t0,tf,nnt);
if t0==0, t=t(2:end); nnt=nnt-1; end   % the original bug is fixed here
```

```
n=1:ns+1+nd; alfa=a+(n-1)*pi*1j; beta=-exp(a)*(-1).^n; n=1:nd;
bdif=fliplr(cumsum(gamma(nd+1)./gamma(nd+2-n)./gamma(n)))./2^nd;
beta(ns+2:ns+1+nd)=beta(ns+2:ns+1+nd).*bdif; beta(1)=beta(1)/2;
for kt=1:nnt,
    tt=t(kt); s=alfa/tt; bt=beta/tt;
    U=integral(@(x)interp1(x0,u0,x,'spline').*exp(-s.*x),...
                t0,tf,'ArrayValued',true);
    btF=bt.*eval(FF).*U; y(kt)=sum(real(btF));
end
```

Example 3.40 *If the analytical form of the input function is not known, but instead only a set of sample points is given, the following statements can be given to calculate the output signal. After around 30 s, the numerical solution can be found and it is exactly the same as the one shown in Fig. 3.12. It can be seen from the example that extremely complicated problems can be solved in this way.*

```
>> x0=0:0.5:25; y0=exp(-0.2*x0).*sin(x0);
   [t,y]=num_laplace(G,0,25,200,x0,y0); plot(t,y)
```

3.5 Nonlinear Equation Solutions and Optimization

3.5.1 Solutions of Nonlinear Equations

A MATLAB function fsolve() can be used directly to solve the nonlinear equations described as $f(x) = 0$, where the dimension of $f(\cdot)$ is identical to that of x, that is, the number of equations is the same as the number of variables, with the syntaxes

```
[x,ε,key,c]=fsolve(Fun,x0)                  % simple syntax
[x,ε,key,c]=fsolve(Fun,x0,OPT,p1,···)       % with additional parameters
```

where Fun is the MATLAB description of the equation to be solved. It can be a MATLAB function, an anonymous function or an inline function. The argument x_0 is the initial search point. The argument OPT is the option template provided in the Optimization Toolbox. The returned variable x is the solution of the equation, while ε is the error at point x. The returned variable key indicates the type of the solution, and a positive key means that a successful solution has been found. The returned variable c provides further information, where the iterations member expresses the number of iterations needed, and member funcCount is the number of calls made to the $f(x)$ function.

The options used in the fsolve() function and other optimization problem solvers can be used to control the accuracy and speed of solutions. The option template can be set with the MATLAB function OPT = optimset and the commonly used control options are shown in Table 3.3.

The function OPT = optimset can be used to get the default control template. If we want to change specific members of the template, for example to turn off the "large-scale" algorithm, the following statements can be used to turn the LargeScale property to 'off'.

```
>> OPT=optimset; OPT.LargeScale='off';
```

Table 3.3 Control parameters in equation and optimization solvers.

Member name	Parameter descriptions
Display	indicates the way that the immediate result is displayed, where `'off'` gives no display (default one), `'iter'` gives a display at each iteration and `'notify'` gives prompts only when the function does not converge, and `'final'` display the final results
GradObj	indicates whether gradient information is provided or not, with options `'off'` and `'on'`
LargeScale	indicates whether a large-scale algorithm is used, with options `'on'` and `'off'`
MaxIter	maximum allowed iterations
MaxFunEvals	maximum number of function calls allowed
TolFun	error tolerance control of the error function
TolX	error tolerance of solutions

■ Example 3.41 *Consider the following nonlinear equations*

$$\begin{cases} x^2 + y^2 - 1 = 0 \\ 0.75x^3 - y + 0.9 = 0. \end{cases}$$

There are two unknown variables x and y. MATLAB can only be used to solve the equation with an unknown vector $\boldsymbol{x}$. Variable substitution should be made first to convert the original equation into the standard form. For instance, if we let $x_1 = x, x_2 = y$, then the original equation can be rewritten as

$$\begin{bmatrix} x_1^2 + x_2^2 - 1 \\ 0.75x_1^3 - x_2 + 0.9 \end{bmatrix} = 0.$$

An anonymous function can be used to describe the original equation. Selecting the initial search point $\boldsymbol{x}_0 = [1, 2]^T$, the `fsolve()` function can be used to solve the equation

```
>> f=@(x)[x(1)^2+x(2)^2-1; 0.75*x(1)^3-x(2)+0.9];
   [x,Y,c,d]=fsolve(f,[1; 2]),
```

It can be seen from the returned variable d that six iterations are made, and a solution to the equation is found: $\boldsymbol{x} = [0.3570, 0.9341]^T$. Substituting it back into the original equation, the error is 1.2×10^{-10}.

Selecting another initial search point $\boldsymbol{x}_0 = [-1, 0]^T$, the following statements can be used, and a new solution can be found: $\boldsymbol{x} = [-0.9817, 0.1904]^T$.

```
>> [x,Y,c,d]=fsolve(f,[-1,0]')
```

It can be seen that another solution can be found. In other words, initial value selection is important in solving nonlinear equations. If the initial search point is not well chosen, no solution may be found.

Observing the original equation it can be found that, in the second equation, we can write $y = 0.75x^3 + 0.9$. Substituting it back into the first equation, it can be seen that a 6th-order polynomial equation of x can be obtained, from which six sets of solutions are expected.

With the use of Symbolic Math Toolboxs function `solve()`, *all the solutions can be found with the following statements*

```
>> [x,y]=solve('x^2+y^2-1=0','75*x^3/100-y+9/10=0')
```

If analytical solutions do not exist for the high-order equation, symbolic functions can be used to solve the problems in high precision as

$$
x = \begin{bmatrix} 0.35696997189122287798839037801365 \\ -0.98170264842676789676449828873194 \\ -0.55395176056834560077984413 \pm 0.35471976465080793456863789j \\ 0.86631809883611811016789894 \pm 1.2153712664671427801318378354j \end{bmatrix},
$$

$$
y = \begin{bmatrix} 0.93411585960628007548796029415446 \\ 0.19042035099187730240977756415289 \\ -1.49160640756582231747872169 \pm 0.70588200721402267753918827lj \\ 0.92933830226674362852985276 \pm 0.211438221858959236156233818j \end{bmatrix}.
$$

It can be seen that apart from the two sets of real roots, there are another four sets of complex solutions, in complex conjugates. The complex roots of the equations cannot be obtained with numerical solutions. The following statements can be used to validate the solutions.

```
>> norm([eval('x.^2+y.^2-1') eval('0.75*x.^3-y+0.9')])
```

Substituting the solutions into the original equations, it can be seen that the norm of the error matrix is 7.2118×10^{-31}.

It should be noted that the analytical solution method is not universal. It can only be used in the solutions of equations convertible to polynomial equations. For other types of equations, normally only one solution can be found. Other methods should be used instead.

Example 3.42   *Consider the following simultaneous equations*

$$
\begin{cases} x^2 \sin(y) = x + 1 \\ y = x \cos(0.1x^2 + 3x) + 0.5. \end{cases}
$$

The symbolic equation solver can be used, while only one solution can be found, with $x = -0.678162895875447687794290198072$, $y = 0.775104225339126883127201584614$.

```
>> [x,y]=solve('x^2*sin(y)=x+1','y=x*cos(0.1*x^2+3*x)+0.5')
```

In fact, with a graphical method, the two functions can be drawn with the following statements, as shown in Fig. 3.13. For the purpose of demonstration, one set of curves is manually modified to dash lines. The intersections of the two sets of curves are the solutions of the simultaneous equations.

```
>> ezplot('x^2*sin(y)-x-1'); hold on    % draw the first equation and hold axis
   ezplot('-y+x*cos(0.1*x^2+3*x)+0.5') % draw the second equation on top of it
```

It can be seen that in the area shown in the figure, there are 19 intersections, and all of them are real solutions to the simultaneous equations. If the interested area is enlarged, more solutions can be found. Thus sometimes graphical methods can be adopted.

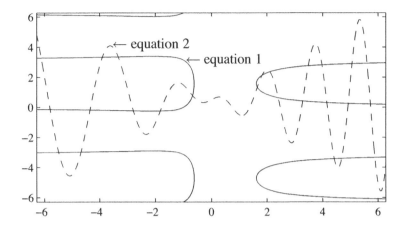

Figure 3.13 Graphical method illustration.

3.5.2 Solutions to Nonlinear Equations with Multiple Solutions

From the results obtained by the graphical method, such as Fig. 3.13, there could exist many solutions. The problem is, how we can find all the solutions? Also, for the typical Riccati equation in Example 3.16, how we can find all the possible solutions. To further extend the question: how can we find the solutions to the modified versions of the Riccati equation given by

$$AX + XD - XBX + C = 0 \qquad (3.77)$$

$$AX + XD - XBX^{\mathrm{T}} + C = 0. \qquad (3.78)$$

The function fsolve() can still be used to find the solutions to these equations, and the initial values can be set randomly. A MATLAB function more_sols() is written as

```
function more_sols(f,X0,A,tol,tlim)
if nargin<=4, tlim=60; end, if nargin<=3, tol=1e-20; end
if nargin<=2, A=1000; end
if nargin<=1, [n,m]=size(f(1)); X0=zeros(n,m,0); end,
ff=optimset; ff.Display='off'; ff1=ff; M=0;
ff.TolX=tol; ff.TolFun=tol; X=X0; [n,m,i]=size(X0); tic
while (1), assignin('base','M',M)
   x0=A*(-0.5+rand(n,m)); [x,a,key]=fsolve(f,x0,ff1); M=M+1;
   t=toc; if t>tlim, break; end
   if key>0, N=size(X,3); % check whether it is a new solution
   for j=1:N, if norm(X(:,:,j)-x)<1e-5; key=0; break; end, end
      if key>0, [x1,a,key]=fsolve(f,x,ff); % if it is, find an accurate one
         if norm(x-x1)<1e-5 & key>0;
            i=i+1, X(:,:,i)=x1; assignin('base','X',X); tic % record it
end, end, end, end
```

with the syntax more_sols$(f, \boldsymbol{X}_0, A, \epsilon, t_{\text{lim}})$, where f is the anonymous description of the original equation, and $\boldsymbol{X}_0$ is the initial solution. A specifies the interested range of solutions, in $(-A/2, A/2)$ intervals, with a default value of 1000, for searching the solutions in a relatively large area. The variable ϵ is the error tolerance (default eps) and t_{lim} is the idle time (default of 60 for 60 s). The function is implemented in an infinite loop structure. If a solution is found, it is validated to see whether it is a new one; if it is, then it should be recorded, otherwise it is abandoned. If there is no new solutions found in t_{lim}, then the function is terminated. Alternatively, the function can be terminated at any time, by pressing **Ctrl-C**. The solutions found are saved in the three-dimensional array $\boldsymbol{X}$, whose $\boldsymbol{X}(:, :, i)$ stores the ith solution.

Example 3.43   *In the equation set in Example 3.42, the equations can be expressed in an anonymous function, then the following statements can be used to find all the solutions in the $(-2\pi, 2\pi)$ area (by selecting $A = 13$); the solutions obtained are shown graphically in Fig. 3.14.*

```
>> f=@(x)[x(1)^2*sin(x(2))-x(1)-1; -x(2)+x(1)*cos(0.1*x(1)^2+3*x(1))+0.5];
   A=13; more_sols(f,zeros(2,1,0),A)
   ezplot('x^2*sin(y)-x-1'); hold on; ezplot('-y+x*cos(0.1*x^2+3*x)+0.5');
   x=X(1,1,:); y=X(2,1,:); plot(x(:),y(:),'o')
```

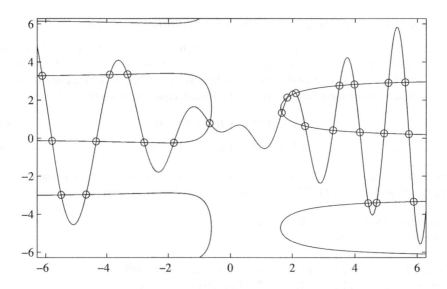

Figure 3.14   Graphical method illustration.

Example 3.44   *The function* fsolve() *can also be used for solving nonlinear matrix equations. For instance, the Riccati equation in Example 3.16 can be solved numerically with the* fsolve() *function, while the* are() *function used can only be used to find one solution. For ease of explanation, the Riccati equation is rewritten here as*

$$A^{\mathrm{T}}X + XA - XBX + C = 0,$$

where

$$A = \begin{bmatrix} -2 & -1 & 0 \\ 1 & 0 & -1 \\ -3 & -2 & -2 \end{bmatrix}, \quad B = \begin{bmatrix} 2 & 2 & -2 \\ -1 & 5 & -2 \\ -1 & 1 & 2 \end{bmatrix}, \quad C = \begin{bmatrix} 5 & -4 & 4 \\ 1 & 0 & 4 \\ 1 & -1 & 5 \end{bmatrix}.$$

The following commands can be used to find as many numerical solutions as possible to the original nonlinear matrix equation.

```
>> A=[-2,1,-3; -1,0,-2; 0,-1,-2]; B=[2,2,-2; -1 5 -2; -1 1 2];
   C=[5 -4 4; 1 0 4; 1 -1 5]; f=@(X)A'*X+X*A-X*B*X+C;
   more_sols(f,zeros(3,3,0))
```

It can be seen that the infinite loop structure is used. The user can interrupt the loop at any time with **Ctrl-C**. *After a while, the following solutions will be found*

$$X_1 = \begin{bmatrix} 1.2213 & -0.4165 & 1.9775 \\ 0.3578 & -0.4894 & -0.8863 \\ -0.7414 & -0.8197 & -2.3560 \end{bmatrix}, \quad X_2 = \begin{bmatrix} -0.7619 & 1.3312 & -0.8400 \\ 1.3183 & -0.3173 & -0.1719 \\ 0.6371 & 0.7885 & -2.1996 \end{bmatrix},$$

$$X_3 = \begin{bmatrix} 0.9874 & -0.7983 & 0.4189 \\ 0.5774 & -0.1308 & 0.5775 \\ -0.2840 & -0.0730 & 0.6924 \end{bmatrix}, \quad X_4 = \begin{bmatrix} 0.6665 & -1.3223 & -1.7200 \\ 0.3120 & -0.5640 & -1.1910 \\ -1.2273 & -1.6129 & -5.5939 \end{bmatrix},$$

$$X_5 = \begin{bmatrix} 0.8878 & -0.9609 & -0.2446 \\ 0.1072 & -0.8984 & -2.5563 \\ -0.0185 & 0.3604 & 2.4620 \end{bmatrix}, \quad X_6 = \begin{bmatrix} -0.1538 & 0.1087 & 0.4623 \\ 2.0277 & -1.7437 & 1.3475 \\ 1.9003 & -1.7513 & 0.5057 \end{bmatrix},$$

$$X_7 = \begin{bmatrix} -2.1032 & 1.2978 & -1.9697 \\ -0.2467 & -0.3563 & -1.4899 \\ -2.1494 & 0.7190 & -4.5465 \end{bmatrix}, \quad X_8 = \begin{bmatrix} 23.9469 & -20.6673 & 2.4529 \\ 30.1460 & -25.9830 & 3.6699 \\ 51.9666 & -44.9108 & 4.6410 \end{bmatrix}.$$

Example 3.45   *Consider a more challenging equation shown in (3.78)*

$$AX + XD - XBX^{\mathrm{T}} + C = 0$$

where

$$A = \begin{bmatrix} 2 & 1 & 9 \\ 9 & 7 & 9 \\ 6 & 5 & 3 \end{bmatrix}, \quad B = \begin{bmatrix} 0 & 3 & 6 \\ 8 & 2 & 0 \\ 8 & 2 & 8 \end{bmatrix}, \quad C = \begin{bmatrix} 7 & 0 & 3 \\ 5 & 6 & 4 \\ 1 & 4 & 4 \end{bmatrix}, \quad D = \begin{bmatrix} 3 & 9 & 5 \\ 1 & 2 & 9 \\ 3 & 3 & 0 \end{bmatrix}.$$

It seems that this type of equation has not been solved anywhere else before, since it is no longer a quadratic equation. With the new `more_sols()` *function, such an equation can be solved directly, where all the 15 solutions can be found and are stored in* **X**.

```
>> A=[2,1,9; 9,7,9; 6,5,3]; B=[0,3,6; 8,2,0; 8,2,8];
   C=[7,0,3; 5,6,4; 1,4,4]; D=[3,9,5; 1,2,9; 3,3,0];
   f=@(X)A*X+X*D-X*B*X.'+C; more_sols(f,zeros(3,3,0))
```

3.5.3   Unconstrained Optimization

Unconstrained optimization problem can be described by

$$\min_{x} f(x), \tag{3.79}$$

where $x = [x_1, x_2, \cdots, x_n]^T$ is referred to as the decision variables. The scalar function $f(x)$ is referred to as the objective function. The mathematical expression means that we want to find a vector x that minimizes the objective function $f(x)$. This problem is also called the minimization problem. In fact, minimization problems can be regarded as the general description of optimization problems, without loss of generality. If maximization problems are involved, we can multiply the objective function by -1, and the maximization problems can then be converted to minimization problems.

Two MATLAB functions fminsearch() and fminunc() are provided in MATLAB and the Optimization Toolbox respectively and they can be used to solve unconstrained optimization problems directly:

$[x, f_{\text{opt}}, \text{key}, c] = \text{fminsearch}(\text{Fun}, x_0, \text{options})$ % MATLAB function
$[x, f_{\text{opt}}, \text{key}, c] = \text{fminunc}(\text{Fun}, x_0, \text{options})$ % Optimization Toolbox
$[x, f_{\text{opt}}, \text{key}, c] = \text{fminsearch}(\text{Fun}, x_0, \text{options}, p_1, \cdots)$ % additional parameters

The definitions of the input and returned arguments are the same as those in function fsolve(). We shall demonstrate the unconstrained optimization problem solutions with some examples.

Example 3.46   *For the objective function $f(x) = 3x_1^2 - 2x_1x_2 + x_2^2 + 4x_1 + 3x_2$, find its minimum value. Here the objective function can be expressed with an anonymous function, and the original problem can be solved with the following statements. The optimal decision variables obtained are $x = [-1.74997728682171, -3.25001058436666]^T$.*

```
>> f=@(x)3*x(1)*x(1)-2*x(1)*x(2)+x(2)*x(2)+4*x(1)+3*x(2);
   [x,f_opt,c,d]=fminsearch(f,[-1.2, 1])
```

In fact, the objective function can be found by solving $\partial f/\partial x_1 = 0$, $\partial f/\partial x_2 = 0$, and it can be found from the following code that the decision variables are $(-7/4, -13/4)$.

```
>> syms x1 x2; f=3*x1^2-2*x1*x2+x2^2+4*x1+3*x2;
   [x1,x2]=solve(diff(f,x1),diff(f,x2))
```

Thus the accuracy of the above obtained numerical solution is not very high. The following MATLAB statements can be used to get a solution of higher accuracy, such that

$$x = [-1.750000022042348, -3.250000020484706]^T.$$

This might be the best possible accuracy which can be achieved under double-precision.

```
>> OPT=optimset; OPT.TolX=1e-20; OPT.LargeScale='off';
   [x,f_opt,c,d]=fminsearch(f,[-1.2, 1],OPT);
```

3.5.4   Linear Programming

Linear programming can be mathematically expressed as

$$\min_{x \text{ s.t. } Ax \leq B} f^{\mathrm{T}} x, \tag{3.80}$$

where the notation s.t. is the abbreviated form of "subject to", meaning that the conditions are to be satisfied, and the conditions are referred to as the constraints. The constraints can further be classified as the linear equations $A_{\mathrm{eq}} x = B_{\mathrm{eq}}$, linear inequality constraints $Ax \leq B$ and the lower and upper bounds of the decision variable x, such that $x_m \leq x \leq x_M$. Thus a more detailed description to the linear programming problem can be written as

$$\min_{x \text{ s.t. } \begin{cases} Ax \leq B \\ A_{\mathrm{eq}} x = B_{\mathrm{eq}} \\ x_m \leq x \leq x_M \end{cases}} f^{\mathrm{T}} x. \tag{3.81}$$

In inequality constraints, the relationship $\leq$ is used. This is again without loss of generality, since the $\geq$ relationship can be converted to $\leq$ problems by multiplying both sides of the inequality by -1.

The function linprog() provided in the Optimization Toolbox can be used to solve linear programming problems, with the syntax

$[x, f_{\mathrm{opt}}, \text{flag}, c] = \text{linprog}(f, A, B, A_{\mathrm{eq}}, B_{\mathrm{eq}}, x_m, x_M, x_0, \text{options})$

where f, A, B, A_{eq}, B_{eq}, x_m and x_M are the same as the ones in the formula. The variable x_0 is the initial search point. If any of the constraints are not available, empty matrices can be used in the function. The options are the optimization control variables. The definitions of the returned variables are the same as in the fminunc() function. An example is used to demonstrate linear programming problem solutions.

Example 3.47   *Consider the linear programming problem given by*

$$\max_{x \text{ s.t. } \begin{cases} x_1 + x_2 + 3x_3 + x_4 = 6 \\ -2x_2 + x_3 + x_4 \leq 3 \\ -x_2 + 6x_3 - x_4 \leq 4 \\ x_1, x_2, x_3, x_4 \geq 0 \end{cases}} [-x_1 + 2x_2 - x_3 + 3x_4].$$

It should first be converted by multiplying the objective function by -1. The new objective function can be written as $x_1 - 2x_2 + x_3 - 3x_4$. Considering the standard form of linear programming problem, the vector f^{T} can be assigned to $[1, -2, 1, -3]$.

The last constraints can be written as $0 \leq x_i \leq \infty$. From the linear equation, it can be seen that $A_{\mathrm{eq}} = [1, 1, 3, 1]$, and $B_{\mathrm{eq}} = 6$. However, the inequality constraints can be written as

$$A = \begin{bmatrix} 0 & -2 & 1 & 1 \\ 0 & -1 & 6 & -1 \end{bmatrix}, \quad B = \begin{bmatrix} 3 \\ 4 \end{bmatrix}.$$

The following commands can then be used to solve the linear programming problem, with the solution $x^* = [0, 1, 0, 5]^T$.

```
>> f=[1,-2,1,-3]; Aeq=[1,1,3,1]; Beq=6;
   A=[0,-2,1,1; 0,-1,6,-1]; B=[3; 4]; xm=zeros(4,1); xM=[];
   [x,f_opt]=linprog(f,A,B,Aeq,Beq,xm,xM)
```

3.5.5   Quadratic Programming

The standard form of quadratic programming is expressed as

$$\min_{x \text{ s.t. } \begin{cases} Ax \leq B \\ A_{eq}x = B_{eq} \\ x_m \leq x \leq x_M \end{cases}} \left(\frac{1}{2} x^T H x + f^T x \right). \tag{3.82}$$

With the `quadprog()` function provided in the Optimization Toolbox, quadratic programming problems can be solved, with the following syntax similar to `linprog()`

$$[x, f_{opt}, \text{flag}, c] = \text{quadprog}(H, f, A, B, A_{eq}, B_{eq}, x_m, x_M, x_0, \text{options})$$

3.5.6   General Nonlinear Programming

The general form of nonlinear constrained optimization problems is expressed as

$$\min_{x \text{ s.t. } G(x) \leq 0} f(x), \tag{3.83}$$

where $x = [x_1, x_2, \cdots, x_n]^T$. The meaning of the mathematical representation is to find a set of x, satisfying all the constraints $G(x) \leq 0$, while minimizing the objective function $f(x)$. Here the constraints can be very complicated, and they can be equalities or inequality constraints.

The constraints can further be classified as linear equalities $A_{eq}x = B_{eq}$, linear inequalities $Ax \leq B$, or lower and upper bounds of the decision vector x_m and x_M such that $x_m \leq x \leq x_M$. Also, nonlinear equations and inequalities can be expressed with a single MATLAB function. Thus the new form of the nonlinear optimization problem can be written as

$$\min_{x \text{ s.t. } \begin{cases} Ax \leq B \\ A_{eq}x = B_{eq} \\ x_m \leq x \leq x_M \\ C(x) \leq 0 \\ C_{eq}(x) = 0 \end{cases}} f(x). \tag{3.84}$$

A MATLAB function `fmincon()` is provided in the Optimization Toolbox, and it can be used directly to solve constrained optimization problems, with the syntax

$$[x, f_{opt}, \text{flag}, c] = \text{fmincon}(F, x_0, A, B, A_{eq}, B_{eq}, \dots$$
$$x_m, x_M, \text{CF}, \text{options}, p_1, p_2, \cdots, p_k)$$

where F is the objective function and CF is the M-function describing the nonlinear constraints. The definitions of other inputs and returned arguments are the same as those in other optimization functions. The arguments p_1, p_2, $\cdots$, p_k are the additional parameters.

Example 3.48   *Consider the following optimization problem*

$$\min_{x \text{ s.t.} \begin{cases} x_1^2 + x_2^2 + x_3^2 - 25 = 0 \\ 8x_1 + 14x_2 + 7x_3 - 56 = 0 \\ x_1, x_2, x_3 \geq 0 \end{cases}} \left[1000 - x_1^2 - 2x_2^2 - x_3^2 - x_1x_2 - x_1x_3 \right].$$

The constraints can be described by a MATLAB function

```
function [c,ceq]=opt_con1(x)
c=[];
ceq=[x(1)*x(1)+x(2)*x(2)+x(3)*x(3)-25; 8*x(1)+14*x(2)+7*x(3)-56];
```

and two arguments are returned, c and ceq, with the former one describing the inequality and the latter describing the nonlinear equations. If certain constraints does not exist, empty matrices should be returned. Since two returned arguments are expected, anonymous functions cannot be used to describe the original problem. M-functions should be used here.

The following statements should be used to solve the original problem, and the optimum point found is $x^* = [3.5121, 0.2170, 3.5522]^{\mathrm{T}}$.

```
>> f=@(x)1000-x(1)*x(1)-2*x(2)*x(2)-x(3)*x(3)-x(1)*x(2)-x(1)*x(3);
   OPT=optimset; OPT.LargeScale='off'; x0=[1;1;1];
   xm=[0;0;0]; xM=[]; A=[]; B=[]; Ae=[]; Be=[];
   [x,f_opt,c,d]=fmincon(f,x0,A,B,Ae,Be,xm,xM,@opt_con1,OPT)
```

Since the second constraint is in fact a linear equation, the nonlinear constraint can further be simplified as

```
function [c,ceq]=opt_con2(x)
ceq=x(1)*x(1)+x(2)*x(2)+x(3)*x(3)-25; c=[];
```

The following statements can be used to describe the original optimization problem

```
>> x0=[1;1;1]; Ae=[8,14,7]; Be=56;
   [x,f_opt,c,d]=fmincon(f,x0,A,B,Ae,Be,xm,xM,@opt_con2,OPT);
```

The same results can be obtained. If the gradients of the objective function are known, the search algorithm is likely to be faster and more accurate. For instance, since the gradients of the objective function to the three variables can be written as

$$\frac{\partial F}{\partial x_1} = -2x_1 - x_2 - x_3, \quad \frac{\partial F}{\partial x_2} = -4x_2 - x_1, \quad \frac{\partial F}{\partial x_3} = -2x_3 - x_1,$$

the new objective function can be rewritten as

```
function [y,Gy]=opt_fun2(x)
y=1000-x(1)*x(1)-2*x(2)*x(2)-x(3)*x(3)-x(1)*x(2)-x(1)*x(3);
Gy=-[2*x(1)+x(2)+x(3); 4*x(2)+x(1); 2*x(3)+x(1)];
```

where Gy *is used to express the gradient.*

The following statements can be used to solve the optimization problem with the gradient information, and the result is the same as the one obtained earlier.

```
>> xm=[0;0;0]; A=[]; B=[]; Ae=[]; Be=[]; ff=optimset; ff.GradObj='on';
   [x,f_opt,c,d]=fmincon(@opt_fun2,x0,A,B,Ae,Be,xm,[],@opt_con1,ff);
```

3.5.7 Global Search Methods in Optimization Problems

In the previously mentioned optimization problem solvers, an initial search point is usually selected, from which optimal solutions may be found. For convex problems such as linear programming and quadratic programming, the search approaches work well. For concavity problems, however, such searching methods may end up with local minima if the initial search point is not properly chosen. Parallel search methods – in particular, evolution type methods – are much better for finding global minima. Among them, the genetic algorithm (GA), particle swarm optimization (PSO) and ant colony algorithms are popular in solving concavity problems. Existing MATLAB functions and toolboxes can be used directly in solving these problems [1, 11].

The Global Optimization Toolbox provides a ga() function to solve optimization problems with a genetic algorithm. The syntax of the function is

$$[x, f_{opt}, \text{flag}, c] = \text{ga}(F, n, A, B, A_{eq}, B_{eq}, x_m, x_M, \text{CF}, \text{OPT})$$

where n is the number of the decision variables, and other arguments are similar to the function fmincon().

3.6 Dynamic Programming and its Applications in Path Planning

The optimization problems discussed so far can be classified as static optimization problems, since the objective functions and constraints are fixed before the solution process begins. In scientific research and engineering, we often meet another category of optimization problems, where the objective functions and other requests can be described as non-static functions. For instance, in making production schedules, the plan of each quarter may depend upon the actual production of the previous year. Thus the objective function is no longer a static function. Dynamic programming techniques have to be employed.

Dynamic programming, which was proposed by Richard Bellman [12] in 1959, is another area of optimization. The theory has many applications, and it is useful in path planning. In this section, the application and solution of dynamic programming problems in optimal path planning of oriented and other graphs are presented.

3.6.1 Matrix Representation of Graphs

Before introducing the representation of graphs, some ideas about graphs are presented. In graph theory, the graphs are constructed by nodes and edges. The edge is the path that connects directly to the nodes. If the edge is one-directional, then the graph is referred to as oriented or directed (also known as a digraph), otherwise it is referred to as an undigraph. There are different ways to represent a graph, and the representation most suitable for computer modeling is the incidence matrix method. If there are n nodes in a graph, it can be described by an $n \times n$ matrix $\boldsymbol{R}$. Assuming that the edge from node i to node j has a weight of k, the matrix element can be expressed by $\boldsymbol{R}(i, j) = k$. Such a matrix is referred to as an incidence matrix. If there is no edge from node i to j, we can assign $\boldsymbol{R}(i, j) = 0$. However, certain algorithms may assign the element value to $\boldsymbol{R}(i, j) = \infty$.

The sparse form representation of incidence matrices can also be used in MATLAB to describe graphs. Suppose that a graph is composed of n nodes and m edges. It is known that the ith edge is from node a_i to node b_i, and the edge has a weighting of w_i, $i = 1, 2, \cdots, m$. Three vectors can be established, and an incidence matrix can be declared in the following matrix

$a = [a_1, a_2, \cdots, a_m, n]$; $b = [b_1, b_2, \cdots, b_m, n]$;   % start and end nodes
$w = [w_1, w_2, \cdots, w_m, 0]$; $R = \mathtt{sparse}(a, b, w)$;   % edge and incidence matrix

Note that the last element in each vector ensures a square incidence matrix $\boldsymbol{R}$. Sparse matrices and ordinary ones can be converted with functions `full()` and `sparse()` respectively.

3.6.2   Optimal Path Planning of Oriented Graphs

The oriented graph representation of optimal path searching can also be encountered in many application areas. With dynamic programming theory, the method of backward derivation is usually adopted from the destination to the starting node. An example will be given to demonstrate the backward derivation methods. Then a computer solution to the same problem will be used with relevant functions in the Bioinformatics Toolbox, which is the MATLAB implementation of the Dijkstra algorithms.

3.6.2.1   Solving Optimal Path Planning Problems of Oriented Graphs

An example of an oriented graph is used to show dynamic programming with applications to the shortest path problem. The problem is to be demonstrated with a manual solution method.

Example 3.49   *Consider the oriented graph problem shown in Fig. 3.15 [13]. The number on top of each edge indicates the distance required to travel from its starting node to the ending node. The shortest path from node ① to node ⑨ has to be calculated.*

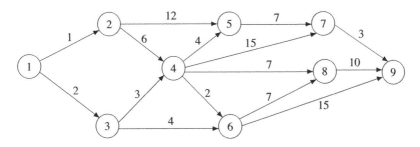

Figure 3.15   Shortest path problem of an oriented graph.

Consider the destination node, node ⑨. Assign the distance at it as 0. Connected to this node, there are three nodes, respectively nodes ⑥, ⑦ and ⑧. Since there is only one edge to travel to the destination node, the shortest distances of these nodes are respectively 15, 3 and 10, that is, the

weights of the edges. From node ⑤ to node ⑦, there is only one edge, thus the shortest distance from node ⑤ is 10, that is, the sum of the distance labeled on node ⑦ plus the weight of the edge. Now let us find the label on node ④. From node ④, there are edges to nodes ⑤, ⑥, ⑦ and ⑧ respectively. Summing the edges separately to the corresponding labels, the sums are respectively 14, 18, 17 and 17, being 14 the smallest. Thus the label on node ④ should be 14. In a similar way, the labels on node ② and ③ can be assigned to 20 and 17, and the label on node ① can be set to 19, the shortest distance expected. From the above, it can be seen that the shortest path is nodes ① → ③ → ④ → ⑤ → ⑦ → ⑨, as shown in Fig. 3.16.

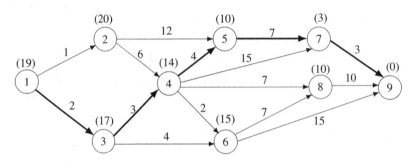

Figure 3.16 Manual solution to the oriented graph problem.

From the above, it can be seen that the method is quite straightforward and easy to understand. However, for large-scale problems, manual derivation is extremely complicated and error-prone. Computer solutions are necessary and will be explained later.

3.6.2.2   Search and Illustration of Oriented Graphs

Some relevant functions are provided in the Bioinformatics Toolbox to solve graphs and shortest path searching problems. For instance, to establish an object for an oriented graph, the function `view()` is used to display the object, and the function `graphshortestpath()` is used to solve directly the shortest path problems. The syntaxes of these functions are

```
P = biograph(R)                      % create an object P
[d,p] = graphshortestpath(P,n₁,n₂)   % find the shortest path
```

where R is the incidence matrix of an oriented graph, expressed as either an ordinary or a sparse matrix. The matrix can be processed by the `biograph()` function to establish an object for the oriented graph. For the oriented graph shown in Fig. 3.15, the matrix element $R(i, j)$ refers to the weight of the edge from node i to node j. Having established the object P, function `graphshortestpath()` can be used to solve the shortest path problem directly. The arguments n_1 and n_2 are respectively the starting end terminal node numbers. The returned variable d is the shortest distance, while vector p returns the node numbers on the shortest path. Other functions can be used to further show the search results graphically.

■ **Example 3.50**   *Considering again the problem in Example 3.49, the MATLAB functions in the Bioinformatics Toolbox can be used to reexamine the original problem.*

From Fig. 3.15, it can be seen that the information of each edge in the graph is summarized as shown in Table 3.4, where the starting node, ending node, and the weight of each edge are obtained. The following commands can be used to enter the incidence matrix R, from which the *graph is generated automatically with the* `view()` *function, and the handle is assigned to* `h`. *The automatically drawn graph is shown in Fig. 3.17(a). Note that when constructing the incidence matrix* $\boldsymbol{R}$*, you have to make sure that it is a square matrix.*

Table 3.4   Edge data.

start	end	weight
1	2	1
1	3	2
2	5	12
2	4	6
3	4	3
3	6	4
4	5	7
4	7	15
4	8	7
4	6	2
5	7	7
6	8	7
6	9	15
7	9	3
8	9	10

```
>> ab=[1 1 2 2 3 3 4 4 4 4 5 6 6 7 8]; bb=[2 3 5 4 4 6 5 7 8 6 7 8 9 9 9];
   w=[1 2 12 6 3 4 4 15 7 2 7 7 15 3 10]; R=sparse(ab,bb,w); R(9,9)=0;
   h=view(biograph(R,[],'ShowWeights','on'))   % show the graph
```

For a given oriented graph object defined in $\boldsymbol{R}$*, the function* `graphshortestpath()` *can be used to find the shortest path, and with the help of the* `view()` *function, the shortest path is shown in red; the result is shown in Fig. 3.17(b). It can be seen that the result obtained is exactly the same as the one obtained manually.*

```
>> [d,p]=graphshortestpath(R,1,9)   % shortest path from node ① to node ⑨
   set(h.Nodes(p),'Color',[1 0.4 0.4])
   edges=getedgesbynodeid(h,get(h.Nodes(p),'ID'));
   set(edges,'LineColor',[1 0 0])    % the path in red is the shortest
```

3.6.3   Optimal Path Planning of Graphs

In practical applications, for instance in route finding in cities, the relevant graphs can also be described by undigraphs, since for nodes A and B, the distance from node A to node B is exactly the same as that from node B to node A. Manipulating undigraphs is also simple. Assume that the edges are assigned first to be one-directional, and the incidence matrix $\boldsymbol{R}_1$ can then be entered. The incidence matrix for the undigraph can be obtained directly from $\boldsymbol{R} = \boldsymbol{R}_1 + \boldsymbol{R}_1^{\mathrm{T}}$. If there are one-way streets in the city, manual modification can be made. For instance, if the edge from node i to j is one way, the element of $\boldsymbol{R}(i, j)$ is retained, and set $\boldsymbol{R}(j, i) = 0$.

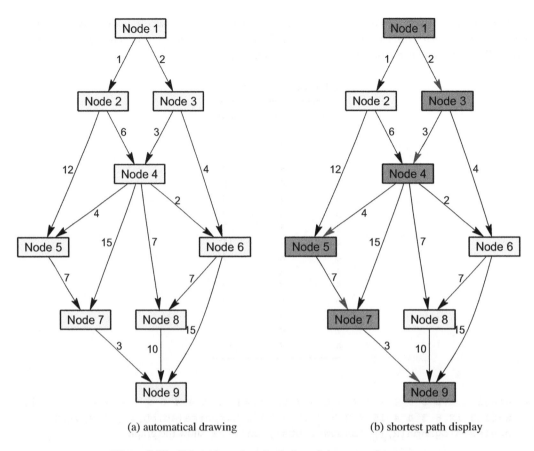

(a) automatical drawing   (b) shortest path display

Figure 3.17   Oriented graph and solution of shortest path program.

For some special undigraphs, the weights from node i to node j may be different from the weight from node j to node i. The elements in matrix R can be modified manually, too, so that even more general problems can be solved with the functions.

3.7   Data Interpolation and Statistical Analysis

3.7.1   Interpolation of One-dimensional Data

Assume that $f(x)$ is an unknown one-dimensional function. Through experiment, a set of n sample points are measured, where the function value $f_1, f_2, \cdots, f_n$ at the independent variable points x_1, $x_2, \cdots, x_n$ are recorded. From the sample points, finding the function values at other values of x is called interpolation of functions.

Some functions, such as `interp1()`, `interp2()` and `polyfit()`, are provided in MATLAB to solve interpolation and curve fitting problems.

The function $y_1 = $ `interp1`$(x, y, x_1, \text{method})$ can be used for one-dimensional data interpolation, where the two vectors x, y are the sample points of independent variables and function values. The argument x_1 stores the x variable values for interpolation, and y_1 is the interpolated results. The available "method" options are `'linear'`, `'cubic'` and `'spline'`, and the last option is usually the best choice.

The function $p = \text{polyfit}(x, y, n)$ can be used to perform polynomial fitting, where x and y are the coordinates of the simple points, and n is the expected degree. The returned argument p contains the coefficients of the polynomial.

Apart from these interpolation functions, more professional interpolation functions are provided in the Spline Toolbox. They can also be used directly for solving interpolation problems.

■ **Example 3.51**   *Assume that the sample points can be generated from a known function* $f(x) = (x^2 - 3x + 5)e^{-5x} \sin x$, *with the following statements. The sample points generated are shown in Fig. 3.18(a).*

```
>> x=0:.12:1; y=(x.^2-3*x+5).*exp(-5*x).*sin(x); plot(x,y,x,y,'o')
```

It can be seen that the "curve" by joining adjacent sample points with lines is not smooth at all. A set of more densely distributed interpolation points should be used, and the interpolation at these points can be obtained with the following statements:

```
>> x1=0:.02:1; y0=(x1.^2-3*x1+5).*exp(-5*x1).*sin(x1);
   y1=interp1(x,y,x1); y2=interp1(x,y,x1,'cubic');
   y3=interp1(x,y,x1,'spline'); plot(x1,[y1',y2',y3'],':',x,y,'o',x1,y0)
```

For different fitting options, the fitting results can be obtained and compared with the theoretical one in Fig. 3.18(b). It can be seen that the default linear interpolation result is not good. The one obtained by the `'spline'` *option is much closer to the theoretical values. In fact, the spline interpolation method is so close to the theoretical one that they cannot be distinguished.*

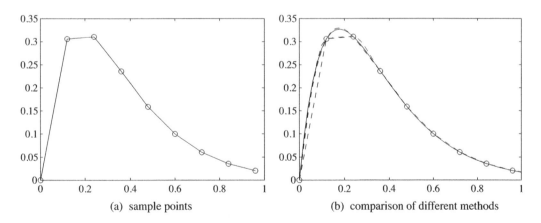

(a) sample points   (b) comparison of different methods

Figure 3.18   Various one-dimensional interpolation results.

The fitting polynomials can be obtained with the generated data

```
>> p1=polyfit(x,y,3), p2=polyfit(x,y,4), p3=polyfit(x,y,5)
```

and the fitting polynomials can be written as

$$P_1(x) = 3.2397x^3 - 5.3334x^2 + 2.1408x + 0.0435$$
$$P_2(x) - 7.1411x^4 + 16.9506x^3 - 13.5159x^2 + 3.6779x + 0.008$$
$$P_3(x) = 10.7256x^5 - 32.8826x^4 + 38.4877x^3 - 20.8059x^2 + 4.5073x + 0.0009$$

The above polynomial can be used to evaluate data fitting results, as shown in Fig. 3.19. It can be seen that the fitting quality of cubic and quartic polynomials is not satisfactory. While the quintic polynomial gives satisfactory fitting.

```
>> y1=polyval(p1,x1); y2=polyval(p2,x1); y3=polyval(p3,x1);
   x0=(x1.^2-3*x1+5).*exp(-5*x1).*sin(x1);
   plot(x1,y0,x1,y1,'--',x1,y2,':',x1,y3,'-.')
```

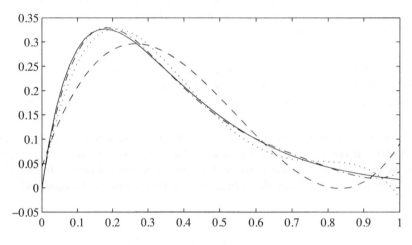

Figure 3.19 Polynomial fitting results.

If the value of x is large, the fitting of the polynomials may not be satisfactory. Thus unless mathematical model requests are made, it is not necessary to use polynomial fitting. Interpolation to the data is sufficient.

3.7.2   Interpolation of Two-dimensional Data

Two kinds of two-dimensional interpolation problems are studied in the section. One is to interpolate the mesh grid data, and the other, scattered points.

The two-dimensional interpolation function `interp2()` has the following syntax

$$z_1 = \text{interp2}(x, y, z, x_1, y_1, \text{method})$$

where x, y and z are the sample points described in a mesh grid form, and x_1, y_1 are the interpolation points. The return variable z_1 stores the interpolated results, and its dimension is the same as that of x_1 and y_1. The same available interpolation methods as in `interp1()` can be used, and `'spline'` is the best choice.

■ **Example 3.52**   *Consider again the function* $z = f(x, y) = (x^2 - 2x)e^{-x^2-y^2-xy}$ *given in Example 2.21. Assume that the sample points are generated in mesh grid form, as shown in Fig. 3.20(a). It can be seen that the linear interpolation surface, as the default form shown in Fig. 3.20(a), is not smooth.*

```
>> [x,y]=meshgrid(-3:.6:3, -2:.4:2); z=(x.^2-2*x).*exp(-x.^2-y.^2-x.*y);
   surf(x,y,z), axis([-3,3,-2,2,-0.7,1.5])
```

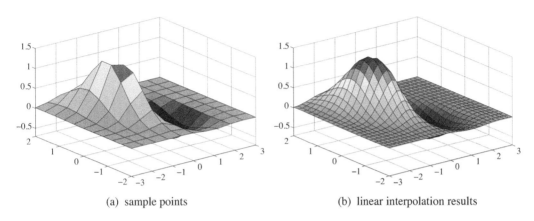

(a) sample points   (b) linear interpolation results

Figure 3.20   Comparison of two-dimensional interpolations.

With more densely distributed interpolating points, the following spline interpolation can be made and the results are shown in Fig. 3.20(b). It can be seen that spline interpolation can achieve accurate interpolation results.

```
>> [x1,y1]=meshgrid(-3:.2:3, -2:.2:2); z1=interp2(x,y,z,x1,y1,'spline');
   surf(x1,y1,z1), axis([-3,3,-2,2,-0.7,1.5])
```

Although the function `interp2()` can be used to solve two-dimensional interpolation problems successfully, this function has its own limitations, since it can only be used to process data given in mesh grid form. In practical applications, the sample data measured are usually given in scattered points form (x_i, y_i, z_i), thus `interp2()` cannot be used. The function `griddata()` should be used instead, with the syntax

$z_1 = $ `griddata(` $x, y, z, x_1, y_1,$ method`)`

where x, y and z are the vectors composed of sample points. The variables x_1 and y_1 are the points to be interpolated. The returned argument is the interpolation values given by variable z_1, with the same size as x_1 and y_1. Here the option is suggested to be `'v4'`.

■ **Example 3.53**   *Consider the prototype function* $z = f(x, y) = (x^2 - 2x)e^{-x^2-y^2-xy}$. *A set of sample points* (x_i, y_i) *is generated randomly in the rectangular domain of* $x \in [-3, 3]$ *and* $y \in [-2, 2]$. *The distribution of the samples is shown in Fig. 3.21(a). It can be seen that the sample points are fairly well distributed. The function values* z_i *can be calculated directly. With the* `griddata()` *function, interpolation can be made and interpolation error is assessed.*

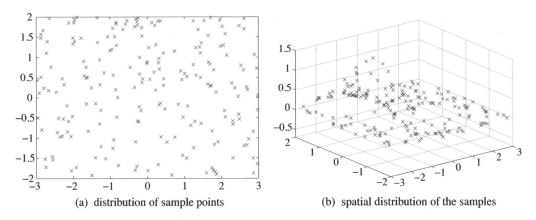

(a) distribution of sample points   (b) spatial distribution of the samples

Figure 3.21   Display of the given samples.

*Here 200 randomly generated sample points are made, and the following statements can be used to construct the vectors **x**, **y** and **z**. Since the points are not mesh grid data, three-dimensional surfaces cannot be drawn from the data; only the spatially distributed points are shown in Fig. 3.21(b).*

```
>> x=-3+6*rand(200,1); y=-2+4*rand(200,1);
   z=(x.^2-2*x).*exp(-x.^2-y.^2-x.*y); plot(x,y,'x') % 2d distribution
   figure, plot3(x,y,z,'x'), axis([-3,3,-2,2,-0.7,1.5])
```

The interpolating points can be selected as mesh grid data, used in Example 2.21. The 'v4' algorithm can be used to perform interpolation on the data, and the interpolated surface is obtained as shown in Fig. 3.22(a), and the error surface is shown in Fig. 3.22(b). It can be seen that two-dimensional interpolation is quite accurate.

```
>> [x1,y1]=meshgrid(-3:.2:3, -2:.2:2); z2=griddata(x,y,z,x1,y1,'v4');
   surf(x1,y1,z2), axis([-3,3,-2,2,-0.7,1.5])
   figure; z0=(x1.^2-2*x1).*exp(-x1.^2-y1.^2-x1.*y1);
   surf(x1,y1,abs(z2-z0))
```

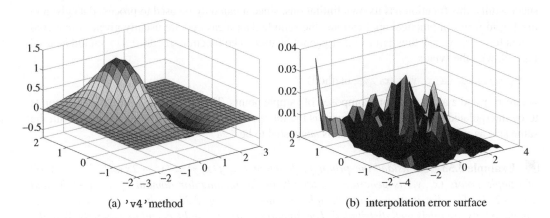

(a) 'v4' method   (b) interpolation error surface

Figure 3.22   Grid data interpolation method.

3.7.3 Least Squares Curve Fitting

Assume that a set of data x_i, y_i, $i = 1, 2, \cdots, N$, is measured. Also it is known that the data satisfy a prototype function $\hat{y}(x) = f(a, x)$, where a is the vector of undetermined coefficients. Least squares fitting is used to minimize the following

$$J = \min_a \sum_{i=1}^{N} [y_i - \hat{y}(x_i)]^2 = \min_a \sum_{i=1}^{N} [y_i - f(a, x_i)]^2, \qquad (3.85)$$

such that J is minimized. The MATLAB function `lsqcurvefit()` can be used to solve least squares curve fitting problems, with the syntax

$$[a, J_m] = \text{lsqcurvefit}(\text{Fun}, a_0, x, y, a_m, a_M, \text{OPT}, p_1, \cdots)$$

where `Fun` is the MATLAB representation of the prototype function, a_0 is the initial search point, and x and y are the vectors of the input and output data. The returned variable a is the undetermined coefficients. The minimized objective function is returned in J_m. The lower and upper bounds of the undetermined variables are assigned in a_m and a_M, and the option is given in variable OPT. This function also allows the use of additional parameters.

■ Example 3.54 *Consider the following statements used to generate a set of sample points x and y.*

```
>> x=0:.1:10; y=0.12*exp(-0.213*x)+0.54*exp(-0.17*x).*sin(1.23*x);
```

The prototype function can be written as $y(x) = a_1 e^{-a_2 x} + a_3 e^{-a_4 x} \sin(a_5 x)$, where a_i, ($i = 1, 2, 3, 4, 5$), are undetermined coefficients. The least squares method can be used to solve this fitting problem. The prototype function can be expressed as an anonymous function, and from it the undetermined coefficients can be found, such that, $a = [0.12, 0.213, 0.54, 0.17, 1.23]^T$, and the fitting error is 1.79×10^{-16}.

```
>> f=@(a,x)a(1)*exp(-a(2)*x)+a(3)*exp(-a(4)*x).*sin(a(5)*x);
   [a,e]=lsqcurvefit(f,[1,1,1,1,1],x,y)
```

3.7.4 Data Sorting

Assume that a set of data is obtained X. Various MATLAB functions can be used to analyze the data. For instance, the maximum and minimum point can be obtained respectively with $[x_M, i] = \max(X)$ and $[x_m, i] = \min(X)$, where x_M and x_m return the maximum and minimum points of the array X, and i is the index of the value. If X is a matrix, the returned variables are vectors, comprising the maximum points and the indexes in each column.

The MATLAB function `sort()` can be used to sort the components in the data vector, and this syntax is exactly the same as the one in the `min()` function. The returned vector is sorted in ascending order. Of course, with proper arrangements, the sorted vector in descending order can also be obtained easily.

3.7.5   Fast Fourier Transform

The Fourier transform of discrete data $x_i, i = 1, 2, \cdots, N$ is the foundation of digital signal processing. Discrete Fourier transformation can be mathematically expressed as

$$X(k) = \sum_{i=1}^{N} x_i e^{-2\pi j(k-1)(i-1)/N}, \quad \text{where } 1 \le k \le N. \tag{3.86}$$

Its inverse transform is defined as

$$x(k) = \frac{1}{N} \sum_{i=1}^{N} X(i) e^{2\pi j(k-1)(i-1)/N}, \quad \text{where } 1 \le k \le N. \tag{3.87}$$

The fast Fourier transform (FFT) technique is the general form of discrete Fourier transforms. In the kernel of MATLAB, a built-in function `fft()` is provided, and it can be used to solve FFT problems with very high efficiency. Another characteristic of the function is that it is not necessary to request the length to satisfy the $2^n - 1$ constraint, although it might be faster if it were.

📋 **Example 3.55**   *Assume that the mathematical function is*

$$x(t) = 12 \sin(2\pi \times 10t + \pi/4) + 5 \cos(2\pi \times 40t).$$

Selecting a step size $h = 0.01$, a set of L time instances t_i can be generated. The function values x_i can be obtained. The corresponding frequencies $f_0 = 1/(ht_f), 2f_0, 3f_0, \cdots$ can be composed, where t_f is the terminating time. The following statements can be used to draw the relationship between frequency and magnitude, as shown in Fig. 3.23(a).

```
>> h=0.01; tf=10; t=0:h:tf; x=12*sin(2*pi*10*t+pi/4)+5*cos(2*pi*40*t);
   X=fft(x); f=t/(h*10); fN=floor(length(f)/2); stem(f(1:fN),abs(X(1:fN)))
```

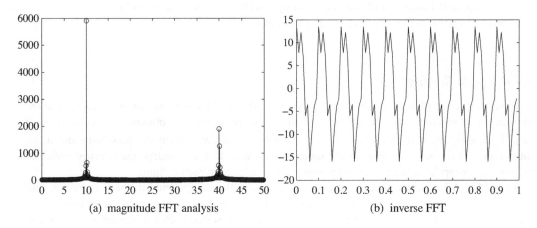

(a) magnitude FFT analysis   (b) inverse FFT

Figure 3.23   FFT analysis of data.

Here only half of the frequency points are used to avoid the aliasing phenomenon. From the frequency analysis results, it can be seen that there are two peak points, at 10 Hz and 40 Hz, respectively. They are the same as in the given function.

The inverse fast Fourier transform can be evaluated with the function ifft()*, and the error occurring when restoring the original function is* 1.1115×10^{-13}. *The inverse FFT result is drawn in Fig. 3.23(b), and it is the same as the original function.*

```
>> ix=real(ifft(X)); ii=1:100; plot(t(ii),x(ii),t(ii),ix(ii),':')
```

MATLAB can also be used to solve two-dimensional and multi-dimensional FFT and inverse FFT problems. The functions are fft2(), ifft2(), fftn() and ifftn().

3.7.6 Data Analysis and Statistics

3.7.6.1 Generation and Validation of Pseudo Random Numbers

Random numbers are often used in system simulation problems. There are two main ways of generating random numbers. One is to generate random numbers electronically with hardware devices, and the other is to generate random numbers mathematically. Random numbers generated in this way are referred to as pseudo random numbers.

There are at least two advantages in using pseudo random numbers in simulation and other applications. One is that the random numbers can be repeated, which makes repeatable experiments possible. The second is that we can generate random signals with any specified properties. For instance, we can choose to generate random numbers satisfying uniform, normal, Poisson or other distributions, as needed. The MATLAB-based pseudo random number generators are discussed in the following.

1) Uniformly distributed random numbers

Reliable uniformly distributed pseudo random numbers in the [0, 1] interval can be generated with the MATLAB function rand(), with $\boldsymbol{x} = \text{rand}(n, m)$, where n and m are the row and column numbers of the matrix.

If one has already obtained a set of uniformly distributed random numbers x_i in the interval [0, 1], the uniformly distributed random numbers in arbitrary interval $[a, b]$ can be generated by the transformation $y_i = a + (b - a)x_i$.

■ **Example 3.56**   *With MATLAB, a set of 30 000 uniformly distributed random numbers can be generated. The mean and variance can be obtained with the following statements, and it returns* $\bar{x} = 0.50264$, $x_{\mathrm{m}} = 0.00001558$, $x_{\mathrm{M}} = 0.999989$. *Also the percentage of random values greater than 0.5 is 50.44667%. It can be seen that the quality of the pseudo random number generated is very high.*

```
>> x=rand(30000,1); y=x(find(x>=0.5)); format long
   [mean(x), min(x), max(x), length(y)/length(x)]
```

2) Normally distributed random numbers

Pseudo random data satisfying standard normal distribution $N(0, 1)$ can be generated with the randn() function, and the syntax is the same as the rand() function. If the pseudo random numbers generated are given in x_i, the random numbers y_i satisfying $N(\mu, \sigma^2)$ can easily be constructed with $y_i = \mu + \sigma x_i$.

3) Poisson distributed random numbers

The probability density function of Poisson distribution is $p_p(x) = \lambda^x e^{-\lambda x}/x!, x = 0, 1, 2, 3, \cdots$. In Statistics Toolbox, many pseudo random number generators are provided. For instance, the `poissrnd()` function can be used to generate Poisson random numbers with the syntax $x = \text{poissrnd}(\lambda, n, m)$, where x is the $n \times m$ random matrix generated, whose elements have a Poisson distribution.

3.7.6.2   Assessing Probability Density from Measured Data

The function `hist()` can be used to assign the vector values into different cells, according to their values. The count in each cell can be obtained. With the vector of the cells, the probability density function can be estimated. Suppose the cells are evenly arranged with the same width Δx, then the function $y = \text{hist}(x, c)$ can be used to get the count in each cell, and the probability density function can be estimated with $y/(\text{length}(x) * \Delta x)$.

Example 3.57   *Generate 30 000 normally distributed random numbers with the* `randn()` *function, and assign 30 cells in the interval* $[-3, 3]$. *The following statements can be used and the approximate probability density function can be estimated from the data.*

```
>> x=randn(30000,1); xx=linspace(-3,3,30);
   y=hist(x,xx); yp=y/(length(x)*(xx(2)-xx(1)));
```

Since the theoretical probability density function is $p(x) = e^{-x^2/2}/\sqrt{2\pi}$, *the estimated one can be obtained as shown in Fig. 3.24, together with the theoretical one, and it can be seen that they agree well.*

```
>> p0=exp(-xx.^2/2)/sqrt(2*pi); bar(xx,yp); hold on; plot(xx,p0)
```

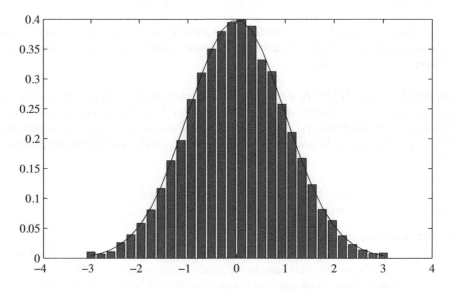

Figure 3.24   Comparisons of probability density function under normal distribution.

The Statistics Toolbox also provides a lot of other functions to theoretically evaluate probability density functions. For instance, the probability density function of normal distribution can be evaluated with $p = $ `normpdf`(x, μ, σ), and they can be used directly.

Experimental data or randomly generated data can be processed easily with the existing MATLAB functions. For instance, the function `mean()` can be used to evaluate the mean of a vector, `std()` can be used to compute the standard deviation, `median()` can be used to get the median value, and `cov()` can be used to get the covariance matrix for the measured data in matrices.

3.7.6.3 Correlation Analysis

Assume that a set of data x_i, y_i, $(i = 1, 2, \cdots, n)$, is measured. The following formula can be used to calculate the correlation coefficients

$$r = \frac{\sqrt{\sum(x_i - \bar{x})(y_i - \bar{y})}}{\sqrt{\sum(x_i - \bar{x})}\sqrt{\sum(y_i - \bar{y})}}. \tag{3.88}$$

In MATLAB, function `corrcoef()` is provided to calculate the correlation the coefficient matrix. In fact, the correlation matrix can be obtained by normalizing the covariance matrix on a column basis.

Example 3.58   *With MATLAB functions, $10\,000 \times 5$ normally distributed pseudo random numbers can be generated. The functions* `mean()` *and* `std()` *can be used to evaluate the mean value and standard deviation of the random numbers. The function* `corrcoef()` *can be used to find the correlation coefficient of the random matrix*

```
>> S=randn(10000,5); M=mean(S), D=std(S), V=corrcoef(S)
```

and it can be found that

$$\bar{S} = [0.0011, 0.0066, 0.0009, 0.0264, 0.0101], \sigma(S) = [1.0011, 1.0036, 1.0049, 1.0058, 1.0061],$$

and the correlation coefficient matrix can be obtained as

$$V = \begin{bmatrix} 1 & 0.0119 & 0.0051 & -0.0114 & -0.0011 \\ 0.0119 & 1 & 0.0093 & -0.0012 & 0.0071 \\ 0.0051 & 0.0093 & 1 & 0.0048 & 0.0095 \\ -0.0114 & -0.0012 & 0.0048 & 1 & -0.0017 \\ -0.0011 & 0.0071 & 0.0095 & -0.0017 & 1 \end{bmatrix}.$$

It can be seen from the results, that the mean and variance of the random numbers generated satisfy well the standard $N(0, 1)$ distribution. Also, it can be seen that the correlation coefficient matrix is very close to an identity matrix, meaning that the random numbers generated are independent.

For the sample points x_i, the autocorrelation function can be evaluated as

$$C_{xx}(k) = \frac{1}{N} \sum_{l=1}^{n-k-1} x(l)x(k+l), \quad 0 \le k \le m - 1, \tag{3.89}$$

where $m < n$. Similarly, the cross-correlation function for two sequences x_i and y_i can be evaluated as

$$C_{xy}(k) = \frac{1}{N} \sum_{l=1}^{n-k-1} x(l)y(k+l), \quad 0 \le k \le m - 1. \tag{3.90}$$

In MATLAB, functions `autocorr()` and `crosscorr()` can be used to calculate and draw autocorrelation functions and cross-correlation functions with the syntaxes

```
[C_xx, m, Bounds] = autocorr(x, n, n_σ);
[C_xy, m, Bounds] = crosscorr(x, y, n, n_σ);
```

where x and y are data vectors. The default value of m is 20. The argument n_σ is related to the standard deviations. The returned arguments C_{xx} and C_{xy} are respectively the autocorrelation and cross-correlation functions. If no argument is returned, correlation functions with trust regions will be drawn automatically.

Example 3.59   *The following MATLAB commands can be used to observe the correlation of the pseudo random numbers generated, with the auto- and cross- correlation functions shown in Figs. 3.25(a) and (b). It can be seen that the statistical behavior of the pseudo random numbers generated is satisfactory.*

```
>> x=randn(1000,1);   % generate 1000 × 1 normally distributed random numbers
   y=randn(800,1);    % generate 800 × 1 numbers, with different seed
   autocorr(x,10);    % autocorrelation function
   figure; crosscorr(x,y); ylim([-0.1,1]) % cross-correlation function
```

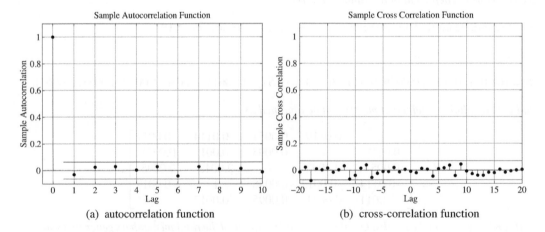

(a) autocorrelation function   (b) cross-correlation function

Figure 3.25   Correlation analysis of random data.

3.7.6.4   Estimation of Power Spectrum Densities

For the discrete sample data vector y, the MATLAB function `psd()` can be used to evaluate its power spectral density. However, it is found that the result obtained is not satisfactory. The Welch transform based estimation algorithm is introduced here instead [14].

Assume that n is the length of the data vector $\mathbf{y}$ to be processed. These data can be divided into $\mathcal{K} = [n/m]$ groups with a vector length of m such that

$$x^{(i)}(k) = y[k + (i - 1)m], \quad 0 < k \le m - 1, \quad 1 \le i \le \mathcal{K}. \tag{3.91}$$

With the Welch algorithm, the following $\mathcal{K}$ equations can be established.

$$J_{\mathrm{m}}^{(i)}(\omega) = \frac{1}{mU} \left| \sum_{k=0}^{m-1} x^{(i)}(k) w(k) e^{-j\omega k} \right|^2, \tag{3.92}$$

where $w(k)$ is the data processing window function. For instance, it can be the Hamming window defined as

$$w(k) = a - (1 - a) \cos\left(\frac{2\pi k}{m - 1} \right), \quad k = 0, \cdots, m - 1. \tag{3.93}$$

If $a = 0.54$, and

$$U = \frac{1}{m} \sum_{k=0}^{m-1} w^2(k), \tag{3.94}$$

the power spectral density of the signal can be estimated with

$$P_{\mathrm{xx}}^{w}(\omega) = \frac{1}{\mathcal{K}} \sum_{j=1}^{\mathcal{K}} J_{\mathrm{m}}^{(i)}(\omega). \tag{3.95}$$

The following procedure can be used to compute the power spectral densities of the signal [8].

- Calculate $X_m^{(i)}(l) = \sum_{k=0}^{m-1} x^{(i)}(k) w(k) e^{-j[2\pi/(m\Delta t)]lk}$ in groups with the `fft()` function.

- In each group, compute $\left| X_m^{(i)}(l) \right|^2$, and the accumulative sum $Y(l) = \sum_{i=1}^{\mathcal{K}} \left| X_m^{(i)}(l) \right|^2$.

- From the following formula, the power spectral density can be obtained

$$P_{\mathrm{xx}}^{w}\left(\frac{2\pi}{m\Delta t} l \right) = \frac{1}{\mathcal{K}mU} Y(l). \tag{3.96}$$

The function `fft()` is used in calculating $X_m^{(i)}$:

1. MATLAB function `fft()` can be used to compute discrete Fourier transforms. The resulting P_{xx}^{w} should be multiplied by Δt.
2. To increase the effectiveness of the algorithm, m should be chosen as $2^k - 1$ for integer k.

A MATLAB function `psd_estm()` can be written to estimate the power spectral density of a given sequence, listed as

```
function [Pxx,f]=psd_estm(y,m,T,a)
if nargin==3, a=0.54; end
k=[0:m-1]; Y=zeros(1,m); m2=floor(m/2); f=k(1:m2)*2*pi/(length(k)*T);
w=a-(1-a)*cos(2*pi*k/(m-1)); K=floor(length(y)/m); U=sum(w.^2)/m;
for i=1:K, xi=y((i-1)*m+k+1)'; Xi=fft(xi.*w); Y=Y+abs(Xi).^2; end
Pxx=Y(1:m2)*T/(K*m*U);
```

The function can be called with the syntax $[P_{xx}, f] = \mathrm{psd\_estm}(y, m, T, a)$. In this function, in order to avoid the aliasing phenomenon, half of the transformation data should be selected. In this function, the definitions of y and m are the same, and T is the sampling interval Δt. The returned arguments f and P_{xx} are respectively the frequency and the power spectral density.

Exercises

3.1 Experience the computation efficiency by taking the inverse of square random matrices of sizes $n = 550$ and $n = 1550$, respectively. Measure the time required and also validate the results.

3.2 Analyze the following matrices, for example, by taking the determinant, trace, rank, norms and characteristic polynomials.

$$A = \begin{bmatrix} 7.5 & 3.5 & 0 & 0 \\ 8 & 33 & 4.1 & 0 \\ 0 & 9 & 103 & -1.5 \\ 0 & 0 & 3.7 & 19.3 \end{bmatrix}, \quad B = \begin{bmatrix} 5 & 7 & 6 & 5 \\ 7 & 10 & 8 & 7 \\ 6 & 8 & 10 & 9 \\ 5 & 7 & 9 & 10 \end{bmatrix},$$

$$C = \begin{bmatrix} 1 & 2 & 3 & 4 \\ 5 & 6 & 7 & 8 \\ 9 & 10 & 11 & 12 \\ 13 & 14 & 15 & 16 \end{bmatrix}, \quad D = \begin{bmatrix} 3 & -3 & -2 & 4 \\ 5 & -5 & 1 & 8 \\ 11 & 8 & 5 & -7 \\ 5 & -1 & -3 & -1 \end{bmatrix}.$$

3.3 Construct a Vandermonde matrix for a vector $V = [1, 2, 3, 4, 5]$. Find the characteristic polynomial, and validate the Cayley–Hamilton theorem. If the error is too large, try to use the `poly1()` function instead, and see whether better results can be achieved.

3.4 For the matrices in Problem (3.2), do a triangular factorization and also find the eigenvalues and eigenvector matrices. For the symmetric matrix B, perform a Cholesky factorization and validate the results.

3.5 Solve the following linear algebraic equations, and validate the solutions. Try to find the analytical solutions to the equations.

$$(a) \begin{bmatrix} 7 & 2 & 1 & -2 \\ 9 & 15 & 3 & -2 \\ -2 & -2 & 11 & 5 \\ 1 & 3 & 2 & 13 \end{bmatrix} X = \begin{bmatrix} 4 \\ 7 \\ -1 \\ 0 \end{bmatrix}, \quad (b) \begin{bmatrix} 1 & 3 & 2 & 13 \\ 7 & 2 & 1 & -2 \\ 9 & 15 & 3 & -2 \\ -2 & -2 & 11 & 5 \end{bmatrix} X = \begin{bmatrix} 9 & 0 \\ 6 & 4 \\ 11 & 7 \\ -2 & -1 \end{bmatrix}.$$

3.6   Analyze the following complex matrix. Find its inverse, rank, eigenvalues and eigenvectors. Validate the Cayley–Hamilton theorem for the complex matrix.

$$A = \begin{bmatrix} 0.2368 & 0.2471 & 0.2568 & 1.2671 \\ 1.1161 & 0.1254 & 0.1397 & 0.1490 \\ 0.1582 & 1.1675 & 0.1768 & 0.1871 \\ 0.1968 & 0.2071 & 0.2168 & 0.2271 \end{bmatrix} + \begin{bmatrix} 0.1345 & 0.1768 & 0.1852 & 1.1161 \\ 1.2671 & 0.2017 & 0.7024 & 0.2721 \\ -0.2836 & -1.1967 & 0.3558 & -0.2078 \\ 0.3536 & -1.2345 & 2.1185 & 0.4773 \end{bmatrix} i.$$

3.7   Find the reduced row echelon form of the following equations

(a) $\begin{cases} 6x_1 + x_2 + 4x_3 - 7x_4 - 3x_5 = 0 \\ -2x_1 - 7x_2 - 8x_3 + 6x_4 = 0 \\ -4x_1 + 5x_2 + x_3 - 6x_4 + 8x_5 = 0 \\ -34x_1 + 36x_2 + 9x_3 - 21x_4 + 49x_5 = 0 \\ -26x_1 - 12x_2 - 27x_3 + 27x_4 + 17x_5 = 0 \end{cases}$ (b) $A = \begin{bmatrix} -1 & 2 & -2 & 1 & 0 \\ 0 & 3 & 2 & 2 & 1 \\ 3 & 1 & 3 & 2 & -1 \end{bmatrix}.$

3.8   Validate for arbitrary 3×3 and 4×4 matrices that the Cayley–Hamilton theorem satisfies.

3.9   Find the rank and Moore–Penrose generalized inverse of the following matrices and check whether the Moore–Penrose generalized inverse conditions are satisfied.

$$A = \begin{bmatrix} 2 & 2 & 3 & 1 \\ 2 & 2 & 3 & 1 \\ 4 & 4 & 6 & 2 \\ 1 & 1 & 1 & 1 \\ -1 & -1 & -1 & 3 \end{bmatrix}, \quad B = \begin{bmatrix} 4 & 1 & 2 & 0 \\ 1 & 1 & 5 & 15 \\ 3 & 1 & 3 & 5 \end{bmatrix}.$$

3.10   For the special matrix A given below, find, using an analytical method, the inverse matrix and eigenvalues of the matrix. Find the state transition matrix e^{At}.

$$A = \begin{bmatrix} -9 & 11 & -21 & 63 & -252 \\ 70 & -69 & 141 & -421 & 1684 \\ -575 & 575 & -1149 & 3451 & -13801 \\ 3891 & -3891 & 7782 & -23345 & 93365 \\ 1024 & -1024 & 2048 & -6144 & 24572 \end{bmatrix}.$$

3.11   For sinusoidal, cosine and logarithmic function, it is known that the Taylor series can be written as shown below. Write a MATLAB code to implement the following matrix functions.

(a) $\cos A = I - \dfrac{1}{2!}A^2 + \dfrac{1}{4!}A^4 - \dfrac{1}{6!}A^6 + \cdots + \dfrac{(-1)^n}{(2n)!}A^{2n} + \cdots,$

(b) $\arcsin A = A + \dfrac{1}{2\cdot3}A^3 + \dfrac{1\cdot3}{2\cdot4\cdot5}A^5 + \dfrac{1\cdot3\cdot5}{2\cdot4\cdot6\cdot7}A^7 + \cdots + \dfrac{(2n)!}{2^{2n}(n!)^2(2n+1)}A^{2n+1} + \cdots,$

(c) $\ln A = A - I - \dfrac{1}{2}(A-I)^2 + \dfrac{1}{3}(A-I)^3 - \dfrac{1}{4}(A-I)^4 + \cdots + \dfrac{(-1)^{n+1}}{n}(A-I)^n + \cdots.$

3.12   For the following measured data, find the numerical derivatives. Also evaluate the numerical integral using the trapezoidal method.

x_i	0	0.1	0.2	0.3	0.4	0.5	0.6	0.7	0.8	0.9	1	1.1	1.2
y_i	0	2.2077	3.2058	3.4435	3.241	2.8164	2.311	1.8101	1.3602	0.9817	0.6791	0.4473	0.2768

3.13 Solve the following limit problems with MATLAB.

(a) $\displaystyle\lim_{x\to\infty} \frac{(x+2)^{x+2}(x+3)^{x+3}}{(x+5)^{2x+5}}$, (b) $\displaystyle\lim_{x\to-1}\lim_{y\to 2} \frac{x^2y + xy^3}{(x+y)^3}$.

3.14 Solve the following calculus problems with MATLAB Symbolic Toolbox.

(a) Indefinite integral $\displaystyle\int \frac{x^3 + 3x^2 - 5}{(x^2 - 2x - 6)(x^3 + x + 1)}\,\mathrm{d}x$.

(b) For function $x^2 \sin(\cos x^2)\cos x$, write out its Taylor series expansion with 20 terms.

(c) $\displaystyle\lim_{n\to\infty}\left(1 + \frac{1}{2} + \cdots + \frac{1}{n} - \ln n\right)$.

3.15 For the function $\displaystyle f(x, y) = \int_0^{xy} \mathrm{e}^{-t^2}\,\mathrm{d}t$, find $\dfrac{x}{y}\dfrac{\partial^2 f}{\partial x^2} - 2\dfrac{\partial^2 f}{\partial x \partial y} + \dfrac{\partial^2 f}{\partial y^2}$.

3.16 For $f(x) = \mathrm{e}^{-5x}\sin\left(3x + \dfrac{\pi}{3}\right)$, please find the integral function $R(t) = \displaystyle\int_0^t f(x)f(t + x)\,\mathrm{d}x$.

3.17 Rewrite the `lorenzeq()` function such that β, σ and ρ can be specified as additional parameters. For different combinations of these parameters, draw the phase space trajectories.

3.18 Consider the well-established Rössoler equation

$$\begin{cases} \dot{x} = y + z \\ \dot{y} = x + ay \\ \dot{z} = b + (x - c)z. \end{cases}$$

For $a = b = 0.2$, $c = 5.7$, draw the three-dimensional phase space plot, and animate the trajectory. Also draw the projection of the curve on the x-y plane.

3.19 Consider the reaction speed equation for a chemical reaction system [15]

$$\begin{cases} \dot{y}_1 = -0.04y_1 + 10^4 y_2 y_3 \\ \dot{y}_2 = 0.04y_1 - 10^4 y_2 y_3 - 3 \times 10^7 y_2^2 \\ \dot{y}_3 = 3 \times 10^7 y_2^2, \end{cases}$$

with initial values $y_1(0) = 1$, $y_2(0) = y_3(0) = 0$. These are often regarded as stiff equations. Try to see whether function `ode45()` can be used to solve the equations. If the method fails, how can we get the correct equation solutions?

3.20 Select a set of state variables to convert the ODE to first-order explicit ones

$$\begin{cases} \ddot{x}\sin\dot{y} + \ddot{y}^2 = -2xy + x\ddot{x}\dot{y} \\ x\ddot{x}\ddot{y} + \cos\ddot{y} = 3y\dot{x}. \end{cases}$$

3.21 Consider the following nonlinear differential equation

$$\begin{cases} \dot{x} = -y + xf\left(\sqrt{x^2 + y^2}\right) \\ \dot{y} = x + yf\left(\sqrt{x^2 + y^2}\right), \end{cases}$$

where $f(r) = r^2 \sin(1/r)$. It was pointed out in [16] that the original ODE has multiple limit cycles, when $r = 1/(n\pi)$, $n = 1, 2, 3, \cdots$. Observe numerically the limit cycle behaviors.

3.22 The Chua's circuit is a famous ODE in chaos studies [17]

$$\begin{cases} \dot{x} = \alpha[y - x - f(x)] \\ \dot{y} = x - y + z \\ \dot{z} = -\beta y - \gamma z, \end{cases}$$

where $f(x)$ is a piecewise diode function $f(x) = bx + (a - b)(|x + 1| - |x - 1|)/2$, and $a < b < 0$. Describe and solve such an ODE with MATLAB, and draw the phase space trajectory for $\alpha = 15$, $\beta = 20$, $\gamma = 0.5$, $a = -120/7$, $b = -75/7$, and initial states $x(0) = -2.121304$, $y(0) = -0.066170$, $z(0) = 2.881090$.

3.23 Lotka–Volterra predator–prey equation is given by

$$\begin{cases} \dot{x}(t) = 4x(t) - 2x(t)y(t) \\ \dot{y}(t) = x(t)y(t) - 3y(t), \end{cases}$$

with initial states $x(0) = 2$, $y(0) = 3$. Solve the equation and draw the phase plane trajectory.

3.24 Solve both analytically and numerically the following ODE and compare the results

$$\begin{cases} \ddot{x}(t) = -2x(t) - 3\dot{x}(t) + e^{-5t}, & x(0) = 1, \dot{x}(0) = 2 \\ \ddot{y}(t) = 2x(t) - 3y(t) - 4\dot{x}(t) - 4\dot{y}(t) - \sin t, & y(0) = 3, \dot{y}(0) = 4. \end{cases}$$

3.25 Consider the following simple linear ODE

$$y^{(4)} + 3y^{(3)} + 3\ddot{y} + 4\dot{y} + 5y = e^{-3t} + e^{-5t}\sin(4t + \pi/3),$$

with initial states $y(0) = 1$, $\dot{y}(0) = \ddot{y}(0) = 1/2$, $y^{(3)}(0) = 0.2$. Find the analytical and numerical solutions and compare the results.

3.26 Assume that the position of an artillery shell is $(x(t), y(t))$ at time t. Their relationship with its velocity $v(t)$ and the angle $\theta(t)$ with horizontal line satisfies the following ODE

$$\begin{cases} \dot{v}(t)\cos\theta(t) - \dot{\theta}(t)v(t)\sin\theta(t) = -kv^2(t)\cos\theta(t) \\ \dot{v}(t)\sin\theta(t) + \dot{\theta}(t)v(t)\cos\theta(t) = -kv^2(t)\sin\theta(t) - mg \\ \dot{x}(t) = v(t)\cos\theta(t) \\ \dot{y}(t) = v(t)\sin\theta(t), \end{cases}$$

where k is a constant, and m is the mass of the shell. The initial conditions are $x(0) = y(0) = 0$, $v(0) = v_0$, $\theta(0) = \theta_0$. Draw the trajectory of the shell.

3.27 The following equations are usually regarded as stiff equations in ODE textbooks. Solve the ODEs with conventional and stiff equation solvers, and compare the results with analytical solutions.

(a) $\begin{cases} \dot{y}_1 = 9y_1 + 24y_2 + 5\cos t - \sin t/3, & y_1(0) = 1/3 \\ \dot{y}_2 = -24y_1 - 51y_2 - 9\cos t + \sin t/3, & y_2(0) = 2/3, \end{cases}$

(b) $\begin{cases} \dot{y}_1 = -0.1y_1 - 49.9y_2, & y_1(0) = 1 \\ \dot{y}_2 = y - 50y_2, & y_2(0) = 2 \\ \dot{y}_3 = 70y_2 - 120y_3, & y_3(0) = 1, \end{cases}$

3.28 A fractional-order plant model is given as

$$G(s) = \frac{(s^{0.4} + 0.4s^{0.2} + 0.5)}{\sqrt{s}(s^{0.2} + 0.02s^{0.1} + 0.6)^{0.4}(s^{0.3} + 0.5)^{0.6}}.$$

Finds its step response and the time domain response under the input $u(t) = e^{-0.3t}\sin t^2$.

3.29 Find all possible solutions to this variation of the Riccati equation and validate the results

$$AX + XD - XBX + C = 0,$$

where $A = \begin{bmatrix} 2 & 1 & 9 \\ 9 & 7 & 9 \\ 6 & 5 & 3 \end{bmatrix}$, $B = \begin{bmatrix} 0 & 3 & 6 \\ 8 & 2 & 0 \\ 8 & 2 & 8 \end{bmatrix}$, $C = \begin{bmatrix} 7 & 0 & 3 \\ 5 & 6 & 4 \\ 1 & 4 & 4 \end{bmatrix}$, $D = \begin{bmatrix} 3 & 9 & 5 \\ 1 & 2 & 9 \\ 3 & 3 & 0 \end{bmatrix}$.

3.30 Find all the possible solutions of $\sin(x - y) = 0$, $\cos(x + y) = 0$ in $-2\pi < x, y < 2\pi$.

3.31 Solve the following optimization problems

(a) $\min\limits_{x \text{ s.t.} \begin{cases} 4x_1^2 + x_2^2 \leq 4 \\ x_1, x_2 \geq 0 \end{cases}} (x_1^2 - 2x_1 + x_2)$, (b) $\max\limits_{x \text{ s.t.} x_1+x_2+5=0} [(x_1 - 1)^2 - (x_2 - 1)^2]$,

(c) $\min\limits_{x \text{ s.t.} \begin{cases} x_1/4 - 60x_2 - x_3/25 + 9x_4 \leq 0 \\ x_1/2 - 90x_2 - x_3/50 + 3x_4 \leq 0 \\ x_3 \leq 1 \\ x_1, x_2, x_3, x_4 \geq 0 \end{cases}} \left(-\frac{3}{4}x_1 + 150x_2 - \frac{1}{50}x_3 + 6x_4\right)$,

where (b) is an optimization problem with two variables. Use graphical methods to explain the results.

3.32 Solve the linear programming problem with double indexes

$$\min \quad 2800(x_{11} + x_{21} + x_{31} + x_{41}) + 4500(x_{12} + x_{22} + x_{32}) + 6000(x_{13} + x_{23}) + 7300x_{14}.$$

$$x \text{ s.t.} \begin{cases} x_{11}+x_{12}+x_{13}+x_{14} \geq 15 \\ x_{12}+x_{13}+x_{14}+x_{21}+x_{22}+x_{23} \geq 10 \\ x_{13}+x_{14}+x_{22}+x_{23}+x_{31}+x_{32} \geq 20 \\ x_{14}+x_{23}+x_{32}+x_{41} \geq 12 \\ x_{ij} \geq 0, (i=1,2,3,4, j=1,2,3,4) \end{cases}$$

3.33 Solve the following quadratic programming problem

$$\min\limits_{x \text{ s.t.} \begin{cases} x_1 + x_2 + x_3 + x_4 \leq 5 \\ 3x_1 + 3x_2 + 2x_3 + x_4 \leq 10 \\ x_1, x_2, x_3, x_4 \geq 0 \end{cases}} \left[(x_1 - 1)^2 + (x_2 - 2)^2 + (x_3 - 3)^2 + (x_4 - 4)^2\right].$$

3.34 Solve the following optimization problems [18]:

(a) $\min k$, q,w,k s.t. $\begin{cases} q_3+9.625q_1w+16q_2w+16w^2+12-4q_1-q_2-78w=0 \\ 16q_1w+44-19q_1-8q_2-q_3-24w=0 \\ 2.25-0.25k \leq q_1 \leq 2.25+0.25k \\ 1.5-0.5k \leq q_2 \leq 1.5+0.5k \\ 1.5-1.5k \leq q_3 \leq 1.5+1.5k \end{cases}$

(b) $\min k$. q,k s.t. $\begin{cases} g(q) \leq 0 \\ 800-800k \leq q_1 \leq 800+800k \\ 4-2k \leq q_2 \leq 4+2k \\ 6-3k \leq q_3 \leq 6+3k \end{cases}$

where $g(q) = 10q_2^2q_3^3 + 10q_2^3q_3^2 + 200q_2^2q_3^2 + 100q_2^3q_3 + q_1q_2q_3^2 + q_1q_2^2q_3 + 1000q_2q_3^3 + 8q_1q_3^2 + 1000q_2^2q_3 + 8q_1q_2^2 + 6q_1q_2q_3 - q_1^2 + 60q_1q_3 + 60q_1q_2 - 200q_1$.

3.35 A group of challenging optimization benchmark problems can be solved with MATLAB directly. Please find the global solutions of them.
(a) De Jong problem [19]

$$J = \min_x x^\mathrm{T} x = \min_x(x_1^2 + x_2^2 + \cdots + x_p^2), \text{ and } x_i \in [-512, 512]$$

where $i = 1, \cdots, p$. The theoretical solutions are $x_1 = \cdots = x_p = 0$.
(b) Griewangk benchmark problem

$$J = \min_x \left(1 + \sum_{i=1}^{p} \frac{x_i^2}{4000} - \prod_{i=1}^{p} \cos \frac{x_i}{\sqrt{i}}\right), \text{ where } x_i \in [-600, 600].$$

(c) Ackley benchmark problem [20]

$$J = \min_x \left[20 + 10^{-20} \exp\left(-0.2\sqrt{\frac{1}{p}\sum_{i=1}^{p} x_i^2}\right) - \exp\left(\frac{1}{p}\sum_{i=1}^{p} \cos 2\pi x_i\right)\right].$$

3.36 In the graphs shown in Figs. 3.26(a) and (b), find the shortest paths from node A to node B.

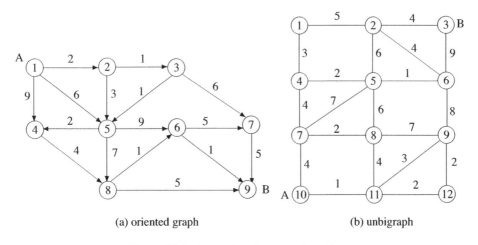

(a) oriented graph   (b) unbigraph

Figure 3.26 Graphs for shortest path problem.

3.37 From the function $f(x, y) = \dfrac{1}{3x^3 + y}e^{-x^2-y^4}\sin\left(xy^2 + x^2 y\right)$, generate a set of sample points. Interpolate these data and see whether the original surface can be reconstructed.

3.38 Generate a sequence of random numbers. Move the sequence leftward 10 places to construct another sequence. Find the cross-correlation function of the two sequences.

3.39 For the set of measured data shown below, draw a smooth curve with interpolation algorithms in $x \in (-2, 4.9)$ and compare the interpolation results.

x_i	−2	−1.7	−1.4	−1.1	−0.8	−0.5	−0.2	0.1	0.4	0.7	1	1.3
y_i	0.1029	0.1174	0.1316	0.1448	0.1566	0.1662	0.1733	0.1775	0.1785	0.1764	0.1711	0.1630
x_i	1.6	1.9	2.2	2.5	2.8	3.1	3.4	3.7	4	4.3	4.6	4.9
y_i	0.1526	0.1402	0.1266	0.1122	0.0977	0.0835	0.0702	0.0579	0.0469	0.0373	0.0291	0.0224

3.40 Assume that a set of measured data are given by

x_i	0.1	0.2	0.3	0.4	0.5	0.6	0.7	0.8	0.9	1
y_i	2.3201	2.6470	2.9707	3.2885	3.6008	3.9090	4.2147	4.5191	4.8232	5.1275

(a) Compare the interpolation results with different algorithms.

(b) Suppose that the prototype function is $y(x) = ax + bx^2 e^{-cx} + d$. Find the least squares solutions to the parameters a, b, c, d.

3.41 Generate 30 000 pseudo random numbers satisfying $N(0.5, 1.4^2)$. Check whether the mean and variance of the generated data are correct or not. Observe by histograms whether the probability distribution agrees well with the theoretical one.

3.42 Observe the pseudo random number generator `randn()` in MATLAB. See whether the seed has any effect on the correlation function and power spectral densities.

3.43 Derive an autocorrelation function of Gaussian distribution, with probability density function given by $f(t) = \dfrac{1}{3\sqrt{2\pi}} e^{-t^2/3^2}$. Generate a set of random numbers and see whether the autocorrelation function of the generated data agrees well with the theoretical curves.

References

[1] D Xue, Y Q Chen. MATLAB solutions to advanced applied mathematical problems. Beijing: Tsinghua University Press, 2nd Edition, 2008. In Chinese

[2] J J Dongarra, J R Bunch, C B Moler, *et al.* LINPACK user's guide. Philadelphia: Society of Industrial and Applied Mathematics (SIAM), 1979

[3] C B Moler, C F Van Loan. Nineteen dubious ways to compute the exponential of a matrix. SIAM Review, 1979, 20:801–836

[4] Wuhan University Press, Shandong University Press. Computational algorithms. Beijing: People's Education Press, 1979. In Chinese

[5] G E Forsythe, M A Malcolm, C B Moler. Computer methods for mathematical computations. Englewood Cliffs: Prentice-Hall, 1977

[6] K J Åström. Introduction to stochastic control theory. London: Academic Press, 1970

[7] E E L Mitchell, J S Gauthier. Advanced continuous simulation language (ACSL) – user's manual. Mitchell & Gauthier Associates, 1987

[8] D Xue. Analysis and computer-aided design of nonlinear systems with Gaussian inputs. PhD thesis, the University of Sussex, 1992

[9]   J Valsa, L Brančik. Approximate formulae for numerical inversion of Laplace transforms. International Journal of Numerical Modelling: Electronic Networks, Devices and Fields, 1998, 11(3):153–166

[10]   Valsa J. Numerical inversion of Laplace transforms in MATLAB, 2011

[11]   The MathWorks Inc. Global optimization toolbox user's guide, 2010

[12]   R Bellman. Dynamic programming. Princeton, NJ: Princeton University Press, 1957

[13]   Y X Lin. Dynamic programming and sequential optimization. Zhengzhou: Henan University Press, 1997. In Chinese

[14]   A V Oppenheim, R W Schafer. Digital signal processing. Englewood Cliffs: Prentice-Hall, 1975

[15]   D G Liu, J G Fei. Numerical simulation algorithms of dynamic systems. Beijing: Science Press, 2001. In Chinese

[16]   R H Enns, G C McGuire. Nonlinear physics with MAPLE for scientists and engineers. Boston: Birkhäuser, 2nd Edition, 2000

[17]   H G Zhang, Z L Wang, W Huang. Control theory of chaotic systems. Shenyang: Northeastern University Press, 2003. In Chinese

[18]   D Henrion. A review of the global optimization toolbox for Maple, 2006

[19]   A Chipperfield, P Fleming. Genetic algorithm toolbox user's guide. Department of Automatic Control and Systems Engineering, University of Sheffield, 1994

[20]   D H Ackley. A connectionist machine for genetic hillclimbing. Boston, USA: Kluwer Academic Publishers, 1987

4

Mathematical Modeling and Simulation with Simulink

Simulink was first released by MathWorks Inc. in 1990. It provides a block diagram based modeling and simulation environment. The previous name for Simulink was "Simulab". However, since it sounded very similar to another famous language, Simula, it was renamed as Simulink the following year. From its new name, two meanings can be seen: "simu" means that it is a simulation tool, while "link" refers to the fact that it can be used to link blocks to form a system to be simulated. Due to its powerful modeling and simulation facilities, Simulink has become the top computer environment in the field of computer simulation and other related fields.

Before the emergence of Simulink, it was very difficult, if not impossible, to model and simulate complicated systems described by block diagrams. Although MATLAB was very powerful for solving numerically simple ODEs, it was extremely difficult to establish a state space model from the whole system in block diagram form. Other simulation languages and tools, such as the ACSL language [1], were employed to describe system models and to perform simulation tasks. Manual programming methods were used to express the original block diagram of the system. The methods then used were not at all straightforward, and it was all too easy to get wrongly described system models and misleading results, and the simulation results are not trustworthy. Also, due to the amount of manual programming involved, the modeling process could involve a great deal of time and effort, so the modeling and simulation are not economical. Worse still, since MATLAB and ACSL are different languages, the argument transfer between them is rather complicated and not very convenient. This disadvantage limited the use of ACSL applications in MATLAB. Partly because of this, most of the ACSL users abandoned it when Simulink appeared, and turned to use Simulink as their major tool for modeling and simulation of complicated systems. Before simulation environments such as Simulink appeared, in order to assess the capabilities of different software to express mathematical models of control systems, some influential benchmark problems were proposed, such as the F-14 aircraft modeling [2]. Once Simulink was widely used, the benchmark problems could easily and straightforwardly be solved.

In Section 4.1, a brief introduction to the various block libraries of Simulink are given, and some of the commonly used blocks are described. In Section 4.2, introductions are made to manipulating blocks such as rotating, connecting and block parameter modification. The establishment of Simulink models, with particular skills, will also be given in this section. In Section 4.3, modeling skills and tools are presented, including the use of the model browser, model printing and simulation parameters

System Simulation Techniques with MATLAB® and Simulink®, First Edition. Dingyü Xue and YangQuan Chen.
© 2014 John Wiley & Sons, Ltd. Published 2014 by John Wiley & Sons, Ltd.

setting. This provides the necessary fundamental knowledge and preparation for more advanced modeling and simulation research to be discussed later. In Section 4.4, modeling and simulation techniques will be demonstrated with some illustrative examples. Linear system modeling and representation methods are presented in Section 4.5 where LTI Viewer based linear system frequency domain analysis and numerical simulation methods are also presented. In Section 4.6, simulation methods for continuous systems driven by stochastic inputs are discussed. Statistical analysis of simulation results are given. Probability density function, correlation and power spectral density of the signals in the systems are all examined with illustrative examples.

4.1   Brief Description of the Simulink Block Library

On entering the command `simulink` in MATLAB command window, or clicking the ▦ icon (or 🪲 icon in earlier versions) in the toolbar of the MATLAB command window, the Simulink model library window will be displayed, as shown in Fig. 4.1.

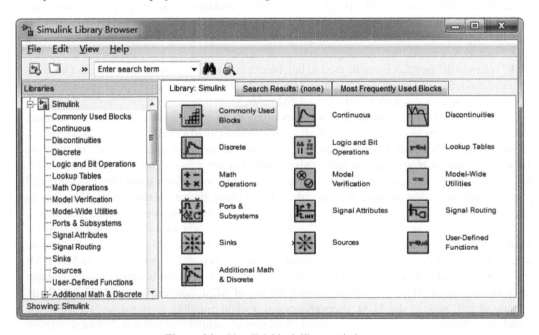

Figure 4.1   Simulink block library window.

Simulink can also be invoked with the command `open_system('simulink')`, and the model library window is displayed as shown in Fig. 4.2. Although the two library windows are displayed differently, their functions are very similar. In order to better describe the block libraries, the latter windows are used in this book. In the new version, the model tree structure is displayed, and users can go to different submodels easily. The lock icon 🔒 can be used to protect the group. To move a block icon in the group, the **Diagram** → **Unlock library** menu should be selected. The » icon in the lower left corner can be used to switch the two model libraries in Figs. 4.1 and 4.2.

From the block library in Fig. 4.2 or the left hand side of the interface shown in Fig. 4.1, it can be seen that the Simulink model library is composed of several groups. Thus the window shown in Fig. 4.1 is also called the Simulink model browser. It can be seen that, in the standard

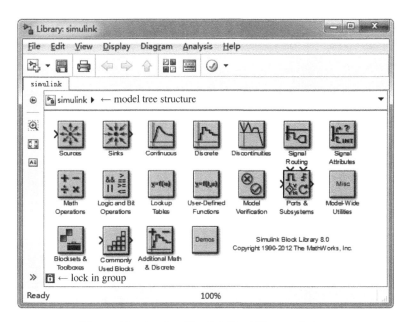

Figure 4.2   Another Simulink block library window.

Simulink library, the groups such as **Sources** (input sources), **Sinks** (for output signal displays), **Continuous**, **Discrete**, **Math Operations**, **Logic and Bit Operations**, **Discontinuities** (for some nonlinearities), **Lookup Tables**, **Signals Routing**, **Signal Attributes**, **User-defined Functions** and **Ports & Subsystems**, are provided. Also, for connection to other toolboxes and blocksets, a lot of other groups are provided in **Blocksets & Toolboxes**. The users can also make their own groups attached to the model browser. **Commonly Used Blocks** group also provides essential blocks for Simulink modeling, and the users are free to add or delete blocks in the group. In this section, some of the frequently used groups and blocks are briefly introduced. Further detailed information will be given later when they are used.

4.1.1   Signal Sources

The **Signal Sources** group provides various of commonly used input and signal source blocks, as shown in Fig. 4.3[1]. The main blocks in the group are as follows:

- **In1** block can be used to represent the input port of the whole system. They can also be used in subsystem modeling and they are also useful in linearization and command-line simulation of the systems.

- **Signal Generator** block can be used as a signal source input in the system, and available signals in the block are square waves, sinusoidal and sawtooth signals.

- **Band-Limited White Noise** block can be used to generate white noise input for continuous and hybrid systems; this block will be addressed later. Random signal generators are provided such as the **Random Number** (for normal distribution) and **Uniform Random Number** blocks. It should

---

[1]Note that the layout of the blocks in the group is rearranged due to the requests in typesetting.

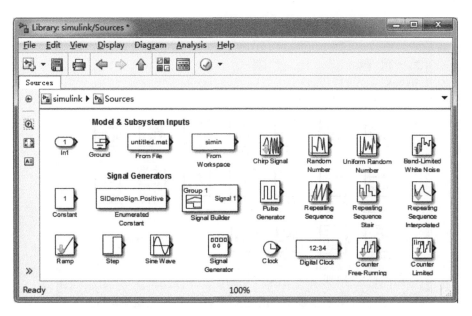

Figure 4.3   Blocks in the **Signal source** group.

be noted that these blocks should not be used to drive continuous systems. The readers are strongly suggested to check Sec. 3.4.6 "Solutions to Linear Stochastic Differential Equations."

- **From File** block and **From Workspace** block can be used to generate input signals from file or MATLAB workspace.
- **Clock** block is used to generate current time t, and it is useful in representing time-dependent quantities. For instance, it is useful in constructing time-varying systems or composing the ITAE criterion of the system.
- **Constant** block generates a constant as input signal.
- **Ground** block is used to generate zero input signal. If a terminal of a block is not connected to any other block, it can be connected to a **Ground** block temporarily, to avoid a warning message.
- Various of other input blocks are provided, such as **Step**, **Pulse Generator**, **Ramp** and **Sine Wave** blocks. Also the **Repeating Sequence** block can be used to generate arbitrary periodic inputs. The **Signal Builder** block is also defined to allow the user to design graphically any waveform.

4.1.2 Continuous Blocks

The blocks in the **Continuous** group are shown in Fig. 4.4, the commonly used ones are:

- Different type of integrator blocks are provided in the group. The **Integrator** block is the most widely used block in modeling continuous systems. Numerical integration of the input signal is performed. It can be used to define the key signals in differential equations. Different kinds of other integrator blocks, such as **Integrator Limited**, **Integrator Second-Order** and **Integrator Second-Order Limited** are provided, and an integrator with reset can also be defined.
- **Derivative** block is used to generate numerically the derivative of the input signal. In real applications, this block without modification should be avoided wherever possible due to high frequency noise amplification.

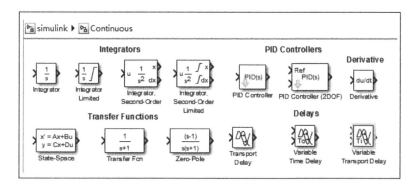

Figure 4.4   Continuous group.

- **State-Space** block of a continuous linear system is described as

$$\begin{cases} \dot{x} = Ax + Bu \\ y = Cx + Du, \end{cases} \tag{4.1}$$

where A is an $n \times n$ square matrix, B is an $n \times p$ matrix, C is a $q \times n$ matrix, while D is a $q \times p$ matrix. They are said to be compatible if the dimensions of the matrices are so defined. In the state space model, the input and output signals are respectively u and y, and they can be in vector form.

- **Transfer Fcn** block is another way to describe linear ODEs with constant coefficients, where Laplace transforms are introduced to map the original linear ODE into an algebraic form. The general form of a transfer function is

$$G(s) = \frac{b_1 s^m + b_2 s^{m-1} + \cdots + b_m s + b_{m+1}}{s^n + a_1 s^{n-1} + a_2 s^{n-2} + \cdots + a_{n-1} s + a_n}, \tag{4.2}$$

where the denominator polynomial is also called the characteristic polynomial of the system, and the highest order of it is called the order of the system. A physically realizable system, is also known as a proper system, $m \le n$. The transfer function of a system can also be regarded as the frequency-dependent gain represented in Laplace form.

- **Pole-Zero** block is another way of describing transfer functions of the system; its mathematical form is

$$G(s) = K \frac{(s - z_1)(s - z_2) \cdots (s - z_m)}{(s - p_1)(s - p_2) \cdots (s - p_n)}, \tag{4.3}$$

where K is called the gain of the system, while $z_i (i = 1, \cdots, m)$ and $p_i (i = 1, \cdots, n)$ are referred to as the zeros and poles of the system. It is obvious that for transfer function models with real coefficients, the poles and zeros are real or in pairs of complex conjugates.

- Three types of time delay blocks are provided. **Transport Delay**, **Variable Time Delay** and **Variable Transport Delay**, which can be used to assign the output signal as the input signal with certain types of delay.

- Two types of PID controllers, **PID controller** and **PID controller (2DOF)**, implement two types of commonly used PID controllers, with the former being the standard PID controller, and with the latter the derivative action is performed in the feedback path. They are probably the most delicate blocks in Simulink, and they can be converted to all possible variations of PID controllers. The PID-type blocks were first introduced with MATLAB R2010a in their present forms.

4.1.3 Discrete-time Blocks

The **Discrete** group provides a lot of discrete-time components and can be used to build discrete models. The blocks of the group are shown in Fig. 4.5, with the following major blocks:

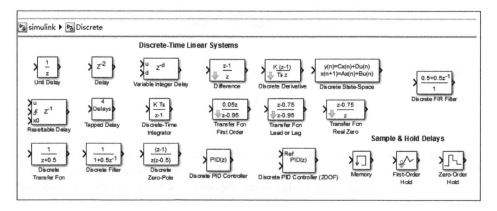

Figure 4.5   Discrete group.

- **Zero-Order Hold** and **First-Order Hold** blocks, the former keeps the input signal the same within a sampling interval, while the latter computes it with linear interpolation.

- **Discrete Zero-Pole**, **Discrete State-Space** and **Discrete Transfer Fcn** blocks are similar to their counterparts in continuous systems, with mathematical descriptions respectively

$$G(z) = K \frac{(z - z_1)(z - z_2) \cdots (z - z_m)}{(z - p_1)(z - p_2) \cdots (z - p_n)}, \tag{4.4}$$

$$G(z) = \frac{b_0 z^m + b_1 z^{m-1} + \cdots + b_{m-1} z + b_m}{z^n + a_1 z^{n-1} + a_2 z^{n-2} + \cdots + a_{n-1} z + a_n}, \tag{4.5}$$

$$\begin{cases} x[(k + 1)T] = Ax(kT) + Bu(kT) \\ y(kT) = Cx(kT) + Du(kT), \end{cases} \tag{4.6}$$

where T is the sampling interval. **Filter, Unit Delay, Integer Delay, Differences, Discrete Derivative, Discrete-Time Integrator, Transfer Fcn Lead or Lag** blocks are all special cases of the discrete transfer function block.

- **Discrete Filter** block is a variation of the transfer function block given by

$$G\left(z^{-1}\right) = \frac{b_0 + b_1 z^{-1} + \cdots + b_{m-1} z^{-m+1} + b_m z^{-m}}{a_0 + a_1 z^{-1} + a_2 z^{-2} + \cdots + a_{n-1} z^{-n+1} + a_n z^{-n}}. \tag{4.7}$$

Also, **Discrete FIR Filter** is the finite impulse response filter defined as

$$G(z) = b_0 + b_1 z^{-1} + \cdots + b_{m-1} z^{-m+1} + b_m z^{-m}. \tag{4.8}$$

- **Memory** block returns its input signal at the previous sampling interval.
- Discrete PID controller blocks, **Discrete PID Controller** and **Discrete PID Controller (2DOF)**, are useful blocks in discrete control systems.

4.1.4   Lookup Table Blocks

The **Lookup table** group blocks implement one-, two- and multi-dimensional lookup tables shown in Fig. 4.6. The main blocks in the group are

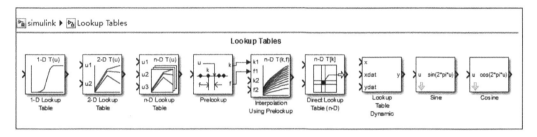

Figure 4.6   Lookup Table group.

- **1-D Lookup Table** block computes the output signal with linear interpolation to the input signal according to the table.
- **2-D Lookup Table** block implements two-dimensional linear interpolation.
- Other lookup table blocks are also provided in the group with **n-D Lookup Table**, **Direct Lookup Table (n-D)**, **Lookup Table Dynamic** and **Interpolation Using Prelookup**.

4.1.5   User-defined Functions

The blocks in the **User-defined Functions** group are shown in Fig. 4.7. Some of the blocks allow MATLAB programming and other programming languages in certain formats. Theoretically

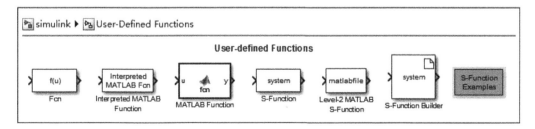

Figure 4.7   User-defined Functions group.

speaking, systems with arbitrary complexities can be modeled in this way. The main blocks in the group are

- **Fcn** block performs function operation on the input signal to calculate the output signal. The formula may contain basic arithmetic and other simple functions.
- User-defined MATLAB function blocks, such as **Interpreted MATLAB Function** (in earlier versions the **MATLAB Fcn**) block and **MATLAB Function** (the former **Embedded MATLAB Function**) block, allow the user to express static nonlinearities with MATLAB functions.
- **S-function** block allows the user to specify a dynamic nonlinear block, that is, the state space equation, using languages such as MATLAB, C/C++, Fortran and Ada. It is not exaggerating to say that with S-function blocks, systems with any complexity can be modeled with Simulink. An automatic code generation interface block, **S-function Builder**, is also provided to construct S-function frameworks. **Level-2 MATLAB S-function** block is also provided to accept more advanced S-function modeling.

4.1.6   Math Blocks

The **Math Operations** group provides various mathematical manipulation blocks, as shown in Fig. 4.8. The commonly used blocks in the group are:

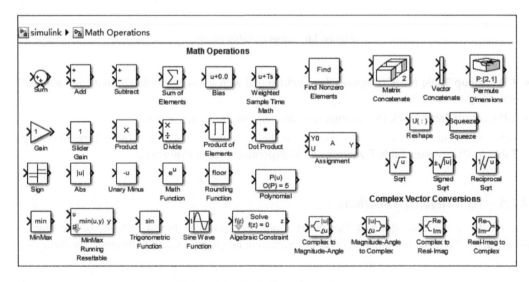

Figure 4.8   Math Operations group.

- **Gain** block can be used to compute the output signals through a gain operation on the inputs. More generally, the block can be used for matrix as well as dot multiplication of the input signals.
- **Sum, Subtract, Abs**, and **Product** blocks compute the output signal with arithmetic operations to the input signals. It is necessary for modeling feedback control systems. The **Sum of Elements** and **Product of Elements** blocks allow an arbitrary number of input signals to be added or multiplied together to form the output signal.

- **Algebraic Constraint** block can be used to solve algebraic equations within a Simulink model, so as to make the input signal zero. This block is useful in modeling DAEs (differential algebraic equations) and other forms of constraints, while the **Polynomial** block can be used for solving polynomial equations.

- **Complex to Real-Imag** and **Complex to Magnitude-Angle** blocks can be used for conversion of different types of complex signals.

- Ordinary math operations blocks, such as **Abs**, **Rounding Function**, **Trigonometric Function** and **Sign**, are also provided in the group to perform basic math operations to the input signals.

4.1.7   Logic and Bit Operation Blocks

The **Logic and Bit Operation** group contains the blocks shown in Fig. 4.9. The commonly used blocks are

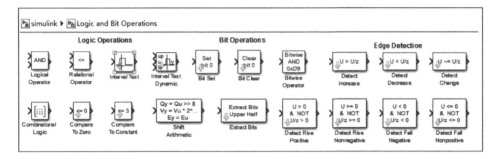

Figure 4.9   Logic and Bit Operation group.

- **Logic Operator** block allows logic operations such as "and", "or", "not" and so on to be performed on the input signals. The logical relationship can be set by its dialog box.

- **Combinatorial Logic** block is used to define truth tables.

- **Relational Operator** is to used for comparisons between the input signals, while **Compare To Constant** and **Compare To Zero** blocks are used respectively to test whether the input signal is a constant or zero. **Interval Test** block is used to test whether the input signal is within the required interval or not.

4.1.8   Nonlinearity Blocks

The **Discontinuity** group contains the commonly used piecewise nonlinear blocks shown in Fig. 4.10. The main blocks included are:

- **Coulomb & Viscous Friction** block can be used to compute the friction outputs.

- **Backlash** block is the same as hysteresis nonlinearity in control.

- **Hit Crossing** block is used to detect accurately the zero-crossing points of its input signals. **Wrap To Zero** block sets the output to zero when the input signal is larger than the threshold.

- There are a lot of piecewise linear static nonlinearity blocks such as **Saturation**, **Dead Zone**, **Quantizer**, **Relay** and **Rate Limiter**. In fact, all of these blocks can also be implemented by

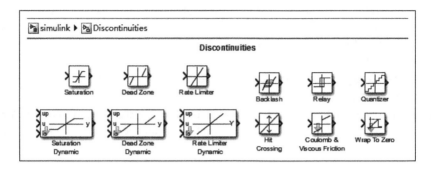

Figure 4.10   Discontinuity group.

one-dimensional lookup table blocks. Moreover, saturation, dead zone and rate limit blocks with dynamic parameters are also supported.

4.1.9   Output Blocks

The **Sinks** group provides necessary output devices to display and return simulation results. The contents of the group are shown in Fig. 4.11, and the major blocks in the group are

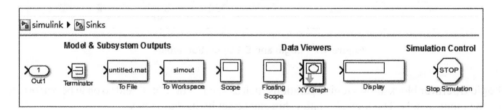

Figure 4.11   Sinks group.

- **Out1** block is used to represent an output port. This block is useful in a subsystem modeling and linearization process. Also, the simulation results can be returned to MATLAB workspace as variable yout.

- **Scope** block can be used to display signals with scopes. Apart from ordinary scopes, **Floating Scope** and **x-y Graph** can also be used to display signal in scopes. Also, the users are allowed to set observation points on any signal.

- **To Workspace** block writes the signal to MATLAB workspace. The default data type is structured array, and it can be set to arrays.

- **To File** block is used to output the signals to data files.

- **Display** block can be used to display the signal as digits.

- **Stop Simulation** block is used to terminate the whole simulation process, when its input signal is zero. Such a block can be used in a Simulink model to control a simulation process.

- **Terminator** block can be used to connect suspended or idle output ports in the blocks to prevent warning messages.

4.1.10   Signal Related Blocks

The **Signals Routing** group is shown in Fig. 4.12. The blocks included in the group are

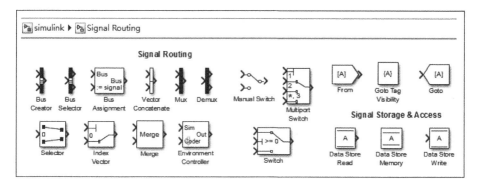

Figure 4.12   Signal Routing group.

- **Mux** and **Demux** blocks are used to represent multiplexer and demux of signals, that is, to combine several individual signals into a vector one or vice versa.
- **Model Info** block is used to display model information.
- **Selector** block is used to select the expected signals from a vector signal and rearrange their order.
- Various switch blocks, such as **Switch**, **Manual Switch** and **Multiport Switch**, can be used to assign the input channels.
- Signal transfer blocks such as **From** and **Goto** can be used to relay signals, to avoid unnecessary crossing of signal lines.

The **Signal Attributes** group provides blocks setting the properties of the signals. The contents of the group are shown in Fig. 4.13, and the commonly used blocks are:

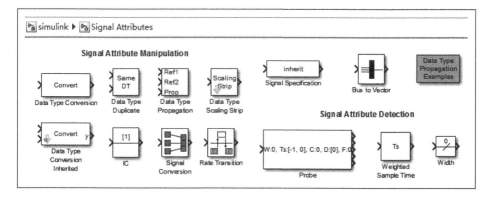

Figure 4.13   Signal Attributes group.

- **Convert** block can be used to convert the data type of the input signal, for example to convert the double-precision input to a logical signal. The data type of the output signal can be specified through its dialog box.

- **IC** block is used to set initial values for the input signals.
- **Data Type Duplicate** block can be used to force the input signals to have exactly the same data type.
- **Width** block measures the number of individual signals in the input signal vector.

4.1.11   Ports and Subsystem Blocks

The **Ports & Subsystem** group contains various subsystem structures as shown in Fig. 4.14. Most commonly used blocks and ports are:

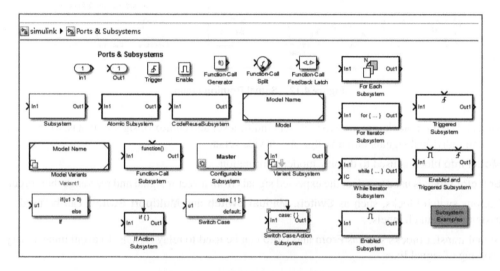

Figure 4.14   Ports & Subsystem group.

- **In1, Out1, Trigger** and **Enable** blocks are used to manipulate input, output and control ports in subsystems.
- **Atomic Subsystem** block establish a blank submodel, and the user can add components to the model, and add or remove control ports.
- **Triggered Subsystem** block will start working when the trigger control signal is acting. The available trigger types are rising edge, falling edge and either of them.
- **Enabled Subsystem** block will be enabled when the enable control signal is acting. The trigger signal and the enabled signal can be used together to control the behavior of the subsystem.
- **Flow control** subsystems include the `for` loop, `while` loop, `if` clause and the `switch` structure. These blocks can be better described with finite state machines by Stateflow.
- **Model** block can be used to embed another existing Simulink model file or model window into the current system.

4.1.12   Commonly Used Blocks

In order to provide easy access to the block library, the commonly used blocks can be put into a group. The default group is shown in Fig. 4.15. One can add or remove blocks to and from the group,

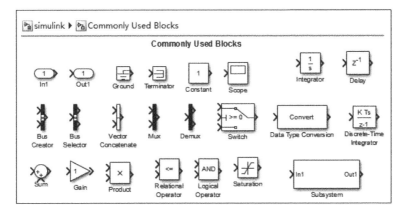

Figure 4.15   Commonly used block group.

so that next time the group is opened, most of the blocks to be used in modeling are already in the window.

4.1.13   Other Toolboxes and Blocksets

Apart from the commonly used blocks discussed earlier, there are a large number of other professional blocks available, to give access to the MATLAB toolboxes and blocksets. These blocks may provide interfaces to certain toolboxes.

Some of the existing toolboxes and blockset groups are shown in Fig. 4.16. Some of the commonly used ones are summarized below:

- Aerospace Blockset has a large number of blocks for aviation and aerospace engineering. It includes the subgroups **Equations of Motion**, **Aerodynamics**, **Propulsion**, **GNC** and **Animation**.

- Comm System Toolbox blockset provides tools and blocks in the simulation of communication systems. The groups in the blockset include **Modulation**, **Synchronization**, **Sequence Operations**, **Interleaving** and **Source Coding**.

- DSP System Toolbox blockset provides some blocks useful in digital signal processing. These blocks include those for measuring power spectral density and autocorrelation function. The groups include **Filtering**, **Transforms**, **Estimation**, **Quantizers** and **Statistics**.

- Gauges blockset provides many visual display devices, normally supported by ActiveX techniques, to exchange data between MATLAB and Simulink, and to display signals using dials and gauges, similar to those used in the process control industry. The groups under the blockset include **Angular Gauges**, **Linear Gauges**, **Numeric Displays**, **LEDs** and **On Off Gauges**.

- Blocksets and toolboxes used for interfacing the existing MATLAB toolboxes. For example, the Control System Toolbox blockset provides an LTI block to load the linear time-invariant model into Simulink. Fuzzy Logic Toolbox blockset can be used to provide access to a fuzzy inference system. Neural Network Blockset is used to embed a neural network block into a Simulink model. MPC Toolbox blockset provides an interface with Model Predictive Control Toolbox. System Identification Toolbox is used for interfacing the toolbox with Simulink.

- Simscape and its Foundation Library provide an easy modeling strategy for multi-domain physical system modeling techniques. The groups in the library include **Electrical**, **Hydraulic**, **Mechanical**,

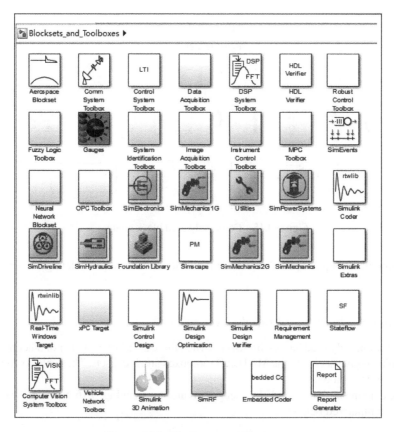

Figure 4.16   Toolboxes and blocksets.

Magnetic, **Thermal**, **Pneumatic** and other engineering blocks. It allows the user to construct engineering simulation systems using a building block method, just like assembling practical hardware systems.

- Professional engineering simulation blocksets include SimPowerSystems, SimHydraulics, Sim-Mechanics, SimDriveline, SimRF and SimElectronics. With the use of these blocksets, engineering systems can be established graphically and the mathematical models behind them can be generated automatically.

- SimEvents provides the necessary blocks for block diagram modeling of event driven or discrete event dynamical systems.

- Simulink 3D Animation blockset, which used to be called the Virtual Reality Toolbox, gives access to virtual reality inputs and three-dimensional display of simulation results.

- Real-time control blocksets. For instance, Real-Time Workshop can be used to translate Simulink models into C code to speed up the simulation process. The xPC blockset can be used in research on hardware-in-the-loop simulation.

- Computer Vision System Toolbox blockset (the former Video and Image Processing blockset) includes image and video loading, edge detection and enhancement, transformation and output blocks. With the blocks, image and video processing systems can easily be constructed. The Image

Acquisition Toolbox allows the acquisition of images and videos from video devices, such that real-time image and video processing can be implemented.

4.2   Simulink Modeling

4.2.1   Establishing a Model Window

Under the Simulink environment, the normal procedures of model creation and editing of a Simulink model are that a blank model window should be opened first, then copy the blocks in the block library to the blank window, and connect the blocks as needed, and then modify the parameters of the model. The whole model can then be used in the simulation.

There are a few methods of opening a blank model window in Simulink:

- With the **File** → **New** → **Model** menu item from MATLAB command window.
- Click the 🔡 icon (or in old versions, the ⬚ icon) in the Simulink window toolbar.
- Select menu item **File** → **New** → **Model** in the Simulink window.
- With the MATLAB function use `new_system()` to declare a new logic model, and use the function `open_system()` to open it. This topic will be further explored in Chapter 6.

The above methods are virtually equivalent. A blank model window is shown in Fig. 4.17. The blank model can be used for accepting various blocks from the Simulink library. The methods of model editing, processing and Simulink methods are presented in the following subsections.

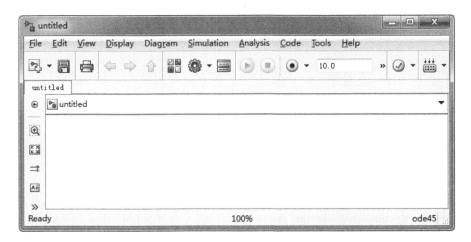

Figure 4.17   Blank model window in Simulink.

4.2.2   Connecting and Simple Manipulation of Blocks

Connecting two blocks in a Simulink window is quite easy. In the Simulink blocks, the output port is indicated with a triangle leaving the block, while the input port is indicated by an input sign > entering the block. To connect the two blocks together, simply click the output port, then drag the mouse to the input port of the other block and release the mouse button. The two blocks will be automatically connected. To connect two blocks together more quickly, click the source block, hold down the **Ctrl** key, and then click the target block; the two blocks will then be connected together automatically.

When the two blocks are correctly connected, a solid line with an arrow is shown, joining the two blocks together as in Fig. 4.18(a). If the two blocks are not connected correctly, a broken line will be shown as in Fig. 4.18(b).

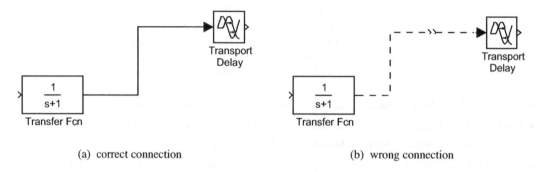

(a) correct connection   (b) wrong connection

Figure 4.18   Connections of two blocks.

Sometimes, we may need to rotate or flip blocks to make the layout look nice. Selecting a block or blocks should be made first. To select a single block, simply click it. The selected block is indicated by four dots around it. For instance, the **Transfer Fcn** block in Fig. 4.19(a). To select a few blocks together, the easiest way is to drag a box around the blocks then release the mouse button. The selected blocks are all indicated in the same way, as shown in Fig. 4.19(b). Also you can hold down **Ctrl**, and select as many blocks as you like.

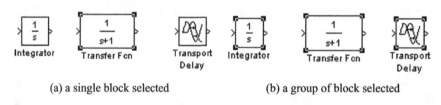

(a) a single block selected   (b) a group of block selected

Figure 4.19   Block selection results.

The **Diagram** menu is the main menu in system modeling. Most of the model and block manipulation work can be completed with this menu. In particular, the **Format** submenu item (in earlier versions **Format** menu is the equivalent menu) shown in Fig. 4.20(a) is displayed. The submenu items **Rotate & Flip** can be selected as shown in Fig. 4.20(b). The **Format** menu in old versions of Simulink is shown in Fig. 4.20(c).

After blocks are selected, they can then be manipulated. For instance, to flip a block, the **Rotate & Flip** menu in the Simulink window can be selected, then the menu item **Flip Block** can be used to flip the block. The flipped blocks are shown in Fig. 4.21(a).

If the connections of the blocks are already made, there will be problems when the blocks are directly flipped; unexpected results will be obtained, as shown in Fig. 4.21(b). After flipping, the connections are still correct, but the layout does not look nice. Connections should be redrawn. It is suggested that rotations should be performed before connections are made.

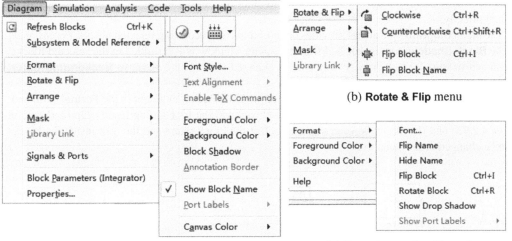

(a) **Diagram menu** for block manipulations

(b) **Rotate & Flip** menu

(c) **Format** menu in old version

Figure 4.20   Simulink **Diagram** → **Format** menu.

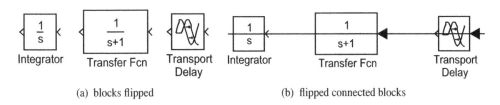

(a) blocks flipped   (b) flipped connected blocks

Figure 4.21   Block flipping.

There is a trick in block layout and connections in Simulink. To insert a block C in between two existing connected blocks A and B, the easiest way is to move block C in between the two blocks, on top of the connection line, and release the mouse button. Block C will be inserted directly and automatic rotation or flipping is performed, if necessary.

To keep connections neat, it is sometimes a good idea to rotate a block 90° clockwise. This can be done, by selecting the **Counter Clockwise** item from the **Rotate & Flip** menu, and the rotated block is as shown in Fig. 4.22(a). The menu item can be used consecutively, different angles of the block can be shown. It is seen from Fig. 4.22(a) that the name of the block has moved to the left hand side of the block. To move the block name to the right of the block, the menu item **Flip Name** can be selected, and the result is shown in Fig. 4.22(b). If you do not want to display the name of the block, the option **Hide Name** can be selected. To show the name again, the menu item **Show Name** can be

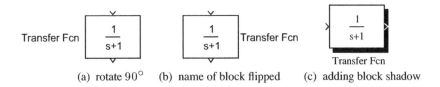

(a) rotate $90°$ (b) name of block flipped   (c) adding block shadow

Figure 4.22   Simple manipulation of blocks.

selected. Also, clicking the name of a block will enter the editing mode, so the block name can be edited directly.

The menu **Diagram** → **Format** provides other options for decorating selected blocks. For instance, if the **Block Shadow** item is selected from the menu, shadow will be added to the selected blocks, as shown in Fig. 4.22(c). Also, colors can be changed with the menu items **Background Color**, **Foreground Color** and **Canvas Color**.

Fonts of selected blocks can easily be modified with the **Font** menu item in the **Format** submenu. The standard font setting dialog box can be displayed and the font setting results can be shown in Figs. 4.23(a) and (b). It can be seen that the font of the block name and the font inside the blocks can be changed at the same time.

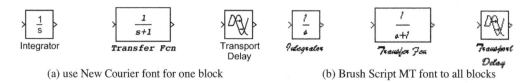

(a) use New Courier font for one block   (b) Brush Script MT font to all blocks

Figure 4.23   Font modifications.

4.2.3   Parameter Modification in Blocks

When drawing a Simulink model, the parameters of the blocks used are usually different from those expected. Thus it is necessary to modify the parameters in the blocks. For instance, the transfer function block in the model library is $G = 1/(s + 1)$ by default, and if we need a transfer function of $G(s) = (s^3 + 7s^2 + 24s + 24)/(s^4 + 10s^3 + 35s^2 + 50s + 24)$ in the Simulink model, we can double click the transfer function block. A dialog box shown in Fig. 4.24(a) is displayed. In the parameter dialog box, the numerator and denominator coefficients can be changed separately. Since transfer functions can be expressed as the ratio of the numerator polynomial to the denominator polynomial, the presentation of the polynomial is important. In MATLAB and Simulink, polynomials can be expressed as its coefficient vector in descending order of s. In the transfer function quoted, the numerator polynomial $s^3 + 7s^2 + 24s + 24$ can be expressed by the vector $[1, 7, 24]$, and the denominator polynomial can be expressed as $[1, 10, 24, 35, 50]$. One can fill in the two vectors in

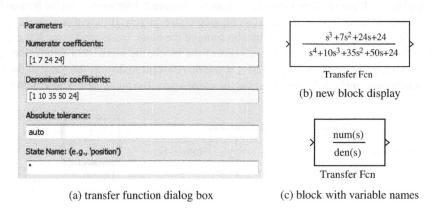

(a) transfer function dialog box   (b) new block display

                                                     (c) block with variable names

Figure 4.24   Transfer function block modification.

the dialog box, as shown in Fig. 4.24(a). Click the **OK** button to accept the new parameters, and the new block is shown in Fig. 4.24(b).

If the numerator and denominator are expressed by the variables num and den in the dialog box, the variables num and den in MATLAB workspace can be used, and the new block is shown in Fig. 4.24(c). It should be noted that, the variables thus used should be assigned in MATLAB workspace first, otherwise the simulation cannot proceed.

The integrator block is an interesting block, since it has many variations. Double click the integrator block, the dialog shown in Fig. 4.25 is given. Through proper setting, different variations of integrators can be constructed.

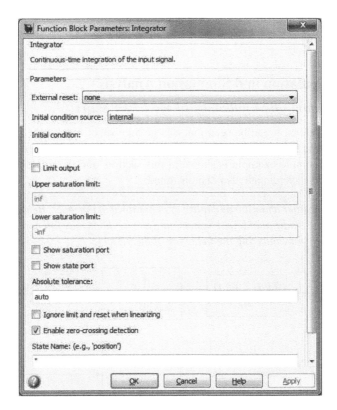

Figure 4.25   Dialog box of an integrator block.

The initial conditions of the integrator block can be set first, by filling in the edit box prompted by **Initial Condition**. It can either be a constant or a variable name. And if from the listbox labeled **Initial condition source**, the item **external** is selected, an extra input port shown in Fig. 4.26(a) will be displayed. The external signal connected to the port can be used to provide initial conditions to the integrator.

Integrators can accept external signals to trigger its reset facilities. This can be done by setting the listbox prompted using **External reset** to **rising**, and the new integrator is shown in Fig. 4.26(b). In this case, the rising edge of the external signal can be used to trigger the reset facilities in the integrator. Of course, other options can be selected from the listbox to assign the reset triggers.

The integrator can be followed by a saturation nonlinear element. This is done by checking the **Limit output** box, and then the upper and lower bounds can be set with specific values in the **Upper**

saturation limit and **Lower saturation limit** edit boxes. The integrator with saturation is shown in Fig. 4.26(c). All these options can be selected together and the new integrator block is constructed as shown in Fig. 4.26(d). It can be seen that the options can make the integrator block more versatile and flexible for different applications.

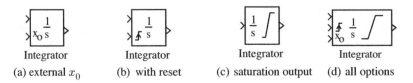

Integrator	Integrator	Integrator	Integrator
(a) external x_0	(b) with reset	(c) saturation output	(d) all options

Figure 4.26   Different variations of integrators.

4.3   Model Manipulation and Simulation Analysis

4.3.1   Model Creation and Fundamental Modeling Skills

From the above illustrations, it can be seen that a large number of blocks are already provided in Simulink. The blocks can be used to construct complicated block diagrams to represent a dynamic system under study. A simple example is shown in this section, and the modeling procedures and simulation analysis of the system are also demonstrated.

Example 4.1   *Consider a nonlinear control system shown in Fig. 4.27. Because of the existence of the nonlinear element, traditional linear system analysis methods may fail completely. Simulation is the only plausible way for analyzing the behavior of the system.*

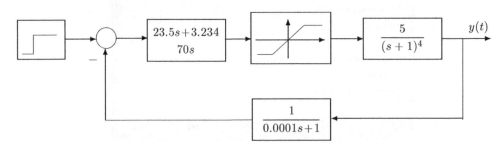

Figure 4.27   Block diagram of a nonlinear control system.

*It can be seen that three **Transfer Fcn** blocks are needed, and also a **Step** input block, a **Sum** block and a nonlinear **Saturation** block. Also, to observe the output of the system, a **Scope** block is also needed.*

*In the Simulink model browser window, the menu **File → New → Model** can be used to create and open a blank window, and the blocks listed above can be copied to the model window, as shown in Fig. 4.28 where, for simplicity, the blocks are numbered.*

*Comparing the block layout and the one in the original block diagram, it can be seen that the input and output ports **Transfer Fcn2** block, labeled as ⑦ should be interchanged. This can be done by selecting it first, then using the **Diagram → Rotate & Flip → Flip Block** menu.*

*The simplest way to connect the blocks is to click the first block, labeled ①, then hold down the **Ctrl** key and in turn click the blocks ②, ③, ④, ⑤ and ⑥; thus the blocks on the forward path can be*

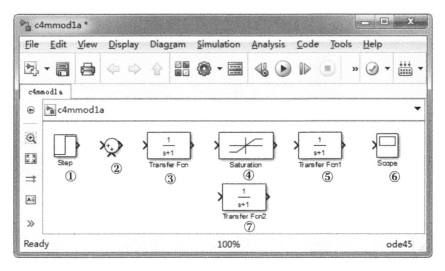

Figure 4.28 Blocks copied into the new model.

correctly connected. The connection of the feedback block labeled ⑦ is rather complicated, because we have to flip it manually first, then the connection lines can be drawn. But we can use a simpler way. Move the cursor to the line between blocks ⑤ and ⑥, hold down **Ctrl** *again, and drag the mouse to the second input port of block ② and release the mouse, the connection of the feedback path is then established, without block ⑦. Now drag block ⑦ on top of the above connection line and release button; the block will automatically embedded itself in the proper place, and the block will be flipped automatically, no matter what the original orientation of it was. The system model shown in Fig. 4.29 can be established. One may save the model to a file by selecting the* **File** → **Save** *menu, and a standard file-save dialog box will be opened. The file can be saved as c4mmod1.slx (in old versions, the suffix should be .mdl).*

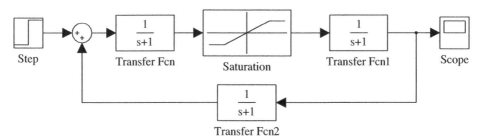

Figure 4.29 Simulink model for the nonlinear system (model file: c4mmod1).

4.3.2 Model Explorer

Double clicking a block, the parameter dialog box can be displayed. One can modify the parameters in the model in the dialog box. If there are many blocks to be modified, the modification method of double clicking the blocks, may be rather tedious. Model explorer can be used to handle this problem easily. We can use the **View** → **Model Explorer** menu in the model window, or click the 🔲 button in the toolbar to launch the model explorer, as shown in Fig. 4.30.

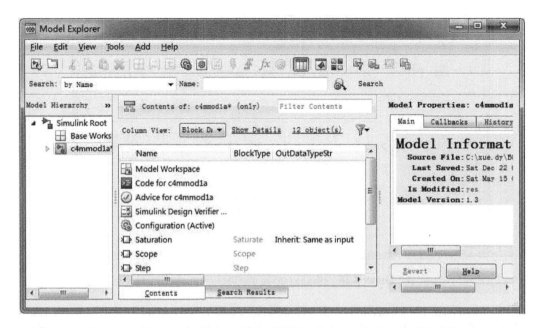

Figure 4.30   Model explorer.

Example 4.2   *Consider again the problem in Example 4.1. In the model explorer window, there are three parts. The left part shows the level of the model in the Simulink environment. For the example, model c4mmod1 lies in the next level of the **Simulink Root**. Below it, there are lower-level objects and settings. The middle part of the model explorer shows in alphabetical order the lower level blocks under it. It can be seen that the **Saturation**, **Scope** and **Step** are listed. To modify the parameters in a block, double click its icon in the middle, and the model parameter edit boxes are shown on the right, and the format displayed in this way is exactly the same as that shown in the dialog boxes. For instance, **Step** block can be assigned as shown in Fig. 4.31(a), and the transfer function block in ③ is displayed as shown in Fig. 4.31(b).*

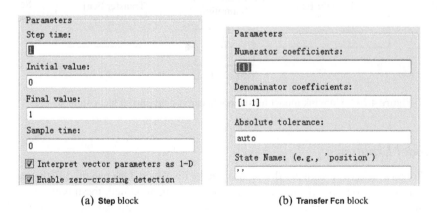

(a) **Step** block   (b) **Transfer Fcn** block

Figure 4.31   Model parameter modification dialog boxes.

After modification, the block diagram of the system can be finalized as shown in Fig. 4.32. It can be seen that the model constructed is the same as the original block diagram. If there are errors in Simulink modeling, the problem can be found easily.

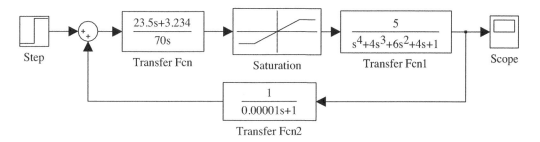

Figure 4.32   Finalized simulation model (model name: c4mmod1c).

4.3.3   On-line Help System in Simulink

As with MATLAB, Simulink also provides its own on-line help system. The menu **Help** → **Help on the selected block** can be selected, to display the on-line help window as shown in Fig. 4.33. The help window can also be invoked from the shortcut menu obtained by right clicking mouse button, and select **Help** menu item, or simply by pressing **F1** key.

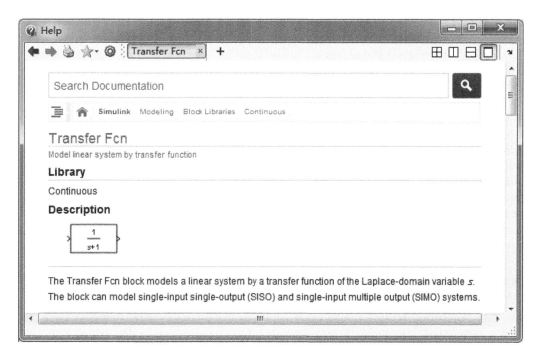

Figure 4.33   On-line help window of Simulink.

4.3.4   Output and Printing of Simulink Models

In the Simulink model window, when the menu **File** → **Print...** is selected, a standard print dialog box will be displayed. By clicking **OK**, the Simulink model can be printed directly to the printer. Alternatively, the **Properties** button can be clicked to select other printing options. For instance, **Orientation** can be assigned to **Landscape** or **Portrait**. Also these properties can be assigned with the **File** → **Page Setup** menu.

We can also print or save the model with the `print` command, with the syntax

`print -s -doptions file_name`

where "options" can be used to select the picture file format. The available picture file format can be listed with the `help print`. If the `print -s -deps myfile` command is used, the model can be printed to the file myfile.eps in encapsulated PostScript format. The option `-s` is used to indicate the Simulink model. If this option is missing, the picture in the graphics window is saved instead.

As with other Windows programs, the **Edit** → **Copy Model to Clipboard** menu in the Simulink model window can copy the whole model into the Windows' clipboard. It can then be pasted to other programs, such as Microsoft Word.

4.3.5   Simulink Environment Setting

Once the Simulink model has been established, we are ready to start a simulation session. The simplest way to start the simulation process is to click the button ▶ in the toolbar of the Simulink model window. Alternatively, it can be invoked by the **Simulation** → **Start** menu item in the model window. Simulation control parameters can be set with the **Simulation** → **Model Configuration Parameters** menu item, or by clicking the ⚙ icon in the toolbar of the model window. The dialog box in Fig. 4.34 can be displayed. The user can fill in the necessary data to control the simulation process.

It can be seen from Fig. 4.34 that on the left hand side of the dialog box, there are many panes, with the default one **Solver** for the control parameter setting for ODE solution algorithm and parameter selections.

1) **Solver** pane

If the **Solver** pane is selected, in the right part of the dialog box, solver-related control parameters can be set. For instance, in the **Solver options** group, the **Type** listbox can be assigned to **Fixed step** and **Variable step**, while the **Solver** listbox can be assigned to the default **ode45**. For stiff equations, the option can be set to **ode15s**. For discrete event systems, the option to choose is **discrete (no continuous states)**. It should be noted that, when there exists continuous element, the **discrete (no continuous states)** cannot be used. The step size in the fixed step scheme can be assigned in the **Fixed step size** edit box, or it can be set to **auto**. In the variable step scheme, however, it is best to assign the step size to **auto**. In real-time simulation, the fixed step scheme must be used. Other parameters in the group should also be set:

- **Simulation interval setting**: The initial and terminating time can be specified in the dialog box. Also, the **STOP** block in **Sinks** group can be used to force termination of the simulation process.

- **Simulation accuracy setting**: Similar to the solutions of ODEs, the **Relative tolerance** property is also an important factor in determining the accuracy of simulation results. The default setting of the error tolerance is 10^{-3}, which is usually too large. In practical simulation processes, a smaller value, such as 10^{-8}, should be used instead. Also the simulation results should be validated as discussed previously.

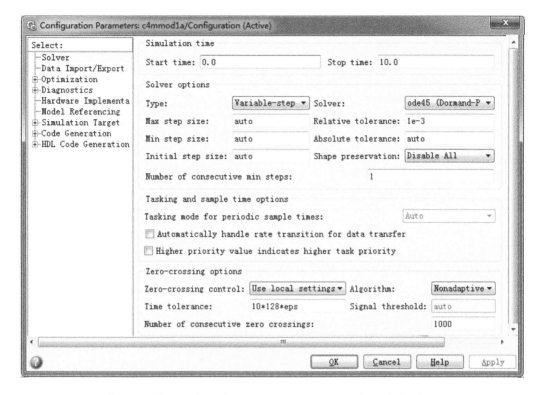

Figure 4.34 Configuration parameters setting dialog box of Simulink.

- **Accurate processing of output signals**: Since the variable step scheme is often adopted in a simulation, sometimes the results obtained are not smooth. If more accurate results are needed, select the option **Refine output** from the **Output options** listbox, and assign the value of **Refine factor** to a number larger than 1.

- **Zero-crossing detection**: Simulation signals around zero may affect simulation results and accuracies. Accurate detection of zero-crossing points is a very important step and it should not be neglected. The zero-crossing detecting algorithm provided in Simulink may be used to effectively solve zero-crossing detection problems, while the speed of computation may be reduced down. From numerical accuracy point of view, **Zero-crossing options** should be set to **Enable all**. However, from simulation speed point of view, the option **Disable all** should be selected. The latter case can be regarded as an approximate or quick simulation.

2) **Data Import/Export** pane

If the **Data Import/Export** pane is selected, the dialog box is shown in Fig. 4.35. It can be seen from the list that, with the default setting, the **Time** and **Output** checkboxes are selected and the relevant signals will be written back to MATLAB workspace, with variable names tout and yout, respectively. It is best to keep these settings. If the states are also expected, the checkbox **States** should be clicked.

In the interface, the maximum length of output signal can be selected, with a default value of 1000, to retain only the last 1000 data items. Earlier simulation data are discarded. In practical simulation, if the step size is too small, the amount of data may be too large. To retain all the data, deselect the **Limit data points to last** checkbox.

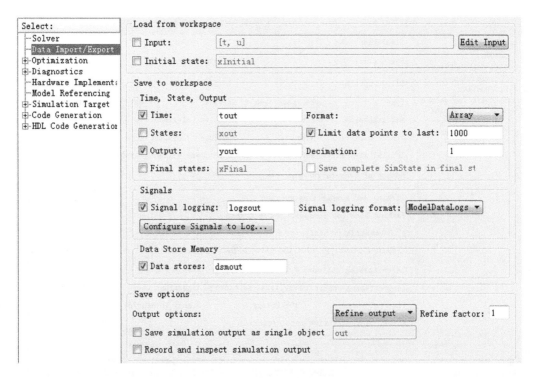

Figure 4.35   Data Import/Export pane.

3) **Diagnostics** pane

If some kinds of problem occur while running the Simulink model, warning or error messages will be given. The error and warning messages are always predefined. If the **Diagnostics** pane is selected, the dialog box shown in Fig. 4.36 will be displayed. The error or warning setting can be modified in the dialog box. In the warning and error setting related to the **Solver** subpane, the commonly used ones are:

- **Algebraic loop**: Algebraic loop is formed such that, within a loop, the output signal can be used as input signals of the blocks, and it propagates to the input port in an algebraic way. If the algebraic loop was formed only with linear blocks, equivalent simplification can be made first, so that the algebraic loop can be eliminated. If there are nonlinear elements, a low-pass filter block can be used and examples will be given later. There are three options for algebraic loop diagnosis: **none**, **warning** and **error**.

- **Minimizing algebraic loop**: Automatic algebraic loop removal methods are provided in Simulink. If such a method fails, the diagnosis level can be set to **none**, **warning** or **error**.

- **Min step size violation**: when the actual step size is chosen to be smaller than the minimum step size allowed, an error or warning message will result.

Other diagnosis functions are also provided such as **Sampling Time**, **Data Conversion** and **Connectivity**. The following commonly used diagnoses are:

- In the modeling of discrete-time systems, the sampling intervals of some of the blocks can be set to −1, indicating that the sampling interval of the block inherits the sampling interval of its input

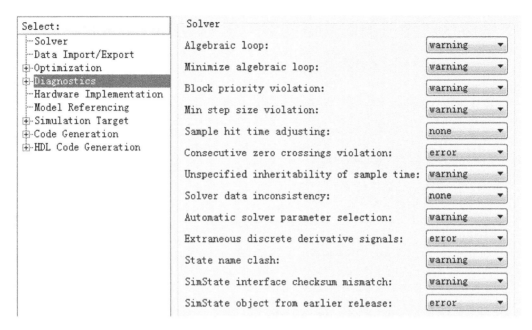

Figure 4.36   Error diagnostics dialog box.

signal. If the sampling interval of the input blocks are assigned to -1, errors may happen, such that the sampling intervals of the blocks cannot be determined. Thus sampling interval diagnosis has an option **Source block specify** -1 **sample time**.

- In the connection diagnosis interface, the option **Unconnected block output** diagnoses whether there are suspended input and output ports. If there is an unconnected output port, it can be connected to the **Terminator** block in the **Sinks** group. Similarly, **Unconnected block input**, **Unconnected lines**, etc, can be diagnosed. In the former, a **Ground** block can be connected first, and in the latter, any unnecessary connections can be deleted. In fact, suspended port may not affect the simulation results. Thus the diagnosis of a suspended port can be canceled.

- Data type related diagnosis options include **int32 to float conversion**, **Unnecessary conversion** and **Vector/matrix conversion**.

4.3.6   Debugging Tools of Simulink Models

Various model testing and debugging tools are now provided in Simulink. The practical ones include **Model Advisor** (with icon ▨) and **Simulink Debugger** (with icon ✾). These tools are useful in modeling, testing and debugging of Simulink models.

The Simulink model can be tested by using the **Analysis** → **Performance Tools** → **Performance Advisor** menu, and the model advisor interface is displayed as shown in Fig. 4.37.

Apart from this tool, there are a lot of other tools, such as the **Analysis** → **Model Advisor** and the **Analysis** → **Model Dependencies** menu items. These tools can be used to detect possible problems in the models, such as connection problems, or suggestions about block combinations. The testing tasks shown on the left hand side of the interface can be selected, then a button labeled **Run This Check** will appear. The user can press it to starting testing. If testing is successful, a

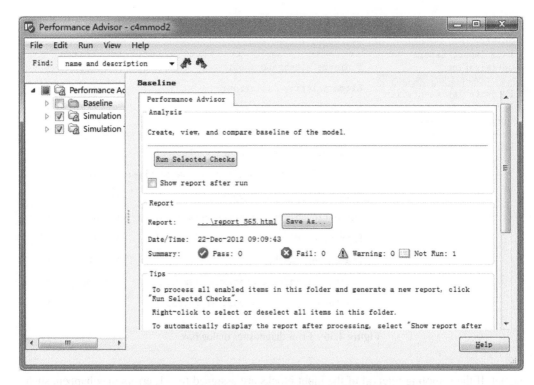

Figure 4.37 Model performance advisor interface.

Passed message will be displayed, otherwise, a testing report will be generated automatically. For small-scale problems, such testing facilities are no use, but it is useful for dealing with large-scale modeling problems.

If in a Simulink model window, the menu **Simulation → Debug → Debug Model** is selected, or if the ❀ button in the toolbar is clicked, the Simulink debugging interface shown in Fig. 4.38 will be displayed. Breakpoints can be set in the Simulink model, or breakpoints can be set automatically when **Zero crossing**, **Step size limited by state**, **Solver Error**, **NAN values** and other similar events occur. The user can also set breakpoints at particular times with **Break at time**. The breakpoints can be used to debug Simulink models. For instance, if in the simulation process, it is found that the system is unstable, an error message will appear, indicating that there is NaN or Inf result at a particular time. A breakpoint can be assigned before that time, to observe the problem happening during this period.

4.4 Illustrative Examples of Simulink Modeling

Several typical examples are presented in this section to demonstrate the establishment of Simulink models as well as simulation analysis of these systems. In the first example, the well-known Van der Pol equation is used to show how to draw block diagrams for ODEs, and some useful conclusions can be drawn. In the second example, a complicated linear system representation of a DC motor drive system is given, and simulation is performed with different parameters. In the third and fourth examples, a nonlinear system and a discrete-time system are demonstrated, and simulation sessions are performed for different parameter changes.

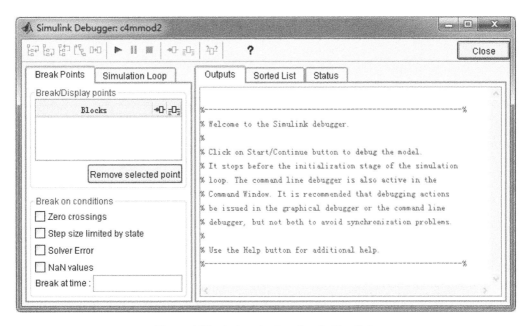

Figure 4.38   Debugging interface in Simulink.

Example 4.3   *Consider the Van der Pol equation studied in Example 3.28*

$$\ddot{y} + \mu(y^2 - 1)\dot{y} + y = 0.$$

To model a differential equation, the key signals in the equation should be constructed first. For instance, the key signals in this equation are $y(t)$, $\dot{y}(t)$ and $\ddot{y}(t)$, and the relationships between them can be expressed as integrators. Suppose that two integrators are drawn first, as shown in Fig. 4.39, with the three key signals labeled.

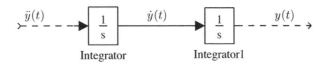

Figure 4.39   Key signal specification with Simulink.

If we rewrite the original Van der Pol equation as $\ddot{y} = -\mu(y^2 - 1)\dot{y} - y$, it can be seen that the right hand side of the equation can easily be constructed from the key signals. The signals can be linked to the $\ddot{y}(t)$ port to complete the block diagram modeling. The Simulink model shown in Fig. 4.40 can be constructed to describe the original ODE.

It can be seen that in the system model, there are several text descriptions. Adding text descriptions to a Simulink model is simple. One can double click on a blank area in the model window and type in some text. After the text input process is completed, it can be moved to the desired place with mouse.

From the example, it can be seen that ODEs can be modeled with Simulink in a graphical way. Based on this idea, more complicated systems can be modeled.

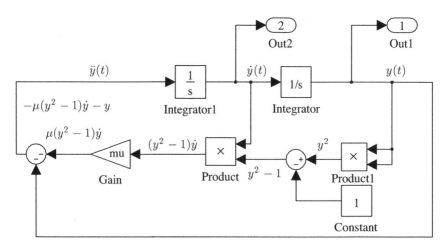

Figure 4.40   Simulink representation of the Van der Pol equation (model name: c4mvdp).

In this model, some parameters should be assigned in the MATLAB workspace: the value of mu *and the initial values of the integrators* x01 *and* x02. *These parameters can be assigned to the blocks by double clicking them and filling in the variable names in the dialog boxes. Also in the* **Sum** *block, double click it and in the dialog box fill in* | --, *where the* | *character means to leave the input port undisplayed. The two following - signs mean to have negative input ports.*

After specifying the parameters in the blocks, we should also assign their values in the MATLAB workspace, with the following MATLAB statements

```
>> mu=1; x01=1; x02=-2;
```

The simulation process can then be initiated. Clicking the ▶ *sign in the toolbar of the Simulink model window, or selecting the* **Simulation → Start** *menu, the simulation process is invoked. After simulation, the results can be assigned to the MATLAB workspace, with the names* tout *and* yout. *The following commands can be used to draw simulation results, as shown in Figs. 4.41(a) and*

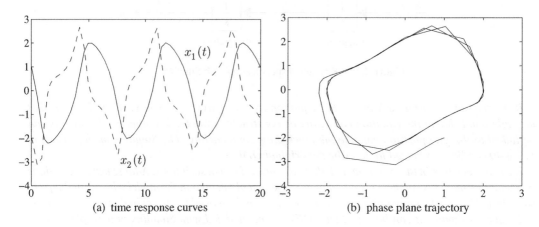

(a) time response curves   (b) phase plane trajectory

Figure 4.41   Simulation results of Van der Pol equations.

(b). The curves are time responses and a phase plane trajectory. It is also seen that in the curves obtained, some parts are smooth, while others may be rough. Simulation results validation and accuracy improvement will be shown later.

```
>> plot(tout,yout), figure, plot(yout(:,1),yout(:,2))
```

*The output format can also be changed. For instance, connect the two signals $y(t)$ and $\dot{y}(t)$ to an **XY Graph** block, as shown in Fig. 4.42. Double click the **XY Graph** block, and a dialog box as shown in Fig. 4.43(a) is displayed. The expected ranges of x and y axes should be provided, as shown in the dialog box. Start the simulation process, and the phase plane trajectory can immediately be obtained as shown in Fig. 4.43(b). It can be seen that the result obtained is exactly the same as that in Fig. 3.28.*

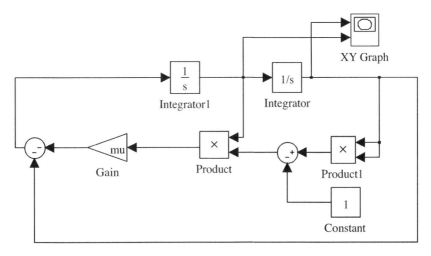

Figure 4.42   Simulink model with an x-y scope. (model name: c4mvdp1).

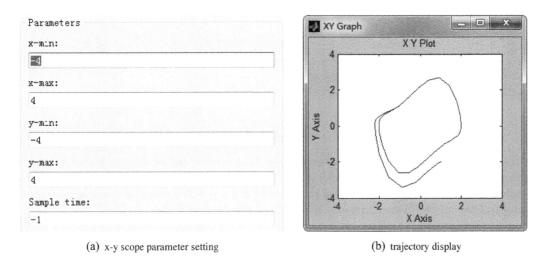

(a) x-y scope parameter setting   (b) trajectory display

Figure 4.43   Phase plane trajectories in x-y scope.

*Also the conventional **Scope** block can be used to show the two signals on the same axis. This can be done by combining the two output signals $y(t)$ and $\dot{y}(t)$ into a vector one with the **Mux** block in the **Signal Routing** group, then connect the vector signal to the **Scope** block directly, as shown in Fig. 4.44.*

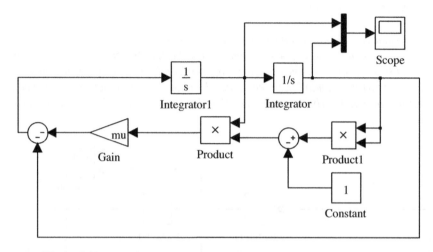

Figure 4.44   Simulink model with vector outputs (model name: c4mvdp2).

After simulation, the result can be obtained as shown in Fig. 4.45(a). The ranges of the axes were assigned automatically. For instance, if the range of the y axis is not satisfactory, the original plot can be processed with the zooming facilities in the toolbar by the button 🔍.

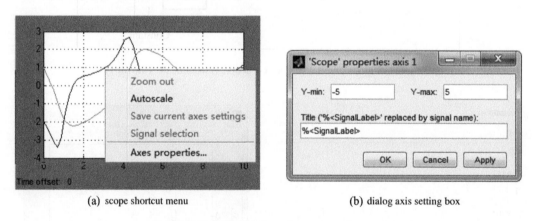

(a) scope shortcut menu   (b) dialog axis setting box

Figure 4.45   Modification of parameters in axes.

*If only the range of the y axis is to be changed, the shortcut menu can be displayed when the plot is clicked with right mouse button, as shown in Fig. 4.45(a). The **Axes properties** menu item can be selected, and the dialog box in Fig. 4.45(b) can be displayed. The range of y can be changed manually. Other toolbar buttons can also be used to change the axis settings.*

Almost all ODEs can be represented as Simulink models. However, compared with the model descriptions with MATLAB statements presented in Chapter 3, sometimes the block diagram based description may seem to be more complicated and less intuitive. Thus for simple ODEs, the best way to solve them is to use MATLAB ODE solvers, rather than block diagrams. If the ODE is only one portion of a system, and its input signal comes from another part of the system, Simulink modeling techniques should be used. So it is very important to learn how to represent an ODE by block diagrams. In later presentations, a simplified way to describe ODEs will also be discussed.

Example 4.4 *Consider the DC motor drive system given in block diagram form [3], as shown in Fig. 4.46. It can be seen that the* **Transfer Fcn** *blocks and* **Sum** *blocks are used extensively in the system. Also,* **Gain** *block,* **Step** *block and* **Scope** *block must also be used. A Simulink model can be constructed as shown in Fig. 4.47.*

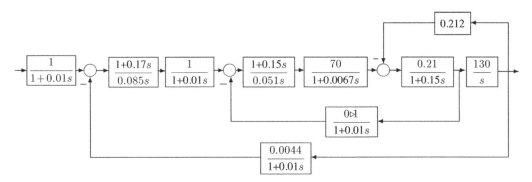

Figure 4.46   Block diagram of DC motor drive systems.

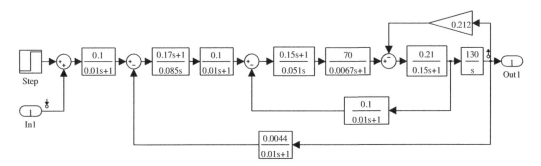

Figure 4.47   Simulink model of DC motor drive system (model name: c4mex2).

From the established Simulink models, simulation analysis can be invoked by the **Simulation** → **Start** *menu item, and simulation results can be returned to the variables* tout *and* yout *in the MATLAB workspace. The command* plot(tout,yout) *can be used to draw the step response of the output signal, as shown in Fig. 4.48(a).*

It can be seen that the output curve is not quite satisfactory. If the outer PI controller is changed to $(\alpha s + 1)/0.085s$, *and we try different values of* $\alpha = 0.17, 0.5, 1, 1.5$, *the output signals under different controller parameters can be obtained as shown in Fig. 4.48(b). It can be seen that, with the new controller* $(1.5s + 1)/0.085s$, *satisfactory results can be achieved. Optimum selection of controller parameters will be presented in Chapter 6.*

It can also be seen that if parameters in a Simulink model are changed, the simulation result with the new parameters can be obtained immediately, and the system behavior can be assessed. Parameter changes can be made by double clicking relevant blocks, and modifying all of the parameters through dialog boxes, or through the parameters with MATLAB commands. The latter way will also be discussed in Chapter 6.

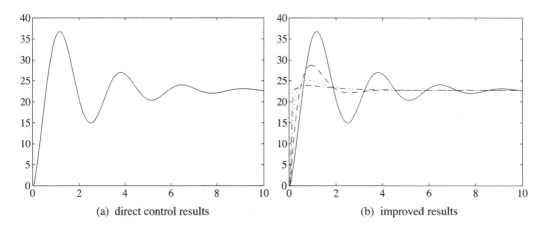

(a) direct control results   (b) improved results

Figure 4.48   Step responses of the system.

Example 4.5   *The block diagram of a control system with hysteresis is shown in Fig. 4.49(a), where the hysteresis component is shown in Fig. 4.49(b).*

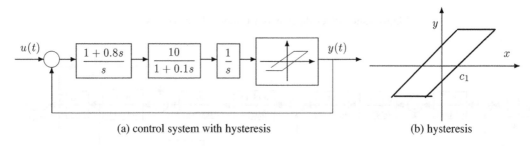

(a) control system with hysteresis   (b) hysteresis

Figure 4.49   Block diagram of a nonlinear system.

*This nonlinear element can be expressed by the **Backlash** block in the **Discontinuity** group. The Simulink model can then be established, as shown in Fig. 4.50. In the hysteresis block, the width is assigned to a variable c_1. Thus the value of c_1 should be assigned in MATLAB before simulation, for example with $c_1 = 1$. The termination time of the simulation is assigned to 3. After simulation, two variables, tout and yout, are returned to the MATLAB workspace, and the response of the system can be obtained as shown in Fig. 4.51(a).*

```
>> c1=1; plot(tout,yout)
```

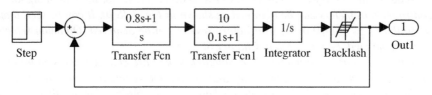

Figure 4.50   Simulink model (model name: c4mex3).

For different values of c_1's, Simulink results can be obtained using the same Simulink model, and the results can be obtained with MATLAB statements, as shown in Fig. 4.51(b). The command hold on *should be used to keep the responses with the same axis setting, so that they can be compared easily.*

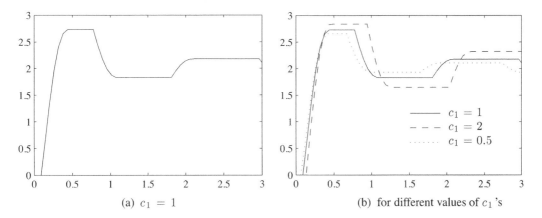

(a) $c_1 = 1$ (b) for different values of c_1's

Figure 4.51   Step responses for the system with hysteresis.

For linear systems, if the amplitude of the input signal is changed, the output waveform is kept unchanged, while the amplitude of it changes with that of the input. For nonlinear systems, the above superposition property no longer holds. For instance, when $c_1 = 1$, the output signals with magnitudes of 2 and 6 are simulated, and the results are compared in Fig. 4.52. It can be seen that the waveforms are also different. This makes the analysis of nonlinear systems far more complicated than their linear counterparts. However, with Simulink, the simulation studies are always possible and it is very easy and straightforward.

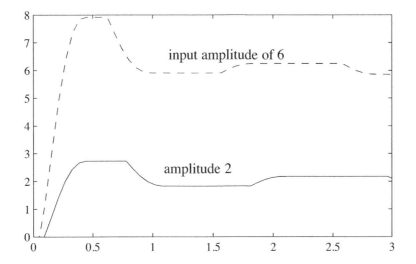

Figure 4.52   Simulation results with different step input amplitudes.

Example 4.6 *The block diagram of a sampled-data system is shown in Fig. 4.53. The zero-order hold (ZOH) model can be represented directly with the* **Zero-Order Hold** *block in the* **Discrete** *group, and the sampling interval is set to 0.1s. A Simulink model of such a system can be constructed as shown in Fig. 4.54(a). In the model, the continuous transfer function is represented by two blocks* g1 *and* g2 *in series connection. After simulation, the output of the system can be drawn with the MATLAB command* stairs(tout,yout), *as shown in Fig. 4.54(b).*

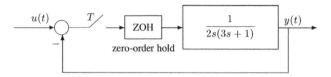

Figure 4.53 Block diagram of a sample-data system.

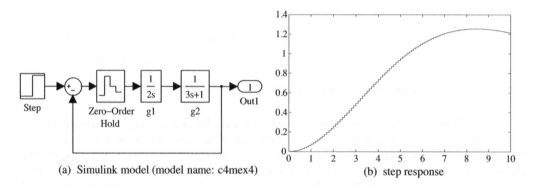

(a) Simulink model (model name: c4mex4)   (b) step response

Figure 4.54 Simulink model and step responses.

4.5   Modeling, Simulation and Analysis of Linear Systems

4.5.1   Modeling of Linear Systems

Various of linear system blocks are provided in the standard Simulink block library, for example the transfer function block, the state space block and the zero-pole-gain block. These blocks are provided as continuous blocks as well as discrete ones. In the Control System Toolbox of MATLAB, various linear time-invariant (LTI) objects, such as transfer function object, state space object and zero-pole-gain object are also provided to represent linear system blocks, with the following syntaxes:

$$G = \mathtt{tf}(\boldsymbol{n}, \boldsymbol{d}); G = \mathtt{ss}(\boldsymbol{A}, \boldsymbol{B}, \boldsymbol{C}, \boldsymbol{D}); G = \mathtt{zpk}(\boldsymbol{z}, \boldsymbol{p}, \boldsymbol{K})$$

where $\boldsymbol{n}$ and $\boldsymbol{d}$ are respectively the coefficient vectors of the numerator and denominator in descending orders of s. With the numerator and denominator polynomials, a transfer function object can be represented in a single LTI object variable G in the MATLAB workspace. Also, if matrices $\boldsymbol{A}$, $\boldsymbol{B}$, $\boldsymbol{C}$ and $\boldsymbol{D}$ are entered, a state space object can be constructed with the second statement. And if the zeros, the poles and the gain of the system are represented by $\boldsymbol{z}$, $\boldsymbol{p}$ and K, respectively, the zero-pole-gain object can be represented by the third statement. For single variable systems, $\boldsymbol{z}$ and $\boldsymbol{p}$ should be presented as column vectors.

Sampled-data systems can also be represented in the same way, and the sampling interval T can also be assigned to the system with the command $G.\mathtt{Ts} = T$.

LTI objects with no time delays can be represented by the **LTI System** block provided in the Control Systems Blockset. One can simply copy the block first to a model window, as shown in Fig. 4.55(a), then double click it to display the dialog box as shown in Fig. 4.55(b). In the edit box labeled as **LTI system variable**, the LTI object G can be filled in directly. The initial state vector can also be assigned to the block in the edit box labeled **Initial states**, if the state space model is given.

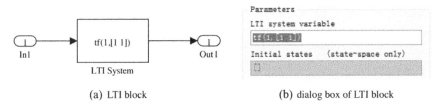

(a) LTI block   (b) dialog box of LTI block

Figure 4.55   Linear model input.

The linear system models with initial conditions can be modeled by the blocks in the **Simulink Extras** group in the **Toolboxes and Blocksets** category. The relevant blocks are shown in Fig. 4.56, and they can be used in modeling nonzero initial condition systems.

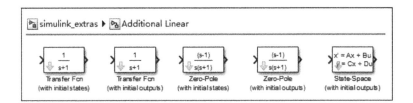

Figure 4.56   Linear continuous blocks with initial conditions.

Example 4.7   *Consider the transfer function model*

$$G(s) = \frac{s^2 + 5}{s^2(s+1)^2((s+2)^2 + 9)}.$$

It might be complicated to enter such a model with the direct use of the tf() *function presented earlier, since manual expansion of the denominator polynomial should be made first. A simpler way of using a transfer function representation will be shown. We can declare the Laplace operator s by* s = tf('s')*, then the following commands can be entered to MATLAB, and the LTI object G can be created in the MATLAB workspace. The variable G can be specified in Fig. 4.55(b) directly.*

```
>> s=tf('s'); G=(s^2+5)/s^2/(s+1)^2/((s+2)^2+9);
```

Example 4.8   *Transfer function matrices of multivariable systems can also be represented by an LTI object. Consider a 4 by 4 transfer function matrix [4]*

$$\boldsymbol{G}(s) = \begin{bmatrix} 1/(1+4s) & 0.7/(1+5s) & 0.3/(1+5s) & 0.2/(1+5s) \\ 0.6/(1+5s) & 1/(1+4s) & 0.4/(1+5s) & 0.35/(1+5s) \\ 0.35/(1+5s) & 0.4/(1+5s) & 1/(1+4s) & 0.6/(1+5s) \\ 0.2/(1+5s) & 0.3/(1+5s) & 0.7/(1+5s) & 1/(1+4s) \end{bmatrix}.$$

It can be seen that modeling a multivariable system using low-level blocks is very difficult and complicated. However, with the LTI block, the system can easily be described in Simulink. One can simply fill the variable name G in the dialog box shown in Fig. 4.55(b).

```
>> h1=tf(1,[4 1]); h2=tf(1,[5 1]);
   h11=h1; h12=0.7*h2; h13=0.5*h2; h14=0.2*h2;
   h21=0.6*h2; h22=h1; h23=0.4*h2; h24=0.35*h2;
   h31=h24; h32=h23; h33=h1; h34=h21; h41=h14; h42=h13; h43=h12; h44=h1;
   G=[h11,h12,h13,h14; h21,h22,h23,h24; h31,h32,h33,h34; h41,h42,h43,h44];
```

Example 4.9   *Consider the discrete state space model given by*

$$x(k+1) = \begin{bmatrix} 0 & -2 & -2 & -1.1 \\ 0.5 & 1.8 & 0.8 & 0.5 \\ 0.5 & 0.8 & 1.8 & 0.5 \\ -0.5 & -0.8 & -0.7 & 0.4 \end{bmatrix} x(k) + \begin{bmatrix} 0.1 & 0.1 \\ 0.2 & 0.1 \\ 0.3 & 0.1 \\ 0.1 & 0 \end{bmatrix} u(k), y(k) = x_1(k)$$

with a sampling interval of $T = 0.1s$. A variable G can be constructed in the MATLAB workspace, and the LTI dialog box can be filled in with the value of G to represent the transfer function model in Simulink.

```
>> F=[0,-2,-2,-1.1; 0.5,1.8,0.8,0.5; 0.5,0.8,1.8,0.5; -0.5,-0.8,-0.7,0.4];
   G=[0.1,0.1; 0.2,0.1; 0.3,0.1; 0.1,0]; C=[1 0 0 0]; G=ss(F,G,C,0,0.1)
```

4.5.2   Analysis Interface for Linear Systems

Many linear system analysis functions are provided in the MATLAB Control System Toolbox. For a given LTI object G, the MATLAB function `bode(G)` can be used to draw a Bode diagram of the system, while `nyquist(G)` and `nichols(G)` can be used respectively to draw its Nyquist plot and Nichols chart. The functions `step(G)` and `impulse(G)` can be used to draw the unit step response and impulse response of the system G. LTI object G can be used to draw the root locus of the system.

Unlike the cases in the old versions, the **In** and **Out** blocks can no longer be used to describe the input and output ports of the system model that is to be linearized or analyzed. We should select the relevant signal line, and from the right-mouse button, in the **Linear Analysis Points** menu, select **Input Point** and **Output Point**.

Linear systems and linearized nonlinear systems can also be analyzed directly with the facilities provided in Simulink. For instance, we can open the linear DC motor drive system described by the c4mex2 model first, then use the **Linear Analysis Points** menu to specify the input and output points, as shown in Fig. 4.57. Select **Analysis** → **Control Design** → **Linear Analysis** menu item (in earlier versions, under the **Tools** menu), the linear system analysis and manipulation interface will be displayed as shown in Fig. 4.58. The default step response of the system, from the input point to the output point, can be displayed automatically in the interface.

One can also click the **Linear Analysis** pane in the toolbar, and the new toolbar is changed to the form in Fig. 4.59. Different analysis tasks can be selected from it by clicking the **Plot Select** icon, as

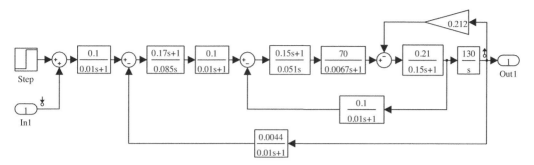

Figure 4.57   Modified c4mex2 model with input and output points (model name: c4mex2x).

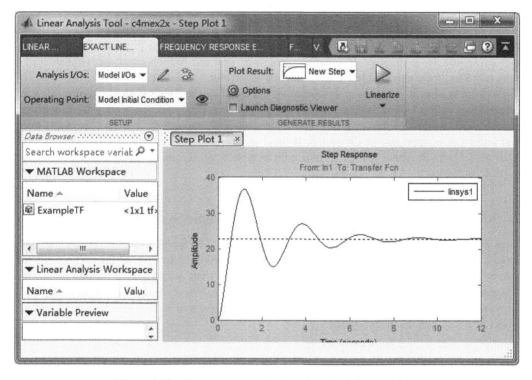

Figure 4.58   Linear system analysis and manipulation interface.

Figure 4.59   Toolbar in linear system analyzer.

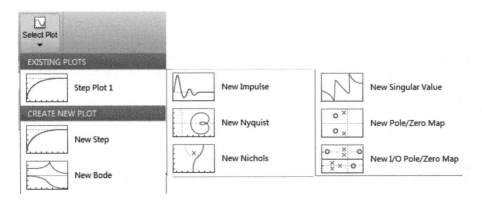

Figure 4.60   Available linear system analysis tasks.

shown in Fig. 4.60. It can be seen that apart from the default step response plot, other plots such as impulse response, Bode diagram, Nyquist plot, pole-zero plot and Nichols charts can be displayed, if selected. Also, for multivariable systems, singular value plots can also be displayed.

If there are nonlinear elements in the Simulink model, click the **Exact Linearization** pane, and the **Operating Points** listbox will be selected to get the operating points. Then click the **Linearize** button to start the linearization process. For the linearization problem, refer to the command line based linearization approach in Chapter 6. If there are no nonlinear elements in the system, the overall model from the input to the output can be retrieved, and the above methods can be used to draw different plots.

4.6   Simulation of Continuous Nonlinear Stochastic Systems

As was pointed out in Chapter 3, for continuous systems driven by white noise, the random number generator cannot be used directly. Also the approach discussed in Chapter 3 can only be applied to linear systems. For nonlinear systems, the method cannot be used. Although the method proposed in [5] can be used to work with some nonlinear systems, it is difficult to extend it to solve general nonlinear system simulation problems. In this section, an effective way is discussed, and the statistical analysis of simulation results is studied.

4.6.1   Simulation of Random Signals in Simulink

The **Band-Limited White Noise** block provided in the Simulink **Sources** group can be used to simulate white noise input with a specific intensity. This block can be used for driving continuous system. In fact, this block acts in a similar way to adding a gain of $1/\sqrt{\Delta t}$ after a random number generator, where Δt is the step size.

Example 4.10   *Consider again the system model shown in Example 3.36. Approximate simulation methods can be used, and the Simulink model is constructed in Fig. 4.61(a). Selecting $T = 0.1s$, the menu item **Simulation → Simulation Parameters** can be selected. A dialog box shown in Fig. 4.34 is displayed. The fixed fourth-order Runge–Kutta algorithm can be selected, with a step size of 0.1s. If the terminating time is assigned to $30\,000T$, the **1000 points Workspace I/O** item in the dialog box should be clicked to off. The simulation process can be invoked and the output*

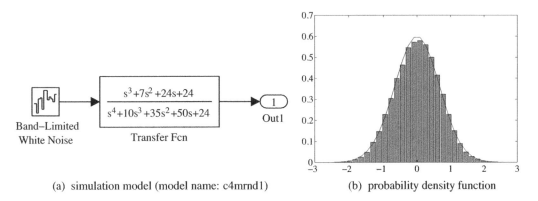

(a) simulation model (model name: c4mrnd1)   (b) probability density function

Figure 4.61   Linear system model driven by random input.

signal can be returned to the MATLAB workspace under the name yout. *The following statements can be used to estimate the probability density function, as shown in Fig. 4.61(b). It can be seen that the results agree well with theoretical results.*

```
>> G=tf([1,7,24,24],[1,10,35,50,24]); v=norm(G,2);
   xx=linspace(-2.5,2.5,30); yy=hist(yout,xx);
   yy=yy/(length(yout)*(xx(2)-xx(1)));
   yp=exp(-xx.^2/(2*v^2))/sqrt(2*pi)/v; bar(xx,yy,'c'),
   hold on; plot(xx,yp)
```

It can be seen from Example 4.10 that the **Band-limited White Noise** block above can be used in simulating continuous systems driven by random inputs.

4.6.2   Statistical Analysis of Simulation Results

Assume that the transfer function model is given by $G(s)$. The autocorrelation function $c_{yy}(\tau)$ of the output signal when driven by a Gaussian white noise can be obtained by a bilateral inverse Laplace transform such that

$$c_{yy}(\tau) = \frac{S}{2\pi \mathrm{j}} \int_{-\mathrm{j}\infty}^{\mathrm{j}\infty} G(s)G(-s)\mathrm{e}^{s\tau}\mathrm{d}s, \tag{4.9}$$

where S is the power spectral density of the white noise signal. With the spectral factorization theorem, $G(s)G(-s)$ can be factorized as

$$SG(s)G(-s) = S\frac{N(s)}{D(s)}\frac{N(-s)}{D(-s)} = \frac{B(s)}{A(s)} + \frac{B(-s)}{A(-s)}, \tag{4.10}$$

where $A(s)$ holds the factors of the denominator of $G(s)G(-s)$, whose poles are on the left-hand side of the s-plane, mathematically denoted by $A(s) = D(s)_-$. $D(s)$ and $N(s)$ are respectively the denominator and numerator polynomials of $G(s)$. Thus the following equations can be obtained

$$B(s)A(-s) + B(-s)A(s) = SN(s)N(-s). \tag{4.11}$$

Assume that the polynomials $A(s)$ and $B(s)$ can be written as

$$A(s) = \sum_{i=0}^{m} \alpha_i s^i, \text{ and } B(s) = \sum_{i=0}^{m-1} \beta_i s^i. \tag{4.12}$$

Then

$$\begin{cases} A(s)B(-s) = \displaystyle\sum_{j=0}^{2m-1} \gamma_j s^j, \gamma_j = \sum_{k+l=j} (-1)^l \alpha_k \beta_l \\[3mm] A(-s)B(s) = \displaystyle\sum_{j=0}^{2m-1} \delta_j s^j, \delta_j = \sum_{k+l=j} (-1)^k \alpha_k \beta_l. \end{cases} \tag{4.13}$$

It follows that

$$A(s)B(-s) + A(-s)B(s) = \sum_{k+l=j} (-1)^l \left[1 + (-1)^{k-l} \right] \alpha_k \beta_l$$

$$= \begin{cases} 2 \displaystyle\sum_{k+l=j} (-1)^l \alpha_k \beta_l, & j \text{ is even} \\[3mm] 0, & j \text{ is odd} . \end{cases} \tag{4.14}$$

Denote $SN(s)N(-s) = f_{m-1}s^{2m-2} + f_{m-2}s^{2m-4} + \cdots + f_1 s^2 + f_0$. The unknowns β_i can be solved with the following linear algebraic equation

$$\begin{bmatrix} (-1)^{m-1}\alpha_{m-1} & (-1)^{m-2}\alpha_m & 0 & \cdots & 0 \\ (-1)^{m-1}\alpha_{m-3} & (-1)^{m-2}\alpha_{m-2} & (-1)^{m-1}\alpha_{m-1} & \cdots & 0 \\ (-1)^{m-1}\alpha_{m-5} & (-1)^{m-2}\alpha_{m-4} & (-1)^{m-1}\alpha_{m-3} & \cdots & 0 \\ \vdots & \vdots & \vdots & \ddots & \vdots \\ 0 & 0 & 0 & \cdots & \alpha_0 \end{bmatrix} \begin{bmatrix} \beta_{m-1} \\ \beta_{m-2} \\ \beta_{m-3} \\ \vdots \\ \beta_0 \end{bmatrix} = \frac{S}{2} \begin{bmatrix} f_{m-1} \\ f_{m-2} \\ f_{m-3} \\ \vdots \\ f_0 \end{bmatrix}, \tag{4.15}$$

and the autocorrelation function of the output signal can be evaluated from

$$c_{yy}(\tau) = \mathscr{L}^{-1} \left[\frac{B(s)}{A(s)} \right], \text{ when } \tau > 0 \tag{4.16}$$

and $c_{yy}(-\tau)$, when $\tau \leq 0$. The autocorrelation function $c_{yy}(\tau)$ can be regarded as the impulse response of $B(s)/A(s)$.

Based on the above algorithm, the following MATLAB function can be developed for computing power spectral density factorizations

```
function [B,A]=spec_fac(num,den)
m=length(den)-1; k=0; NN=conv(num,(-1).^[length(num)-1:-1:0].*num);
X=NN(1:2:end)';   X=0.5*[zeros(m-length(X),1); X];
p=roots(den); ii=find(p>0); p(ii)=-p(ii); A=poly(p);
if m>1, Xx=[(-1)^(m-1)*A(2)  (-1)^m*A(1)  zeros(1,m-2)];
else, Xx=[A(1)]; end, V0=Xx;
```

```
for i=2:m, V0=[0 0 V0(1:m-2)]; k=k+2;
   if k<m+1, V0(2)=(-1)^m*A(k+1);
      if k<m, V0(1)=(-1)^(m-1)*A(k+2); end, end
   Xx=[Xx; V0];
end
B=[inv(Xx)*X]';
```

With this function, the numerator and denominator coefficient vectors num and den, satisfying the spectral factorization, can be obtained in vectors **B** and **A**.

For nonlinear systems, the evaluation of autocorrelation functions is not as simple as this. Simulation studies should be made first and, based on the data, the correlation function analysis method discussed in Chapter 3 can be used to evaluate the autocorrelation functions and cross-correlation functions [6].

■ Example 4.11 *Consider again* $G(s) = \dfrac{s^3 + 7s^2 + 24s + 24}{s^4 + 10s^3 + 35s^2 + 50s + 24}$. *Assume that the variance of the white noise input signal is 1, and the computation step size is 0.1s. The following commands can be used to calculate the theoretical and numerical results for the autocorrelation functions and the results are compared in Fig. 4.62. It can be seen that the two curves agree well.*

```
>> num=[1 7 24 24]; den=[1 10 35 50 24]; G=tf(num,den);
   [B,A]=spec_fac(num,den); G1=tf(B,A); T=0.1;
   [y0,t]=impulse(G1,30000*T);    % impulse response data
   t=[-t(end:-1:2); t]; y0=[y0(end:-1:2); y0]/max(y0); % normalization
   [Cyy,f]=crosscorr(yout,yout); % find autocorrelation function
   f=f*T; plot(f,Cyy,t,y0,':'), xlim([-3 3]) % theoretical in dash lines
```

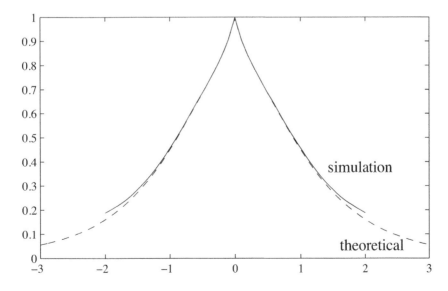

Figure 4.62 Autocorrelation function of the output signal.

The power spectral density function of linear systems is a function of frequency ω and it can be evaluated from

$$G_{\mathrm{p}}(\omega) = G(\mathrm{j}\omega)G(-\mathrm{j}\omega)P, \tag{4.17}$$

where P is the power spectral density of the input signal. If simulation results are obtained, the function `psd_estm()` in Section 3.7 can be used to estimate the power spectral density of the simulated signals.

Example 4.12 *Consider again the simulation results obtained above. The following statements can be used to estimate the power spectral density of the output signal, which is plotted, together with the theoretical result, in Fig. 4.63. It can be found that the two results agree well, which validates the reliability of the simulation method.*

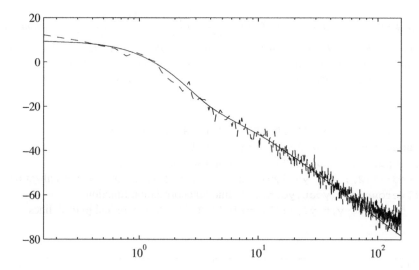

Figure 4.63 Comparison of power spectral density of the output signal.

```
>> [Pxx,f]=psd_estm(yout,2028,T);
   num=G.num{1}.*(-1).^[length(G.num{1})-1:-1:0];
   den=G.den{1}.*(-1).^[length(G.den{1})-1:-1:0]; GG=G*tf(num,den);
   [mag,p]=bode(GG,f);% compute G̃(s) = PG(s)G(-s) and G̃(jω)
   semilogx(f,20*log10(mag(:)),f,20*log10(Pxx))
```

Exercises

4.1 The blocks in the standard Simulink are well organized and classified. Observe each group and get acquainted with them, so that relevant blocks can be found easily whenever necessary.

4.2 In Example 3.27, a numerical solution method is presented for Lorenz equations, and the Apollo satellite trajectory problem is demonstrated in Example 3.33. Construct Simulink models for the two systems, and compare the results with those given in the examples.

4.3 Construct a Simulink model for the nonlinear system [7] shown in Fig. 4.64. Observe the unit step response of the output signal and the error signal. In the nonlinear system, there are two blocks in series connection. Are these two blocks interchangeable? Explain your conclusion with simulation results.

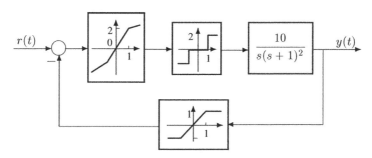

Figure 4.64   Problem (4.3).

4.4 For the two block diagrams of feedback control systems [8] shown in Figs. 4.65(a) and (b), create Simulink models, and find the system responses to unit step inputs. Also discuss the effect of nonlinear elements on the behaviors of the nonlinear systems for various nonlinearity parameters.

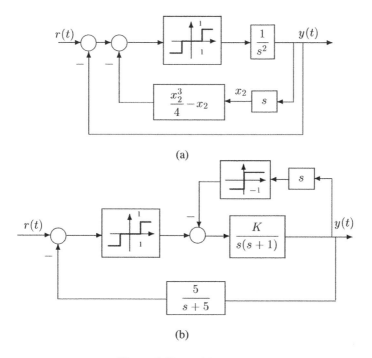

(a)

(b)

Figure 4.65   Problem (4.4).

4.5 The Rössoler equation has been studied in Chapter 3

$$\begin{cases} \dot{x} = y + z \\ \dot{y} = x + ay \\ \dot{z} = b + (x - c)z. \end{cases}$$

Selecting $a = b = 0.2, c = 5.7$, and $x(0) = y(0) = z(0) = 0$, construct a Simulink model and perform the simulation analysis. Compare the simulation results with the ones in Chapter 3.

4.6 Construct a Simulink model for the nonlinear system [9] in Fig. 4.66. Observe the output and error signals under the unit step input signal.

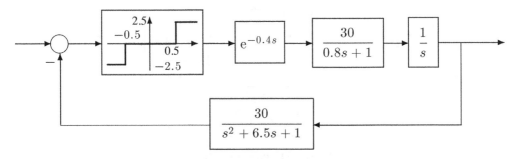

Figure 4.66   Problem (4.6).

4.7 If a Simulink model of a system is as shown in Fig. 4.67, write out the mathematical model of the system.

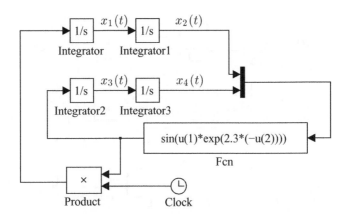

Figure 4.67   Simulink model in Problem (4.7).

4.8 Assume that the linear models are given by

(a) $G(s) = \dfrac{s^2 + 5s + 2}{(s+4)^4 + 4s + 4}$,   (b) $H(z) = \dfrac{z^2 + 0.568}{(z-1)(z^2 - 0.2z + 0.99)} z^{-5}$, $T = 0.05$ s.

Enter the models in the MATLAB workspace and find the analytical and numerical solutions of the systems when driven respectively by impulse, step and ramp inputs. Validate the simulation results.

4.9 Assume that the system is given by the following ODE

$$\begin{cases} \dot{x}_1(t) = -x_1(t) + x_2(t) \\ \dot{x}_2(t) = -x_2(t) - 3x_3(t) + u_1(t) \\ \dot{x}_3(t) = -x_1(t) - 5x_2(t) - 3x_3(t) + u_2(t), \end{cases} \quad \text{and } y = -x_2(t) + u_1(t) - 5u_2(t),$$

and there are two inputs $u_1(t)$ and $u_2(t)$. Establish Simulink models by two different methods.

4.10 Suppose that the difference equation of a system is given by

$$y(k+2) + y(k+1) + 0.16y(k) = u(k-1) + 2u(k-2),$$

and assume that the sampling interval is $T = 0.1$s. Draw the Simulink model for the system in two methods, and perform both time and frequency domain analysis to the system.

4.11 For the DC motor drive system in Example 4.14, draw the open-loop frequency response of the system as a Bode diagram, a Nyquist plot and a Nichols chart.

4.12 For the linear transfer function model $G(s) = \dfrac{6(s+1)}{(s+2)(s+3)(s+4)}$, if the input signal is white noise $\gamma(t)$, find the probability density function of the output signal, and compare the output signals. Perform autocorrelation analysis and power spectral analysis and see whether they agree with the theoretical results.

4.13 Construct a Simulink model for the multivariable transfer function matrix

$$G(s) = \begin{bmatrix} \dfrac{0.806s + 0.264}{s^2 + 1.15s + 0.202} & \dfrac{-15s - 1.42}{s^3 + 12.8s^2 + 13.6s + 2.36} \\[2mm] \dfrac{1.95s^2 + 2.12s + 0.49}{s^3 + 9.15s^2 + 9.39s + 1.62} & \dfrac{7.15s^2 + 25.8s + 9.35}{s^4 + 20.8s^3 + 116.4s^2 + 111.6s + 18.8} \end{bmatrix}.$$

4.14 For the nonlinear system shown in Fig. 4.68. If the input signal $\gamma(t)$ is a Gaussian white noise, with zero mean and a variance of 4. Find the probability density function of the error signal $e(t)$ using simulation results.

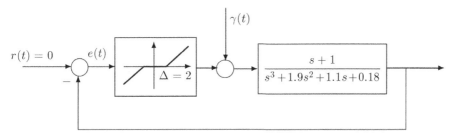

Figure 4.68 Problem (4.14).

References

[1] E E L Mitchell, J S Gauthier. Advanced continuous simulation language (ACSL) – user's manual. Mitchell & Gauthier Associates, 1987

[2] D K Frederick, M Rimer. Benchmark problem for CACSD packages. Abstracts of the second IEEE symposium on computer-aided control system design, 1985. Santa Barbara, USA

[3] X Q Ren. Computer simulation and CAD of control systems. Shenyang: Northeastern University Press, 1986. In Chinese

[4] H H Rosenbrock. Computer-aided control system design. New York: Academic Press, 1974

[5] D Xue, D P Atherton. Simulation analysis of continuous systems driven by Gaussian white noise. M Jamshidi, C J Herget, Eds, Recent advances in computer-aided control systems engineering. Elsevier Science Publishers BV, 1992, 431–452

[6] D Xue. Analysis and computer-aided design of nonlinear systems with Gaussian inputs. Ph.D. thesis, the University of Sussex, 1992

[7] W L Wang. Automatic control principles. Beijing: Science Press, 2001. In Chinese

[8] D P Atherton. Nonlinear control engineering – describing function analysis and design. London: Van Nostrand Reinhold, 1975

[9] D G Liu, J G Fei. Numerical simulation algorithms of dynamic systems. Beijing: Science Press, 2001. In Chinese

5

Commonly Used Blocks and Intermediate-level Modeling Skills

In Chapter 4, some basic Simulink modeling and simulation methods were discussed. This chapter focuses on intermediate-level techniques with some commonly used Simulink blocks to learn modeling skills. In Section 5.1, a simple example is used to further demonstrate the model representation and modeling skills such as including vectorized block modeling and model decoration techniques. Important problems such as the concept of algebraic loops and their elimination, and also the zero-crossing detection method are discussed. In Section 5.2, Simulink modeling of linear multivariable systems is illustrated, where the LTI block in the Control System Toolbox is recommended for simplifying the modeling process. In Section 5.3, the application of blocks such as the lookup table and various switches are explored. General methods in constructing piecewise linear nonlinearities are introduced, whether or not the nonlinearities have memories. In Section 5.4, Simulink modeling techniques for various kinds of differential equations are demonstrated. These equations include ordinary differential equations, differential algebraic equations, delay differential equations, switching differential equations and fractional-order differential equations. In Section 5.5, various visualization output blocks are presented, such as scope output, workspace variable output and gauges output. In Section 5.6, more advanced output visualization methods are presented, including three-dimensional animation methods with virtual reality techniques. Fundamental world modeling with VRML (*Virtual Reality Modeling Language*) is briefly introduced, and the VRML models driven by MATLAB and Simulink output are discussed. In Section 5.7, subsystem modeling is introduced using subsystem masking techniques. An illustrative example of Simulink modeling of a complicated system is presented in detail.

5.1 Commonly Used Blocks and Modeling Skills

5.1.1 *Examples of Vectorized Blocks*

In the current version of Simulink, many blocks support single-channel signal input as well as vectorized inputs, that is, several input signals can be combined together through a multiplexer (**Mux**) block into a vectorized signal. Examples will be given to show the vectorized blocks and their applications.

System Simulation Techniques with MATLAB® and Simulink®, First Edition. Dingyü Xue and YangQuan Chen.
© 2014 John Wiley & Sons, Ltd. Published 2014 by John Wiley & Sons, Ltd.

Example 5.1   *In Example 4.3, Simulink modeling of the Van der Pol equation was provided. The vectorized form of the state space equation can be rewritten as*

$$\begin{bmatrix} \dot{x}_1 \\ \dot{x}_2 \end{bmatrix} = \begin{bmatrix} x_2 \\ -\mu(x_1^2 - 1)x_2 - x_1 \end{bmatrix}.$$

*A single integrator block can be used to implement the vectorized modeling process, and the new model is constructed as shown in Fig. 5.1. Note that, the **Demux** block can be used to extract the original signals from the vectorized ones. The block **Mux** is used to combine the individual signals back to vectorized ones to feed the integrator.*

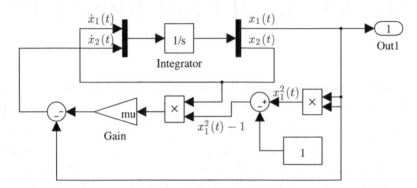

Figure 5.1   Simulink description of Van der Pol in vectorized form (model name: c5mvdp1).

*For such a system, if the initial values in the integrator are assigned to a column vector, simulation results obtained are identical to the results obtained in Example 4.3. The model can be further simplified. For instance, the **Fcn** block can be used to generate the nonlinear operation in the second equation, as shown in Fig. 5.2(a), where the dialog box of the **Fcn** block, can be filled in with* -mu*(u[1]*u[1]-1)*u[2]-u[1]. *Unfortunately, the block **Fcn** does not support vector output, otherwise the block diagram would have been simpler. Later, the static nonlinear function with blocks will be presented.*

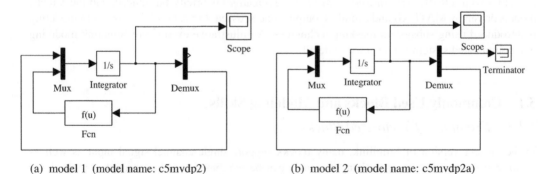

(a) model 1 (model name: c5mvdp2)   (b) model 2 (model name: c5mvdp2a)

Figure 5.2   Simpler forms of Van der Pol equations.

*When the Simulink model is executed, a warning message "Warning: Output port 1 of block 'c5mvdp2/Demux' is not connected." will be displayed, indicating that the first output port in block **Demux** is not connected to any block. The execution can be carried out, and no other block will* *be affected. If the warning is considered a distraction, the menu item* **Simulation** → **Configuration Parameters** *can be used to open the parameter setting dialog box. In the* **Diagnostics** *pane, the* **Unconnected block outport** *item should be set to* **None**. *To avoid changing the setting, an alternative way is to connect the suspended output terminal to a* **Terminator** *block, as shown in Fig. 5.2(b).*

5.1.2   Signals Labeling in Simulink Models

In order to increase the readability of the Simulink model so that users may understand better the internal structures of the model, many decoration options are provided in Simulink to change the properties of blocks and signals. The **Display** menu shown in Fig. 5.3 can be used (or, in earlier versions, the **Format** menu):

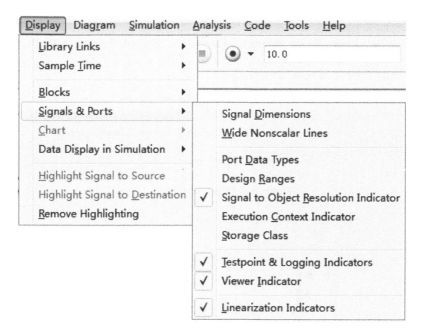

Figure 5.3   Display menu.

- **Display** → **Port/Signal Display** menu: The submenu **Wide Nonscalar Lines** can be used to display the vector signals in wider lines. The submenu **Signal Dimensions** can be used to display the number of signals in vector ones. The Simulink model in Fig. 5.1 can be displayed in the form shown in Fig. 5.4, when the above two menu items are checked. The submenu **Port Data Types** can be used to display the data types of all the signals. For example, **double** is used to indicate the double-precision data type in signals, while **double(2)** can be used to indicate two channel vector signals in double-precision data type.

- **Display** → **Blocks** menu: The menu is shown in Fig. 5.5. If the menu **Sorted Execution Order** is selected, all the blocks are sorted in the execution order and the sequence numbers are displayed on the corners of the blocks. Normally, the input source blocks and integrator blocks, since initial values are known, are sorted first. The other blocks are sorted after these blocks, in the sequence of signal generation. The Simulink model with sorted blocks is shown in Fig. 5.6. Also,

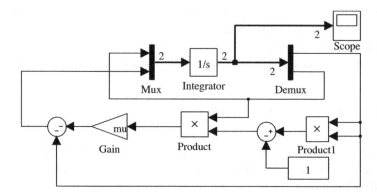

Figure 5.4   Simulink model with two options checked (model name: c5mvdp3).

if the submenu **Block Version for Referenced Models** is checked, the versions of the blocks are displayed.

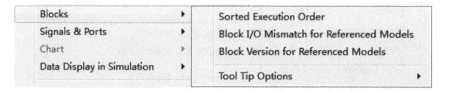

Figure 5.5   The **Block** menu.

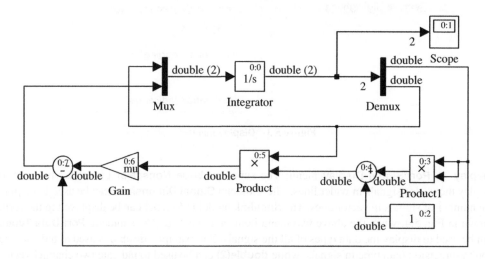

Figure 5.6   The other two options are selected.

- **Display** → **Library Link Display** submenu: The contents of the menu are shown in Fig. 5.7(a). The relationship between the current blocks and their base ones in the model library is assigned. If the **None** option is selected, the relationship between the blocks with the base ones is broken. When the current models are selected, the original base blocks are not affected.

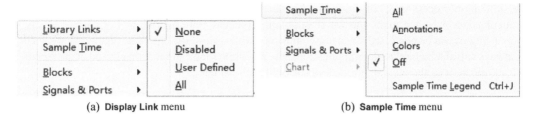

(a) **Display Link** menu   (b) **Sample Time** menu

Figure 5.7   **Library Links** and **Sample Time** menus.

- **Display** → **Sample Time Display** menu: The submenu **Color** can be selected to allow blocks with different sampling intervals to be displayed in different colors.

5.1.3   Algebraic Loop and its Elimination in Simulink Models

The following example can be used to introduce the algebraic loop problems in simulation models, and then several methods to eliminate algebraic loops will be presented, from which an effective method is recommended.

Example 5.2   *Consider the simple linear feedback system shown in Fig. 5.8(a). A Simulink model can easily be constructed as shown in Fig. 5.8(b). Performing simulation with such a system, the following warning messages will be given*

```
Warning: Block diagram 'untitled' contains 1 algebraic loop(s).
Found algebraic loop containing block(s):
 'untitled/Transfer Fcn'
 'untitled/Sum' (algebraic variable)
```

*which indicates that in the loop constructed by the blocks **Transfer Fcn** and **Sum**, there is an algebraic loop. Analyzing the Simulink model given in Fig. 5.8(b), it can be seen that the input signal to the transfer function block is $v(t) = u(t) - y(t)$, and the signals on the right hand side of the above formula involve the signal $y(t)$, while exactly the same signal is needed in computing the input of the block. A loop is then formed and such a loop is referred to as an algebraic loop of the system.*

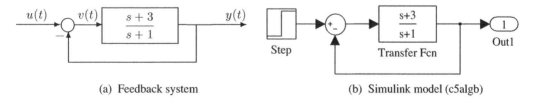

(a) Feedback system   (b) Simulink model (c5algb)

Figure 5.8   Simple feedback control system with an algebraic loop.

The algebraic loop in the system in Example 5.2 is formed with one block, and it is very easy to spot. In some other applications, the algebraic loops may be composed of several blocks and it might be very difficult to actually find the algebraic loops indicated by the warning messages.

The simulation process may still go on, and in each simulation step, algebraic equation solutions are involved, which may make the whole simulation process extremely slow. For some systems, algebraic loops can be neglected since although simulation is slow, the algebraic equations involved are always solvable, while in other applications, the algebraic loop cannot be neglected at all. In this case, elimination methods of algebraic loops should be considered.

Example 5.3   *Although there is an algebraic loop in the previously presented feedback system, the time response can still be obtained as shown in Fig. 5.9(a). Despite the warning messages, the results obtained can be validated, which means that Simulink can handle the algebraic loop problem in this example successfully.*

To avoid the algebraic loop, the whole system can be simplified manually. The transfer function $(s + 3)/(2s + 4)$ can be used to replace the original feedback system, as shown in Fig. 5.9(b). The algebraic loop can be avoided, and correct system responses can be obtained. If the Control System Toolbox is used and the LTI object is obtained, the new Simulink model shown in Fig. 5.9(c) can be constructed.

```
>> Go=tf([1 3],[1 1]); Gc=feedback(Go,1)
```

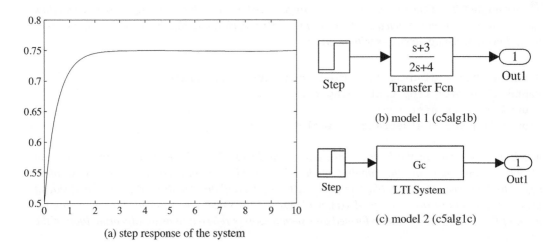

(a) step response of the system

(b) model 1 (c5alg1b)

(c) model 2 (c5alg1c)

Figure 5.9   Simulation results and algebraic loop avoidance.

Since the original system is rather simple, manual equivalent transformation can be used to eliminate algebraic loops. Unfortunately, not all the algebraic loops can be eliminated in this way. In this case, other methods can be employed. We have to use a very small time delay in the output signal, so that the output and input does not happen at the same time. In this way, algebraic loops can be avoided. Similarly, a **Memory** block can be used. Also, a Low-pass Filter block can be used to avoid algebraic loops. These methods are explored in this section, and compared, so that the most effective method can be found.

Example 5.4   *Consider the model with an algebraic loop in Example 5.2. A time delay block with a small delay constant $e^{-\tau s}$ can be added, where τ is a small positive value. The Simulink* *model can be modified as shown in Fig. 5.10(a). Thus, the output signal is y(t) and the input signals become r(t) − y(t − τ). It can be seen that the algebraic loop can be avoided, since the signal y at the input and output are not the values of the same time.*

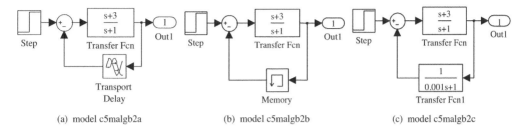

　　(a) model c5malgb2a　　　　　　　(b) model c5malgb2b　　　　　　　(c) model c5malgb2c

Figure 5.10　Three possible solutions to algebraic loops.

Similar to the idea presented above, Simulink models shown in Figs 5.10(b) and (c) can be constructed. The blocks **Memory** *and* **low-pass filter** *can be used to process the output signal. Theoretically, algebraic loops can be avoided with these methods.*

Although the above methods may theoretically avoid algebraic loop problems, new problems can arise. Since the structure of the overall system is changed due to such substitution, the properties of the original system may not be retained. Other questions are how the parameter τ is to be selected, how good the approximation result is and how big is the computational load. These problems should be further explored.

For the small delay block, we can assign the delay constant to be τ = 0.001 and τ = 0.0001, respectively. The simulation results shown in Figs 5.11(a) and (b) will be obtained. It can be seen that the approximation is not satisfactory. The delay solution to algebraic loop problems is not recommended.

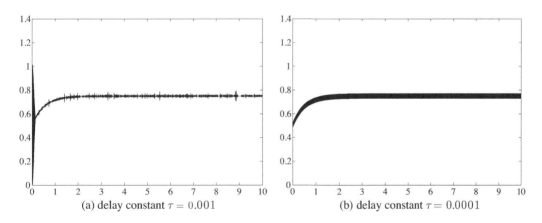

　　(a) delay constant $\tau = 0.001$　　　　　　　　　(b) delay constant $\tau = 0.0001$

Figure 5.11　Solutions when delay term is used.

Now consider the **Memory** *block, and the result in Fig. 5.12(a) can be obtained. It can be seen that the fitting is very poor. The main reason may be the zero initial condition used in the* **Memory** *block. If the initial value is changed to 0.5, the result is as shown in Fig. 5.12(b). It can be seen that although there is an improvement, the approximation is still not good. So, this method is again* *not recommended. If the conditions and simulation algorithms are not properly assigned, the results may be even worse.*

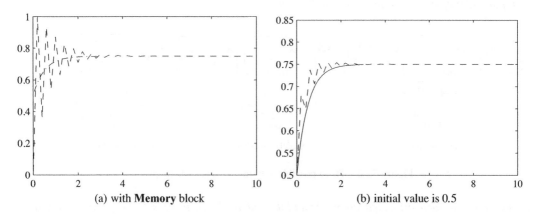

(a) with **Memory** block (b) initial value is 0.5

Figure 5.12 The **Memory** block is used.

Now consider the low-pass filter block $1/(\tau s + 1)$. If the lag constant is selected as $\tau = 0.001$, the response of the system is then obtained as shown in Fig. 5.13. It can be seen that the result is very close to the exact theoretical result. Thus, this scheme can be considered as successful. If the lag constant is selected as 0.01, 0.000 01 and 0.000 001, the same results can be obtained. If the stiff equation solver is used, and τ is assigned to an extremely small value, the computation load may not increase much.

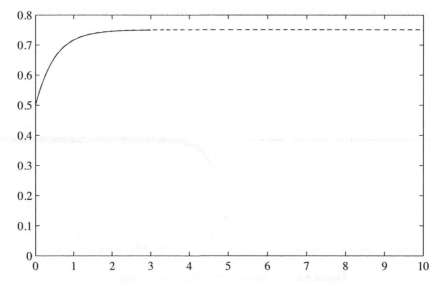

Figure 5.13 Solutions when a low-pass filter block is used.

It can be seen from the above example that if an appropriate time constant τ is selected, the low-pass filter block can be used to avoid the algebraic loop problem successfully. Consistent simulation results can be obtained for different selections of the τ constant. It can be concluded that when the lag constant τ is much smaller than that of the system, algebraic loops can be avoided successfully. Also, stiff equation solvers are suggested to avoid unnecessary oscillations in the solutions.

5.1.4 Zero-crossing Detection and Simulation of Simulink Models

Zero-crossing usually refers to the phenomenon where the sign of a signal changes within one step size of the simulation. Zero-crossing detection means detecting exactly the time when the signal equals zero. If such a detection is neglected, inaccurate simulation results may be obtained, and sometimes incorrect simulation results may even be obtained. In Simulink simulation, the concept of zero-crossing may even be extended. An example is given below to show the existence of zero-crossing point and its effect on the simulation results. Then the method of zero-crossing detection in Simulink is presented.

Example 5.5 *Consider the function evaluation problem of $y = |\sin t^2|$. We can, of course, specify a t vector. A $\boldsymbol{y}$ vector can be generated and the curve can be drawn with the* `plot()` *command. For instance, a step size of $T = 0.02s$ can be selected and the following statements can be issued; the plot in Fig. 5.14(a) is obtained.*

```
>> t=0:0.02:pi; y=abs(sin(t.^2)); plot(t,y)
```

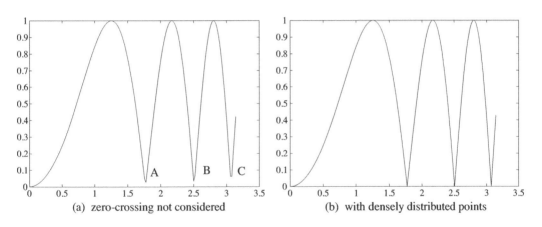

(a) zero-crossing not considered   (b) with densely distributed points

Figure 5.14 The curve of $y = |\sin t^2|$ and demonstration of zero-crossing.

It is obvious that some of the points – for example the points A, B, C close to zero – in the curve may be wrong. The trend should be that the curve should continue decreasing to zero, then increase again, rather than being in the current form. If vector t is much more densely distributed, say using a new step size of $t = 0.001$, the new curve in Fig. 5.14(b) will be obtained. In this case, the process of searching an exact A point at zero is regarded as zero-crossing detection. In drawing function curves, it is possible by merely reducing the step size, but in simulation, more sensitive zero-crossing algorithms are needed.

In fact, zero-crossing detection is the default setting in Simulink, which is done by setting the main **Solver** pane in the Simulink dialog box. Select the **Simulink** → **Configuration Parameters** menu item in the dialog box. The part related to **Zero-crossing options** is shown in Fig. 5.15(a). As a default, the item **Use local settings** is selected. If the item **Enable all** is selected, zero-detection facilities are allowed, while the item **Disable all** will turn the detection off.

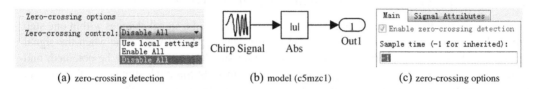

(a) zero-crossing detection   (b) model (c5mzc1)   (c) zero-crossing options

Figure 5.15   Zero-crossing detection setting and demonstration.

No matter whether in MATLAB or Simulink, exact detection of zero-crossing points means that in each simulation step, an algebraic equation solution should be made. This will, of course, slow down the whole simulation process. If the item **Use local settings** is selected, at least when the point is not around the zero points, there is no need to solve the algebraic equation. The overall speed in simulation may thus be increased. Such a selection can be used to ensure the best simulation speed. For specific problems, if solving zero-crossing points is too slow, while the stiff equation solver does not help much, zero-crossing detection should be turned off. Of course, the simulation result should be validated.

Example 5.6   *Consider the problem shown earlier. If Simulink is used, the model shown in Fig. 5.15(b) can be established. The user will be able to see the necessity of zero-crossing detection. Since many blocks are provided with zero-crossing facilities. The detection can be turned on and off in the |U| block with the dialog box shown in Fig. 5.15(c).*

5.2   Modeling and Simulation of Multivariable Linear Systems

As shown in the previous chapter single variable linear models can be presented as low-level blocks such as **Transfer Fcn**, **State Space** and **Pole-zero** blocks. Sometimes, delay terms can be expressed by **Transport Delay**. For discrete-time systems, there are also low-level blocks for the models. In this section, we see how multivariable systems can also be modeled in the Simulink environment, and a few modeling approaches will be presented and suggestions will be made.

5.2.1   Modeling State Space Multivariable Systems

The mathematical form of continuous-time state space model is given by

$$\begin{cases} \dot{x}(t) = Ax(t) + Bu(t) \\ y(t) = Cx(t) + Du(t) \end{cases} \tag{5.1}$$

where the sizes of the A, B, C and D matrices are respectively $n \times n$, $n \times p$, $q \times n$ and $q \times p$. In the system, there are p inputs and q outputs. The order of the system is n. The discrete-time state space model can also be described accordingly. The low-level **State Space** block can be used directly for the modeling of continuous-time multivariable systems.

Example 5.7 *Consider the state space model of a two-input, two-output system*

$$\dot{x} = \begin{bmatrix} 2.25 & -5 & -1.25 & -0.5 \\ 2.25 & -4.25 & -1.25 & -0.25 \\ 0.25 & -0.5 & -1.25 & -1 \\ 1.25 & -1.75 & -0.25 & -0.75 \end{bmatrix} x + \begin{bmatrix} 4 & 6 \\ 2 & 4 \\ 2 & 2 \\ 0 & 2 \end{bmatrix} u, \, y = \begin{bmatrix} 0 & 0 & 0 & 1 \\ 0 & 2 & 0 & 2 \end{bmatrix} x,$$

and the two input signals are respectively $\sin t$ *and* $\cos t$.

With Control System Toolbox commands, the time domain response of the output signals and states can be obtained as shown in Fig. 5.16.

```
>> A=[2.25,-5,-1.25,-0.5; 2.25,-4.25,-1.25,-0.25;
      0.25,-0.5,-1.25,-1; 1.25,-1.75,-0.25,-0.75];
   B=[4,6; 2,4; 2,2; 0,2]; C=[0,0,0,1; 0,2,0,2]; D=[0,0; 0,0];
   G=ss(A,B,C,D); t=0:0.02:10; u=[sin(t); cos(t)];
   [y,t,x]=lsim(G,u,t); plot(t,y), figure; plot(t,x)
```

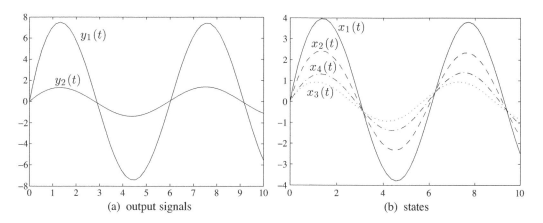

(a) output signals   (b) states

Figure 5.16 Simulation results of two-input, two-output system.

Simulink provides a linear state space model block, but the internal state signals cannot be measured directly. A new Simulink model can be constructed as shown in Fig. 5.17(a), where the

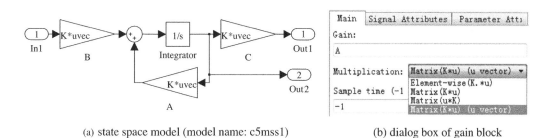

(a) state space model (model name: c5mss1)   (b) dialog box of gain block

Figure 5.17 State space representation with state outputs.

*state signals can be measured. Also, matrix gain rather than scalar gain should be used in the model. By double clicking the gain block, the dialog box shown in Fig. 5.17(b) can be displayed, and then the **Matrix (K* u)** term should be selected. Also, there is no limitation in the modeling process. The model thus established can be used to represent arbitrary linear multivariable continuous-time systems.*

The following method can be used to establish the Simulink model, as shown in Fig. 5.18. Of course, the matrices for the state space model should be entered first.

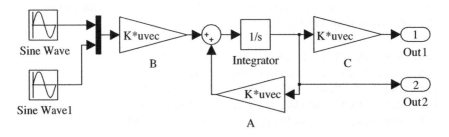

Figure 5.18   Simulink model of multivariable system (model name: c5mss2).

After simulation, the two variables tout *and* yout *can be returned in the MATLAB workspace, where the first two columns of* yout *are the outputs of the system, and the next four columns return the state signals. The following commands can be used to draw the outputs and states, exactly the same as the ones shown in Fig. 5.16.*

```
>> plot(tout,yout(:,1:2));              % system outputs
   figure; plot(tout,yout(:,3:6))       % states of the system
```

If the state space block in Simulink is used, and the states are expected, the matrices C and D should be augmented such that

```
>> C=[C; eye(4)]; D=[D; zeros(4,2)];
```

*The modified state space model should have an output vector of six signals, with the first two being the original output, while the next four are the states. A **Demux** block can be connected to the output vector of the state space block, with its parameter assigned to [2, 4].*

*If the **Selector** block in the **Signal Routing** group is used, the users are allowed to select some signals from an output vector. For instance, if the two output signals and the first and third states are expected from the augmented state space block, the Simulink model can be constructed as shown in Fig. 5.19. On double clicking the **Selector** block, the dialog box shown in Fig. 5.20 will be displayed. The edit box labeled as **Input port size** can be assigned to 6 (the total number of its input, otherwise error messages will be displayed), and the edit box labeled by **Elements** should be assigned to the interested signals to be selected. In this example, it can be set to [1, 2, 3, 5].*

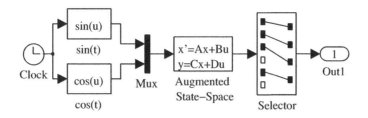

Figure 5.19   Simulink model (model name: c5ex2f).

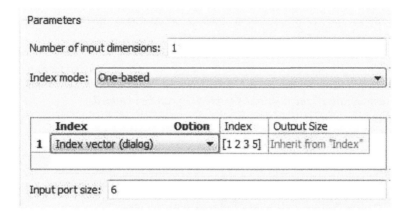

Figure 5.20   Parameters of **Selector** block.

5.2.2   Multivariable System Modeling with Control System Toolbox

Apart from the continuous state space models, linear multivariable system models can also be described by discrete-time state space models, or by transfer function matrices and systems with time delays. With the Control System Toolbox, these models can always be represented by a linear time-invariant (LTI) object G.

A continuous single-variable transfer function model is often given in the form

$$G(s) = \frac{b_1 s^m + b_2 s^{m-1} + \cdots + b_m s + b_{m+1}}{s^n + a_1 s^{n-1} + a_2 s^{n-2} + \cdots + a_{n-1} s + a_n} e^{-\tau s}. \tag{5.2}$$

It can be entered in MATLAB with

```
G = tf(n,d);  G.ioDelay = τ;
or alternatively   G = tf(n,d,'ioDelay', τ)
```

Once all the transfer functions are specified, a transfer function matrix can be entered as an ordinary matrix. Discrete-time systems can also be entered in a similar way.

Multivariable systems can be modeled in Simulink directly with the blockset provided in the Control System Toolbox. An **LTI System** block is provided to represent any LTI object, regardless its particular form: transfer function, state space, continuous, discrete-time or systems with delays. The multivariable system modeling technique is demonstrated for several different system models in the following.

▮ Example 5.8 *Consider again the two-input two-output system studied in Example 5.7. With the following statements, the state space model object G can be established in the MATLAB workspace.*

```
>> A=[2.25, -5, -1.25, -0.5; 2.25, -4.25, -1.25, -0.25;
      0.25, -0.5, -1.25,-1; 1.25, -1.75, -0.25, -0.75];
   B=[4,6;2,4;2,2;0,2]; C=[0,0,0,1;0,2,0,2]; G = ss(A,B,C,0);
```

*Open the Control System Blockset in the Simulink model library, and copy the only block **LTI System** into the Simulink model window. Open the dialog box of the block and fill in the variable G in the block. The Simulink model shown in Fig. 5.21 can be constructed. Perform simulation to the system and the same results as the ones in Example 5.16 can be obtained.*

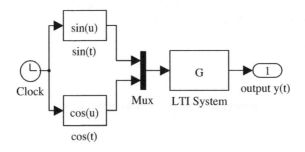

Figure 5.21 State space model (model name: c5ex2d).

*Again using the augmentation method, the matrices **C** and **D** can be augmented such that a new state space model G_1 can be entered into MATLAB workspace.*

```
>> C1=[C; eye(4)]; D1=[D; zeros(4,2)]; G1=ss(A,B,C1,D1);
```

Thus the Simulink model shown in Fig. 5.22 can be constructed. Apart from the output signal, the states can also be returned at an output port. If the model is embedded in the one presented in the previous example, the same simulation results can be obtained.

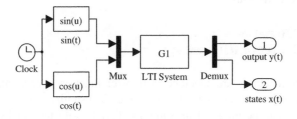

Figure 5.22 State space model with state output (model name: c5ex2e).

■ Example 5.9   *Consider again the 4×4 transfer function matrix discussed in Example 4.8*

$$G(s) = \begin{bmatrix} 1/(1+4s) & 0.7/(1+5s) & 0.3/(1+5s) & 0.2/(1+5s) \\ 0.6/(1+5s) & 1/(1+4s) & 0.4/(1+5s) & 0.35/(1+5s) \\ 0.35/(1+5s) & 0.4/(1+5s) & 1/(1+4s) & 0.6/(1+5s) \\ 0.2/(1+5s) & 0.3/(1+5s) & 0.7/(1+5s) & 1/(1+4s) \end{bmatrix}.$$

The following MATLAB commands can be used to represent a transfer function matrix

```
>> h1=tf(1,[4 1]); h2=tf(1,[5 1]);
   h11=h1; h12=0.7*h2; h13=0.5*h2; h14=0.2*h2;
   h21=0.6*h2; h22=h1; h23=0.4*h2; h24=0.35*h2;
   h31=h24; h32=h23; h33=h1; h34=h21; h41=h14; h42=h13; h43=h12; h44=h1;
     G=[h11,h12,h13,h14;h21,h22,h23,h24;h31,h32,h33,h34;h41,h42,h43,h44];
```

With the model object G, the constructed Simulink model is shown in Fig. 5.23(a). In the model, the first input is assumed to be the unit step input, while the other three inputs are assumed to be zero. All the output signals can be simulated as shown in Fig. 5.23(b). These curves are the outputs of the system driven by just the first unit step input signal.

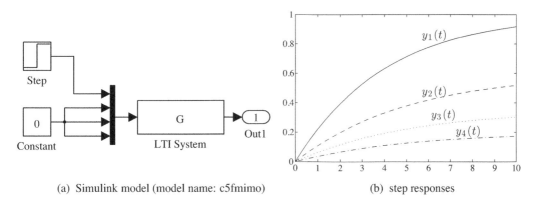

(a) Simulink model (model name: c5fmimo)   (b) step responses

Figure 5.23   Modeling and simulation of multivariable systems.

■ Example 5.10   *Consider the transfer function matrix given below [1]*

$$G(s) = \begin{bmatrix} \dfrac{0.1134e^{-0.72s}}{1.78s^2 + 4.48s + 1} & \dfrac{0.924}{2.07s + 1} \\[3mm] \dfrac{0.3378e^{-0.3s}}{0.361s^2 + 1.09s + 1} & \dfrac{-0.318e^{-1.29s}}{2.93s + 1} \end{bmatrix}.$$

*The transfer function matrix contains time delay terms, and in old versions of the **LTI System** block, the LTI models with delay cannot be expressed. Low-level modeling should be used. From the definition of transfer function matrix, the input and output relationship can be written as*

$$\begin{bmatrix} y_1(s) \\ y_2(s) \end{bmatrix} = \begin{bmatrix} g_{11}(s) & g_{12}(s) \\ g_{21}(s) & g_{22}(s) \end{bmatrix} \begin{bmatrix} u_1(s) \\ u_2(s) \end{bmatrix},$$

where $g_{ij}(s)$ may contain delay terms. The above matrix form can be rewritten in the expanded form as

$$\begin{cases} y_1(s) = g_{11}(s)u_1(s) + g_{12}(s)u_2(s) \\ y_2(s) = g_{21}(s)u_1(s) + g_{22}(s)u_2(s). \end{cases}$$

*With the delay terms represented by the **Transport delay** block, the Simulink model for the multivariable system can be established as shown in Fig. 5.24.*

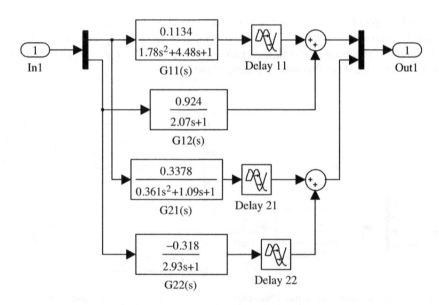

Figure 5.24 Simulink model for the multivariable system with delays (model file: c5mmimo).

*In the new version of Simulink, linear systems with delays and internal delays can directly be described by the **LTI System** block. A new Simulink model can then be established, which is similar to the one in Fig. 5.22. The following statements can be used to enter the LTI object G.*

```
>> g11=tf(0.1134,[1.78 4.48 1],'ioDelay',0.72s);
   g21=tf(0.3378,[0.361 1.09 1],'ioDelay',0.3);
   g22=tf(-0.318,[2.93,1],'ioDelay',1.29);
   g12=tf(0.924,[2.07 1]); G=[g11,g12; g21,g22];
```

5.3   Nonlinear Components with Lookup Table Blocks

Various of nonlinear element blocks are provided in Simulink, and they make nonlinear system modeling and simulation much more simple and straightforward. In this section, several examples are used to demonstrate the use of nonlinear elements and their extensions.

5.3.1   Single-valued Nonlinearities

A wrong impression may be gained from the **Discontinuity** group that Simulink can only be used to handle simple static nonlinearities. In fact, many nonlinearities can be represented by the combination of a few blocks. For instance, the nonlinearity with both dead zone and saturation can be equivalently composed of a saturation and a dead zone block in series connection, as shown in Fig. 5.25 [2].

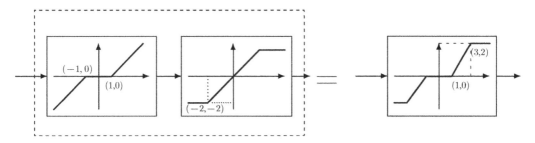

Figure 5.25   Equivalent construction of complicated nonlinear element.

Example 5.11   *Several commonly used nonlinear elements, such as backlash, dead zone and saturation are excited by a sinusoidal signal. Observations are made as to the impact of nonlinearities on the distortion of the sinusoidal waveform. The Simulink model shown in Fig. 5.26(a) can be constructed. The **Amplitude** property of 2 can be assigned in the dialog box. The parameter of the nonlinearities is assigned to* c1, *as shown in Fig. 5.26(b). If we select* c1 = 0.5, *the simulation*

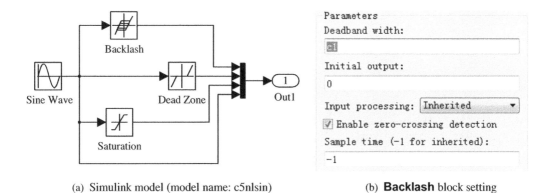

(a) Simulink model (model name: c5nlsin)   (b) **Backlash** block setting

Figure 5.26   Block diagram of nonlinear blocks.

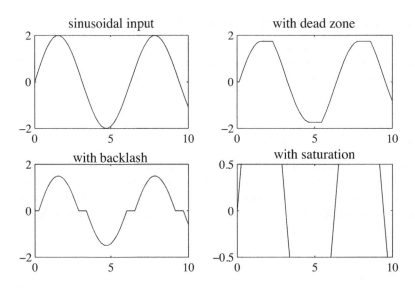

Figure 5.27 Signal distortion with $c_1 = 0.5$.

results can be returned in the variables `tout` and `yout`. The following commands can be used to draw the distorted signals, as in Fig. 5.27.

```
>> t=tout; y=yout;
   subplot(221), plot(t,y(:,4)), subplot(222), plot(t,y(:,1))
   subplot(223), plot(t,y(:,2)), subplot(224), plot(t,y(:,3))
```

If the value of c1 is assigned to 1, and the above process is repeated, the results shown in Fig. 5.28 will be obtained. It can be seen that this kind of problem can be solved easily with Simulink.

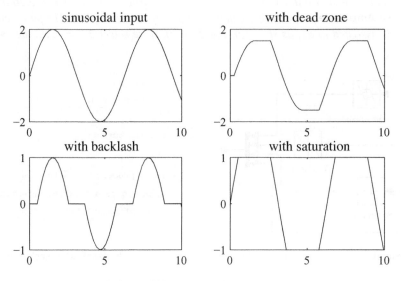

Figure 5.28 Distortion with $c_1 = 1$.

The **1-D Lookup Table** block in the **Lookup Tables** group can be used to construct memoryless piecewise linear nonlinearities of any complexity. The following examples are used to illustrate the modeling methods.

Example 5.12   *Consider the nonlinear element shown in Fig. 5.25. It can be seen that the turning points in the nonlinearity can be obtained as* $(-4, -2)$, $(-3, -2)$, $(-1, 0)$, $(1, 0)$, $(3, 2)$, $(4, 2)$, *thus in the dialog box shown in Fig. 5.29(a), the edit box labeled* **Vector of input values** *can be assigned to* $[-4, -3, -1, 1, 3, 4]$, *while in the edit box of* **Table data**, *the vector* $[-2, -2, 0, 0, 2, 2]$ *can be specified. The nonlinearity can be constructed, with the icon of the block shown in Fig. 5.29(b).*

(a) dialog box   (b) lookup table icon

Figure 5.29   One-dimensional lookup table block (model name: c5mtab1).

Saturation with a dead zone nonlinear element can be modeled in two ways, and a Simulink model can be constructed as shown in Fig. 5.30(a). In the dead zone block, the parameters are set to -1 *and* 1, *while in the saturation element, the parameters are* -2 *and* 2. *The output signals can be obtained as shown in Fig. 5.30(b). The results with the two methods are exactly the same.*

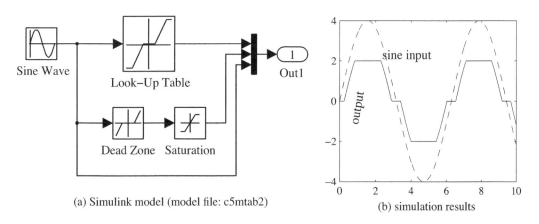

(a) Simulink model (model file: c5mtab2)   (b) simulation results

Figure 5.30   Simulink model of nonlinearities.

5.3.2   *Multi-valued Nonlinearities with Memories*

In the previous example we saw that, with a lookup table method, a memoryless, or single-valued, nonlinear block of any form can be constructed with lookup table blocks. However, if there are multi-valued or hysteresis-type or loop-type nonlinearities, such a method cannot easily be used. Blocks such as switch and memory can therefore be introduced.

Example 5.13   *Consider the loop-type nonlinear element shown in Fig. 5.31, and it is seen that such a nonlinearity is no longer single-valued. It is seen that when the input to the block is increasing, the nonlinearity is a polyline, while when the input is decreasing, another different polyline is used. The loop nonlinearity can be expressed as single-valued functions shown in Fig. 5.32. Of course, the single-valued function is conditional, and it is also controlled by the trend of the input signal.*

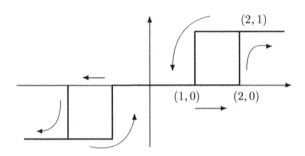

Figure 5.31   Given nonlinearity with loops.

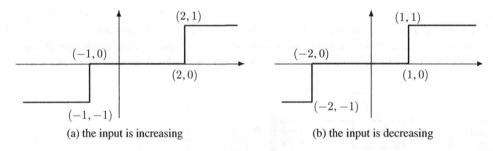

(a) the input is increasing   (b) the input is decreasing

Figure 5.32   Single-value function representation.

A **Memory** *block in the* **Discrete** *group is provided, and it returns the value of the input signal at a previous time instance. This block can be used to construct the Simulink model shown in Fig. 5.33(a). A comparison block is also used to check whether the input signal at the current time is larger than its value at the previous time instance, so that the trend of the input signal is checked. A switch block is also used to select the polylines according to the trend of the input signals. Then the threshold value of the block is set to 0.5.*

   The input and output turning points of the polylines can be specified by

$$x_1 = [-3, -1, -1 + \epsilon, 2, 2 + \epsilon, 3], \ y_1 = [-1, -1, 0, 0, 1, 1],$$
$$x_2 = [-3, -2, -2 + \epsilon, 1, 1 + \epsilon, 3], \ y_2 = [-1, -1, 0, 0, 1, 1],$$

where ϵ can be assigned to a very small value, such as to the eps *constant reserved by MATLAB. If the amplitude of the sinusoidal input signal is 3, the simulation results can be obtained as shown in Fig. 5.33(b).*

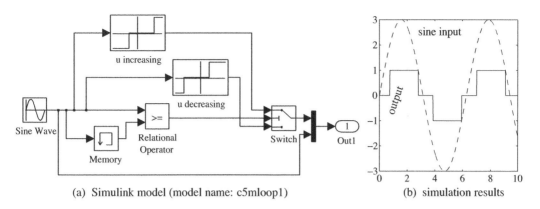

(a) Simulink model (model name: c5mloop1)   (b) simulation results

Figure 5.33   Simulation results for system with sinusoidal input.

If the nonlinear block is changed to the one shown in Fig. 5.34, the Simulink model described above can still be used, and the parameters for the nonlinearity can be redefined as

$$x_1 = [-3, -2, -1, 2, 3, 4], \quad y_1 = [-1, -1, 0, 0, 1, 1],$$
$$x_2 = [-3, -2, -1, 1, 2, 3], \quad y_2 = [-1, -1, 0, 0, 1, 1].$$

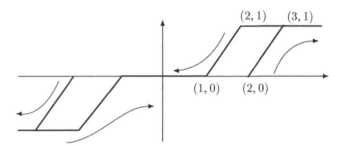

Figure 5.34   Representation of the new nonlinear element.

The new Simulink model can be established as shown in Fig. 5.35(a), and the new simulation result obtained is shown in Fig. 5.35(b).

*Two other switch blocks, **Multiport Switch** and **Manual Switch**, are also provided in Simulink. The former is shown in Fig. 5.36(a), and the input signal at the top is the control signal, with values 1,2,...,n. It can be used to select the channel in other input signals. Double click the block to get the dialog box shown in Fig. 5.36(b). The total number of inputs can be specified in the dialog box.*

The manual switch block shown in Fig. 5.36(c) is also a very useful block in Simulink. When it is double clicked, its status, or the status of the input signal, is toggled. For instance, this block can be used to model autotuning PID controllers.

Example 5.14   *Consider again the problem in Example 5.11. A multiport switch block can be used to select nonlinear elements. Based on this idea, a Simulink model shown in Fig. 5.37(a) can be established. An external input* `key` *is introduced to control the selection of nonlinearities. The value*

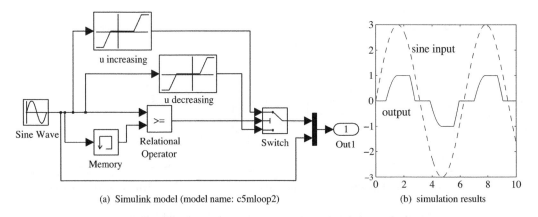

(a) Simulink model (model name: c5mloop2)   (b) simulation results

Figure 5.35   Simulation result of the new loop nonlinearity.

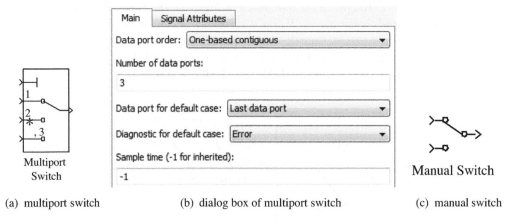

(a) multiport switch   (b) dialog box of multiport switch   (c) manual switch

Figure 5.36   Multiport and manual switches.

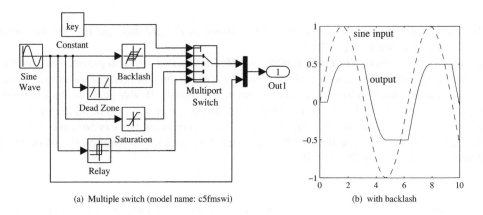

(a) Multiple switch (model name: c5fmswi)   (b) with backlash

Figure 5.37   Multiple switch and simulation results.

of key *should be assigned in MATLAB workspace, for instance,* key $=$ 1. *Also, the parameters in backlash (the width) nonlinear element is assigned as 0.5. Input and output signals in this model can be obtained as shown in Fig. 5.37(b). It can be seen that because of the nonlinearity, the input signal is distorted.*

5.3.3   Multi-dimensional Lookup Table Blocks

Two-dimensional and multi-dimensional lookup table blocks are also provided in Simulink, and the linear interpolation algorithm is used to evaluate the output signal. Two-dimensional lookup table blocks are used as an example, to demonstrate its usages and applications.

■ **Example 5.15**   *Consider the two-dimensional function* $z = f(x, y) = (x^2 - 2x)e^{-x^2 - y^2 - xy}$ *studied in Example 2.21. Assume that a set of sparsely distributed mesh grid points on the x-y plane is given. With MATLAB statements, the height z of the points can be evaluated and the surface can be obtained as shown in Fig. 5.38(a).*

```
>> xx=linspace(-3,3,15); yy=linspace(-2,2,15);
   [x,y]=meshgrid(xx,yy); z=(x.^2-2*x).*exp(-x.^2-y.^2-x.*y);
   surf(xx,yy,z); axis([-3,3,-2,2,-0.6,1.2])
```

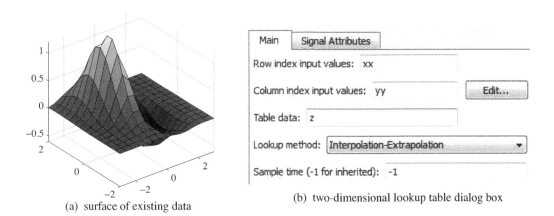

(a) surface of existing data   (b) two-dimensional lookup table dialog box

Figure 5.38   Multi-dimensional lookup table block.

The data are stored in the scale vectors xx, yy *and the height matrix* z, *respectively. Double clicking the* **Look-up Table (2-D)** *block, the dialog box shown in Fig. 5.38(b) will be displayed. The three variables can be set in the dialog box such that a proper lookup table block for the data can be established.*

Now two set of pseudo random numbers, uniformly distributed, can be generated as the input signals **x** *and* **y**. *The Simulink model shown in Fig. 5.39(a) can be constructed. Note that the two pseudo random number generators in the model look the same, but the seed parameters must be set to different numbers. For instance, the seed in the second block can be set to 31242240, while in the first, the default seed is used. Also the ranges of the two signals can be set to* $(-3, 3)$ *and* $(-2, 2)$ *respectively. If a fixed step size simulation scheme is selected, with a step size of 0.001s, and a termination time of 10s, 10,000 points will be simulated. With the simulation results, the* *three-dimensional plot can be obtained with the following statements, and the result is as shown in Fig. 5.39(b). It can be seen that the plot obtained is quite similar to the theoretical surface plot.*

```
>> plot3(yout(:,2),yout(:,1),yout(:,3),'.'); axis([-3,3,-2,2,-0.6,1.2])
```

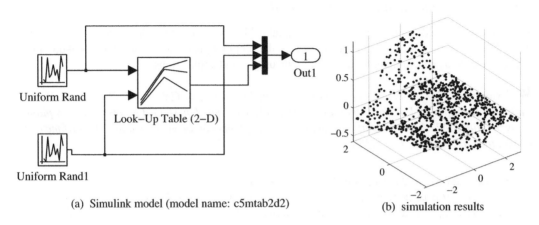

(a) Simulink model (model name: c5mtab2d2)

(b) simulation results

Figure 5.39   Setting and simulation of two-dimensional lookup table block.

Multi-dimensional lookup table blocks are also provided in Simulink, and the linear interpolation algorithm is implemented in the blocks. The specification of sample data points is rather complicated, since the data in mesh grids has to be provided. If the data are given in scattered points, mesh grid data must be obtained with the MATLAB function `griddata()`, so that the relevant lookup table block can be established.

5.3.4   Code Realization of Static Nonlinearities

The so-called static nonlinearity is a function which can be described directly by $y = f(u)$, where u is the input to the block, and y is the output of the block. They can both be vector signals. It has been pointed out that the block **Fcn** provided in Simulink does not support vector output, thus sometimes modeling with such a block may be complicated. Apart from the standard **Fcn** block, custom MATLAB function blocks are also supported. Moreover, embedded MATLAB function blocks are also supported. With these blocks, static nonlinear functions can be easily modeled. We will now look at the use of these two blocks.

1) Interpreted MATLAB function block. The block **Interpreted MATLAB Fcn** block (in earlier versions of Simulink, the **MATLAB Fcn** block) in the **User-defined Functions** group can be copied into the model window. A *.m file should be created for the static nonlinearity, and the filename should be associated to the block with dialog box.

2) MATLAB function block. The block presented above needs an associated *.m file. Sometimes this may not be very convenient for the users. An alternative way is to create such a block with embedded functions. In this case, the **MATLAB Function** block (former **Embedded MATLAB Function** block), in the same group, can be used instead. When the block is copied into the model window, the program editor can be opened automatically. The MATLAB function describing such a static function can be entered into the editor. Unfortunately, while normal Simulink blocks can be run directly, when this block is used, the whole Simulink model has to be compiled (and the user may notice the compilation process) before the simulation can be performed.

Example 5.16   *Again using the Van der Pol equation studied in Example 5.1, a Simulink modeling problem can be demonstrated. In Example 5.1, the **Fcn** block was used and since the block does not support vector output, the structure of the block diagram was complicated. Also the blocks **Mux** and **Demux** should be used to convert signals.*

Consider again the original mathematical model. For $\mu = 2$, the original model can be rewritten as

$$\begin{bmatrix} \dot{x}_1 \\ \dot{x}_2 \end{bmatrix} = \begin{bmatrix} x_2 \\ -2(x_1^2 - 1)x_2 - x_1 \end{bmatrix}.$$

From the right hand side of the equation, expressed in vector form, the MATLAB function can be written as

```
function y=c5fvdp(x), y=[x(2); -2*(x(1)^2-1)*x(2)-x(1)];
```

*Then the block diagram shown in Fig. 5.40(a) can be constructed. Its structure is far simpler than that created in Example 5.1. If the **MATLAB Function** block is used, the Simulink model shown in Fig. 5.40(b) can be established. An editor is opened automatically. The above code can be entered into the editor. The additional parameter μ cannot be used. Specific values should be used instead.*

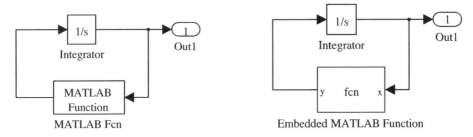

(a) **Interpreted MATLAB Fcn** block (c5mvdpa1)   (b) **MATLAB Function** block (dc5mvdpa2)

Figure 5.40   Simple Simulink modeling of Van der Pol equations.

Note that only static functions can be represented by the two kinds of blocks. No additional parameters can be used in such a function. If dynamic nonlinearities are to be described, or additional parameters are necessary, S-function blocks should be used. S-function programming will be presented in Chapter 6, with its applications.

5.4   Block Diagram Based Solutions of Differential Equations

Numerical solutions to ordinary differential equation with MATLAB have been extensively studied in Chapter 3. These equations can also be solved with block diagram based approach by the use of Simulink. In this section, illustrative examples for solving different kinds of differential equations, including differential algebraic equations, delay differential equations, switching differential equations and fractional-order differential equations, will be presented. It may not be possible to solve some of the equations with MATLAB function calls, but they can be solved easily with block diagram based modeling techniques with Simulink.

5.4.1 Ordinary Differential Equations

Defining key signals is an important stage in Simulink modeling of differential equations. For differential equations, a set of integrators can be used. Assume that the inputs of these integrators are assigned as $\dot{x}_i(t)$, and the output is then $x_i(t)$. With these signals, the Simulink model can be constructed. With the powerful facilities of MATLAB and Simulink, the differential equations can be solved easily.

Example 5.17 *Consider the Lorenz equation given by*

$$\begin{cases} \dot{x}_1(t) = -\beta x_1(t) + x_2(t)x_3(t) \\ \dot{x}_2(t) = -\rho x_2(t) + \rho x_3(t) \\ \dot{x}_3(t) = -x_1(t)x_2(t) + \sigma x_2(t) - x_3(t), \end{cases} \quad \text{and} \quad \begin{cases} \beta = 8/3, \rho = 10, \sigma = 28, \\ x_1(0) = x_2(0) = 0, \\ x_3(0) = 10^{-10}. \end{cases}$$

The differential equation can be constructed in Simulink. Since there are first-order derivatives of the state variables, three integrators are needed. The input ports of the integrators are $\dot{x}_1(t)$, $\dot{x}_2(t)$, $\dot{x}_3(t)$, and the outputs are then $x_1(t)$, $x_2(t)$, $x_3(t)$. Thus, these signals can be used to construct the framework of the Simulink model. Double clicking the integrator block, the initial values of the state variables can be entered into the integrators.

*When constructing the framework for ODE solutions, a vectorized integrator can be used, such that its input and output are denoted respectively as $\dot{x}(t)$ and $x(t)$, with $x(t) = [x_1(t), x_2(t), x_3(t)]^T$. The initial state vector is given by $[0; 0; 1e-10]$. With the **Fcn** blocks, the three equations can be constructed, and the three signals can be vectorized with a **Mux** block into a vector signal, and then this signal can be connected into the input end of the integrator block. Thus the Simulink model shown in Fig. 5.41(a) can be constructed. The signal $x(t)$ can be connected to an output port. The parameters* `beta`, `rho` *and* `sigma` *can be assigned to the MATLAB workspace.*

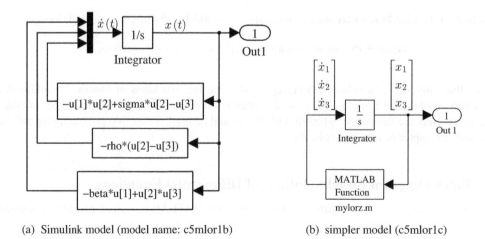

(a) Simulink model (model name: c5mlor1b)   (b) simpler model (c5mlor1c)

Figure 5.41 Simulink modeling of Lorenz equations.

If we use the **Interpreted MATLAB Fcn** block (in earlier versions of Simulink, the **MATLAB Fcn** block), a simpler model can be constructed, as shown in Fig. 5.41(b) to describe the original differential equations. The MATLAB function code can be written as

```
function y=mylorz(x)
y=[-8/3*x(1)+x(2)*x(3);  -10*x(2)+10*x(3);  -x(1)*x(2)+28*x(2)-x(3)];
```

It should be noted that for small-scale problems, the complexity of Simulink modeling is similar to MATLAB's `ode45()` function call. However, for large-scale problems, especially systems with complicated hybrid connections, the Simulink modeling technique is much simpler and more straightforward than the `ode45()` function call. For complicated systems with delays, Simulink models can be used easily, while a MATLAB function approach may fail.

5.4.2   Differential Algebraic Equations

In Section 3.4.5, the concept of differential algebraic equation was presented. Examples were given to solve the equations using MATLAB functions. In fact, differential algebraic equations can also be modeled with Simulink. The **Algebraic Constraints** block in the **Math** group can be used to describe the algebraic constraints.

Example 5.18   *Consider the differential algebraic equation shown in Example 3.34. For simplicity, the original equation is presented here again*

$$\begin{cases} \dot{x}_1 = -0.2x_1 + x_2x_3 + 0.3x_1x_2 \\ \dot{x}_2 = 2x_1x_2 - 5x_2x_3 - 2x_2^2 \\ x_1 + x_2 + x_3 - 1 = 0, \end{cases}$$

and the initial conditions are $x_1(0) = 0.8$, $x_2(0) = x_3(0) = 0.1$. The two signals $x_1(t)$ and $x_2(t)$ can be used to assign as the outputs of two integrators. The last equation can be written as $f(x_3) = x_1 + x_2 + x_3 - 1 = 0$. With the algebraic constraints block, the DAE (Differential Algebraic Equation) model can be established, and with it the function $x_3(t)$ can be designed.

If low-level blocks are used in the modeling, the Simulink model shown in Fig. 5.42 can be established, where the initial values of the two integrators are assigned respectively to 0.8 and 0.1, and the initial value of the algebraic constraint block is set to 0.1.

If a MATLAB function block is used, the Simulink model shown in Fig. 5.43 can be established, where the MATLAB function block can be written as

```
function y=c5fdae2(x)
y=[-0.2*x(1)+x(2)*x(3)+0.3*x(1)*x(2);2*x(1)*x(2)-5*x(2)*x(3)-2*x(2)^2];
```

It can be seen that the results obtained are the same as those obtained in Example 3.34, and yet the complexity involved in solving the equations is low.

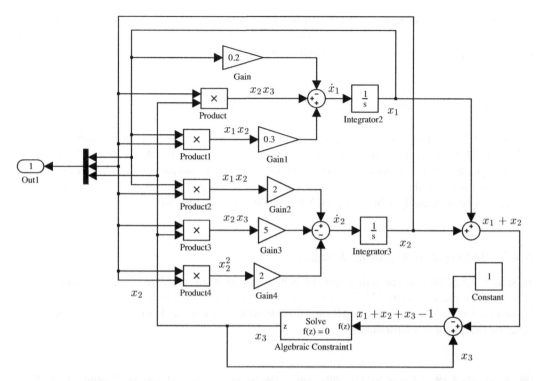

Figure 5.42   Simulink model of the differential algebraic equation (model name: c5fdae).

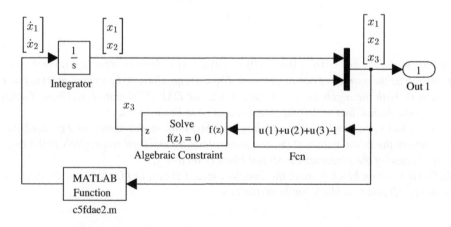

Figure 5.43   Simpler simulation model (model name: c5mdae2).

5.4.3   Delayed Differential Equations

The general form of the delay differential equation is given by

$$\dot{x}(t) = f(t, x(t), x(t - \tau_1), x(t - \tau_2), \ldots, x(t - \tau_n)) \tag{5.3}$$

where $\tau_i > 0$ are the delay constants of the state vector $x(t)$. Such an equation can be solved by the MATLAB function dde23() [3]. However, for neutral-type delay differential equations or equations with variable delays, the function cannot be used. Simulink is a possible way to solve such systems.

Example 5.19   *Consider the delay differential equation*

$$\begin{cases} \dot{x}(t) = 1 - 3x(t) - y(t - 1) - 0.2x^3(t - 0.5) - x(t - 0.5) \\ \ddot{y}(t) + 3\dot{y}(t) + 2y(t) = 4x(t). \end{cases}$$

In the first equation, if the term $-3x(t)$ is moved to the left hand side of the equation, the equation can be converted to

$$\dot{x}(t) + 3x(t) = 1 - y(t - 1) - 0.2x^3(t - 0.5) - x(t - 0.5).$$

From the equation, the signal $x(t)$ can be regarded as the output signal of a transfer function block $1/(s + 3)$, with the input $1 - y(t - 1) - 0.2x^3(t - 0.5) - x(t - 0.5)$. In the second equation, the signal $y(t)$ is the output of the transfer function $4/(s^2 + 3s + 2)$, with input signal $x(t)$. The signals $x(t)$ and $y(t)$ can be connected to **Transport Delay** *blocks to generate the delay signals. From the above analysis, the Simulink model shown in Fig. 5.44(a) can be established. After simulation, the output signal $y(t)$ can be obtained as shown in Fig. 5.44(b).*

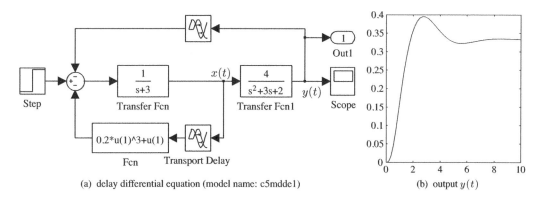

(a) delay differential equation (model name: c5mdde1)   (b) output $y(t)$

Figure 5.44   Simulink model and solution of delay differential equation.

Of course, if one is not used to the transfer function description, integrators can be used by selecting $x_1 = x, x_2 = y, x_3 = \dot{y}$. Thus the original differential equation can be converted to the following first-order explicit differential equation

$$\begin{cases} \dot{x}_1(t) = 1 - x_1(t) - x_2(t - 1) + 0.2x_1^3(t - 0.5) - x_1(t - 0.5) \\ \dot{x}_2(t) = x_3(t) \\ \dot{x}_3(t) = -4x_1(t) - 3x_3(t) - 2x_2(t). \end{cases}$$

Example 5.20 *Consider the delay differential equation described by*

$$\dot{x}(t) = A_1 x(t - \tau_1) + A_2 \dot{x}(t - \tau_2) + B u(t),$$

where $\tau_1 = 0.15$, $\tau_2 = 0.5$, *and*

$$A_1 = \begin{bmatrix} -13 & 3 & -3 \\ 106 & -116 & 62 \\ 207 & -207 & 113 \end{bmatrix}, \quad A_2 = \begin{bmatrix} 0.02 & 0 & 0 \\ 0 & 0.03 & 0 \\ 0 & 0 & 0.04 \end{bmatrix}, \quad B = \begin{bmatrix} 0 \\ 1 \\ 2 \end{bmatrix}.$$

Since the terms $\dot{x}(t)$ and $\dot{x}(t - \tau_2)$ both appear in the same equation, such an equation is referred to as a "neutral-type" delay differential equation. The DDE solver `dde23()` *provided in MATLAB cannot be used to solve this kind of equation. Simulink based modeling techniques have advantages in modeling this kind of equations. Before running the Simulink model, the following statements can be used to enter the matrices into the MATLAB environment*

```
>> A1=[-13,3,-3; 106,-116,62; 207,-207,113];
   A2=diag([0.02,0.03,0.04]); B=[0; 1; 2];
```

Consider again the original differential equation. An integrator can be arranged for the state vector $x(t)$, and its output is assumed to be the vector $x(t)$, then the input is the vector $\dot{x}(t)$. These two vectors can be connected to delay blocks to construct the vectors $x(t - \tau_1)$ and $\dot{x}(t - \tau_2)$. Thus the Simulink model shown in Fig. 5.45(a) can be established, and the solution is obtained as shown in Fig. 5.45(b).

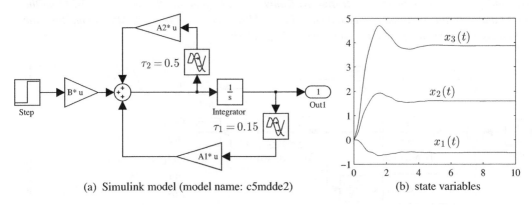

(a) Simulink model (model name: c5mdde2)   (b) state variables

Figure 5.45 Simulink model and solution of the neutral-type delay differential equation.

Example 5.21 *Suppose that the initial conditions of the states are all zero. The differential equation with variable delay is represented as*

$$\begin{cases} \dot{x}_1(t) = -2x_2(t) - 3x_1(t - 0.2|\sin t|) \\ \dot{x}_2(t) = -0.05x_1(t)x_3(t) - 2x_2(t - 0.8) \\ \dot{x}_3(t) = 0.3x_1(t)x_2(t)x_3(t) + \cos(x_1(t)x_2(t)) + 2\sin 0.1t^2. \end{cases}$$

It can be seen that there are variable delays in the system, that is, the delay signal at time $t - 0.2|\sin t|$ in signal x_1. Thus the `dde23()` *cannot be used in solving such equations. Simulink can handle this problem easily.*

*Similar to the modeling procedure in other ODEs, three integrators are required to represent the states x_1, x_2 and x_3 and their derivatives. The Simulink model can be constructed as shown in Fig. 5.46. Note that the variable delay signal can be generated with the **Variable Time Delay** block, with the signal in its second input port generated by $0.2 \sin t$.*

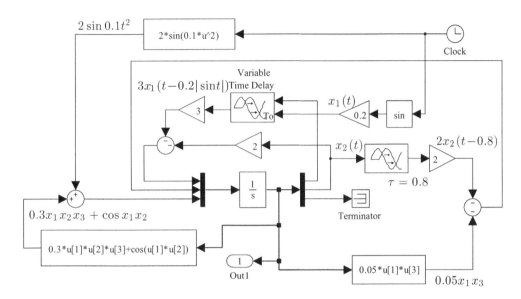

Figure 5.46 Simulink model of system with variable delay (model file: c5mdde3).

Numerical solutions to the system can be obtained as shown in Fig. 5.47. Different control parameters such as relative error tolerance can be used to validate the simulation results.

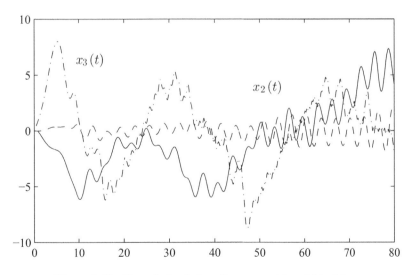

Figure 5.47 Numerical solution of systems with variable delay.

5.4.4   Switching Differential Equations

The switching system is an important branch in control theory [4]. Switching here means that the system is switched, according to a particular rule, among different models. A typical mathematical model of a switching system is

$$\dot{x}(t) = f_i(t, x, u), i = 1, \ldots, m \tag{5.4}$$

and under certain conditions or control laws, the whole system can be switched from one model to another. According to switching system theory, a controller can be designed such that unstable models f_i can be stabilized by suitable switching conditions.

Example 5.22   *Assume that the two models are described as $\dot{x} = A_i x$, where*

$$A_1 = \begin{bmatrix} 0.1 & -1 \\ 2 & 0.1 \end{bmatrix}, \quad A_2 = \begin{bmatrix} 0.1 & -2 \\ 1 & 0.1 \end{bmatrix}.$$

It can be seen that the two models are both unstable. If $x_1 x_2 < 0$, that is, the states lie in the quadrants II or IV, the system is switched to model A_1, while if $x_1 x_2 \geq 0$, that is, in quadrants I or III, the system is switched to model A_2. Assume also that the initial states are $x_1(0) = x_2(0) = 5$.

*The switch block can be used to represent switching conditions, and the Simulink model shown in Fig. 5.48(a) can be established, with the dialog box setting is shown in Fig. 5.48(b). To realize the state-dependent switch conditions, the **Threshold** property of the switch can be set to 0. Also, to get more accurate simulation results, the checkbox **Enable zero-crossing detection** should be checked.*

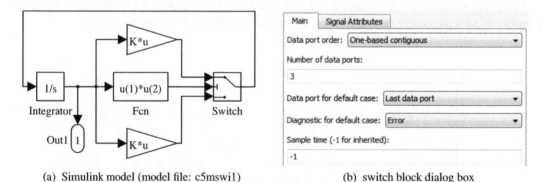

(a) Simulink model (model file: c5mswi1)   (b) switch block dialog box

Figure 5.48   Simulink model of switch differential equation.

After simulation, the results are returned to the MATLAB workspace, and the following commands can be used to draw the states as well as the phase plane trajectory, as shown in Figs 5.49(a) and (b). It can be seen that, under the switching condition, the entire system is stable.

```
>> plot(tout,yout), figure; plot(yout(:,1),yout(:,2)), axis square
```

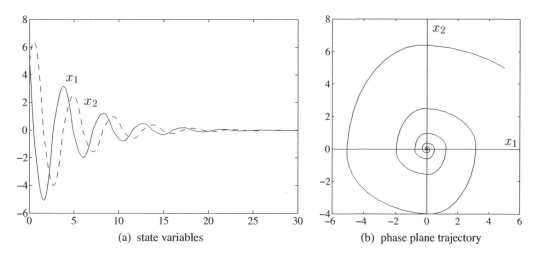

(a) state variables   (b) phase plane trajectory

Figure 5.49 Solutions of switched differential equations.

5.4.5   Fractional-order Differential Equations

Fractional calculus is a direct extension to traditional (integer-order) calculus, where the orders of derivatives can be assigned to non-integers. Thus the algorithms in traditional ODEs cannot be used. Integer-order filters, for instance Oustaloup filters [5], can be used to approximate fractional-order differentiators s^γ. Such a filter can be masked into a Simulink block, and details of block masking will be given in Section 5.7.3.

Example 5.23   *Consider the following nonlinear fractional-order model*

$$\frac{3\mathscr{D}^{0.9}y(t)}{3 + 0.2\mathscr{D}^{0.8}y(t) + 0.9\mathscr{D}^{0.2}y(t)} + \left|2\mathscr{D}^{0.7}y(t)\right|^{1.5} + \frac{4}{3}y(t) = 5\sin(10t),$$

where $\mathscr{D}^{0.9}y(t)$ represents the 0.9th order derivative of the signal $y(t)$. With slight arrangement of the original equation, the explicit equation of signal $y(t)$ can be established

$$y(t) = \frac{3}{4}\left[5\sin(10t) - \frac{3\mathscr{D}^{0.9}y(t)}{3 + 0.2\mathscr{D}^{0.8}y(t) + 0.9\mathscr{D}^{0.2}y(t)} - \left|2\mathscr{D}^{0.7}y(t)\right|^{1.5}\right].$$

The Simulink model for the system is established as shown in Fig. 5.50(a). From the simulation model it can be seen that the fractional-order derivative of the signal is approximated by the filters. Thus simulation accuracy depends upon the quality of fitting with the Oustaloup filter. The order and frequency ranges are important in the approximation. In Fig. 5.50(b), the simulation results are shown with different orders and frequency ranges, and it can be seen that the simulation results are consistent.

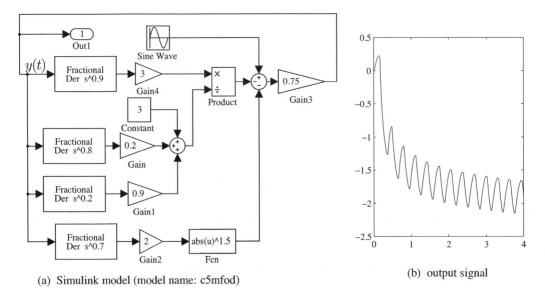

(a) Simulink model (model name: c5mfod)

(b) output signal

Figure 5.50   Description of nonlinear fractional differential equation.

5.5   Output Block Library

Output facilities are very important in simulation, since simulation results can be obtained, and it should be returned to the user in an appropriate form. There are several ways to access simulation results:

- **Scope output**: It has been stated that there are two types of scopes, ordinary scopes and trajectory scopes. The former displays the time response curves, while the latter displays the relationship between two signals. Scopes are flexible ways to display simulation results. There is no need for additional command after simulation. Under default setting, the simulation results obtained cannot be returned to the MATLAB workspace for post-processing, with the powerful graphics facilities in MATLAB.

- **Signal viewer**: To display simply a certain signal in Simulink, the easiest way is to add a signal viewer to the signal. First select a signal, then click the right mouse button to pop up the shortcut menu. Select the **Create & Connect Viewer** → **Simulink** → **Scope** menu item, a signal viewer to the signal, and a sign ❧ will appear on the signal. To display another signal in the same viewer, select the signal, then right click, and select from the shortcut menu the **Connect to Existing Viewer** menu item.

- **Direct numerical display**: In the **Sinks** group, the direct numerical display block **Display** can be used. One can connect it to a signal in a Simulink model. The block can also be used to display signals simultaneously. It should be noted that since simulation in Simulink is not performed in real time, normally the simulation process can be executed very fast. Numerical display flashes by too fast to read.

- **Output port**: As shown earlier, if the output ports are used in the Simulink model, the variables tout and yout can be returned automatically to the MATLAB workspace, representing the time and output port signals respectively. These signals can be post-processed in MATLAB.

- **Return to workspace**: The block **To Workspace**, in the **Sinks** group, can be used to return the interested signal back to the MATLAB workspace for post-processing. The user can freely select the names of MATLAB variables. The default time variable is `tout`.

- **Output to files**: The block **To File**, in the **Sinks** group, can be used to save simulation results directly to data files.

- **Gauge display**: A group of ActiveX components is provided in the Gauges Blockset, and they can be used to display in gauges or dials. This kind of display is similar to the instruments in industrial control fields.

- **Digital analysis and processing of signals**: Many digital signal processing blocks are provided to display correlation functions, power spectral densities and fast Fourier transforms and their inverses.

- **Three-dimensional animation**: Three-dimensional animation and virtual reality techniques can be used to display simulation results.

It should be noted that several output blocks can be connected to the same signal. For instance, in the model shown in Fig. 5.51, the output port **Out1**, **Scope**, gauges **Vacuum** and **To Workspace** blocks can all be connected to the same signal.

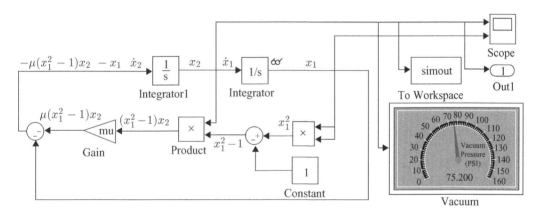

Figure 5.51   Signals connected to many output blocks (model name: c5mmore).

5.5.1   Output Block Group

1) Scopes and output ports

As shown earlier, scopes can be used in a Simulink model. For instance, in the example of Van der Pol equation of Example 4.3, two curves were shown in a scope with the vectorized signal constructed with a **Mux** block. In fact, scope blocks can be used to accept more channels of the input signals. Click the 🖺 button in the toolbar, and a dialog box as shown in Fig. 5.52(a) is displayed. If we input the number 2 in the edit box labeled **Number of axes**, two channels of input ports appear in the scope block. The ports can be connected to the signals x_1 and x_2 in the Van der Pol equation, and the simulation results are displayed in Fig. 5.52(b).

Output port is a practical way of signal outputting. Note that the output port setting is closely related to the dialog box parameter **Workspace I/O** option shown in Fig. 4.35. So, in the **Save to**

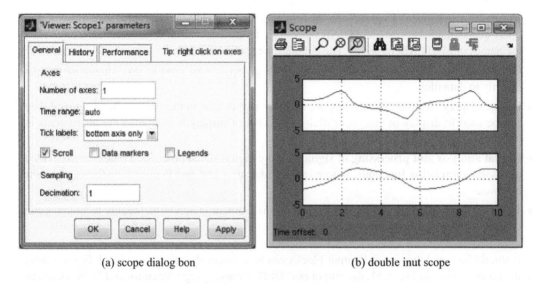

(a) scope dialog bon   (b) double inut scope

Figure 5.52   Scope options and multi-signal display.

Workspace checkboxes, **tout** and **yout** must be checked, otherwise the two variables will not be returned.

Also, if the checkbox **Limit data points to last** is selected, only the last 1000 simulation points will be returned. To retain all the simulation points, the checkbox should be deselected.

2) Workspace and file output

The **To Workspace** output block was the major output device in earlier versions of Simulink. With the powerful output port, the use of **To Workspace** is not as important as it was. Here we still use the Van der Pol equation example to show the use of such a block.

Example 5.24   *From the model shown in Fig. 4.44, the original scope can be replaced by the* **To Workspace** *block, and the new model can be constructed as shown in Fig. 5.53.*

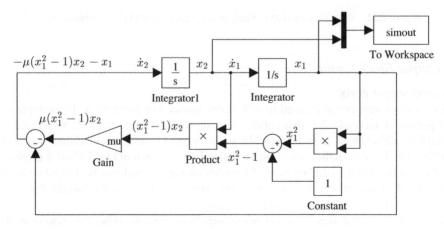

Figure 5.53   Simulink model with **To workspace** block (model name: c5mspc1).

Double click the **To Workspace** *block, and the dialog box shown in Fig. 5.54 will be displayed. The variable name can be specified in the edit box labeled* **Variable name**. *The* **Save format** *listbox is also useful, and it allows three types of data structures:* **Structure** *(the default),* **Array** *and* **Structure with time**.

Figure 5.54   To Workspace dialog box.

Under the default data type, three members time, signals *and* blockName *are returned in the variable* simout, *where* time *member is an empty matrix. If the data type* **Structure with time** *is selected, the* time *member returns a time column vector. The member* signals *is another structured array, including members* values, dimensions *and* label. *The output data can be extracted from* simout.signals.values. *The following statements can be used to draw the simulation results*

```
>> plot(tout,simout.signals.values)
```

It can be seen that the variable format is very tedious. In this case, the variable format **Array** *is much easier to use.*

In fact, the output port block also allows these three types of data formats, as shown in Fig. 4.35, with a default format of **Array** *in the output port block.*

If the **To File** block in the **Sinks** group is used, the simulation results will be saved to ∗.mat file. The saved data can be loaded into the MATLAB workspace for post-processing.

5.5.2   Examples of Output Blocks

Two examples are given here to show the output methods. In the first, the control system with time-varying criteria is shown in the simulation, and the second demonstrates the use of the block **STOP** to control simulation process.

■ **Example 5.25**   *In control system design, sometimes time-related performance criteria are used, such as the integral of time-weighted absolute errors (ITAE) and the integral of squared error (ISE). They are defined as*

$$f(t) = \int_0^t \tau \mid e(\tau)\mid\mathrm{d}\tau, \quad g(t) = \int_0^t e^2(\tau)\mathrm{d}\tau. \tag{5.5}$$

Now consider the control system model shown in Fig. 5.55. The Simulink model can be established with the two criteria in (5.5), as shown in Fig. 5.56.

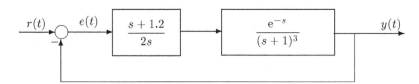

Figure 5.55   Block diagram of a control system.

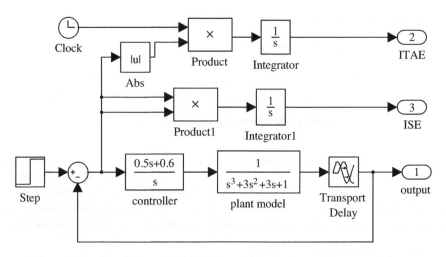

Figure 5.56   Simulink model with ITAE and ISE criteria (model name: c5mitae).

After the Simulink model is constructed, a terminating simulation time of 30 s can be tested. After simulation process, with the following statements, the two criteria can be obtained as shown in Fig. 5.57.

```
>> plot(tout,yout(:,2));        %  draw ITAE curve
   figure; plot(tout,yout(:,3)) %  open a new window to draw ISE criterion
```

■ **Example 5.26**   *Consider the PI control problem presented above. If we assume the terminating time of 30s, the output signal can be obtained as shown in Fig. 5.58. It can be seen that within the*

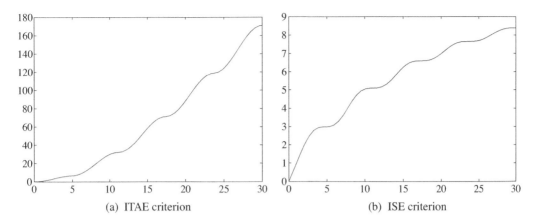

(a) ITAE criterion   (b) ISE criterion

Figure 5.57   Criteria of the control system versus time.

simulation interval, the output of the system has not settled down. In control terminology, the settling time has not been reached. Now the problem is how to automatically select a settling time, so that the whole tuning time can be simulated.

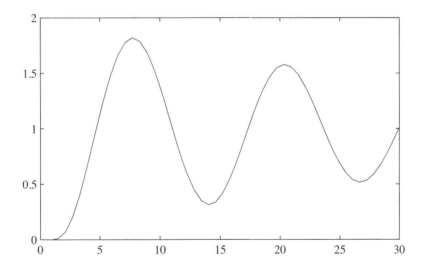

Figure 5.58   Output signal of the process under a PI controller.

*In this example, the use of the **STOP** block is demonstrated. If the input signal to the block is nonzero, then the whole simulation process will be terminated automatically. A Simulink model can be constructed as shown in Fig. 5.59. The upper part of the block diagram is used to check whether the output enters the settling zone long enough. More specifically, it takes the absolute value of the error signal, and if it is larger than 0.02, the output of it can be set to 0, otherwise -1 will be set, and it is assumed that the signal thus generated is $x_1(t)$. It takes the integral to x_1, and uses its rising edge to reset the integrator, as shown in Fig. 5.60(a). The smaller the signal $x_2(t)$, the longer the condition $|e(t)| < 0.02$ lasts. If the time lasted is longer than 10s, the comparative block can then generate a nonzero signal to activate the **STOP** block and terminate the simulation process.*

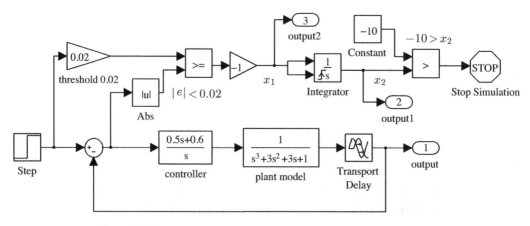

Figure 5.59   Simulink model with **Stop** block (model name: c5mstop).

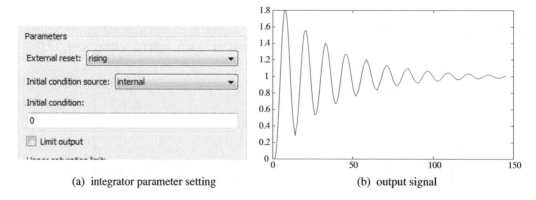

(a) integrator parameter setting   (b) output signal

Figure 5.60   System setting and output signal.

In this case, it is the same as assuming an extremely long termination time, for example to 1000s. If the simulation process is invoked, the output shown in Fig. 5.60(b) can be obtained. It can be seen that the simulation is automatically completed when the output enters the settling area for longer than 10 s.

The two intermediate signals $x_1(t)$ and $x_2(t)$ shown in Fig. 5.61 are the intermediate signals. These signals can also be drawn with the following MATLAB statements

```
>> plot(tout,yout(:,3)), ylim([-1.1,0.1]);
   figure; plot(tout,yout(:,2)), ylim([-10,1]);
```

From the signal $x_2(t)$, the integrator reset action can be easily seen.

It can be seen from the above example that it might be difficult to measure the elapsed time for a condition to have been satisfied. Here a rather complicated structure is constructed to perform the task. With this kind of construction, this sort of problem can be solved successfully. There are of course many other methods for terminating the simulation processes. For instance, finite state machine and Stateflow can be used. Details will be discussed later.

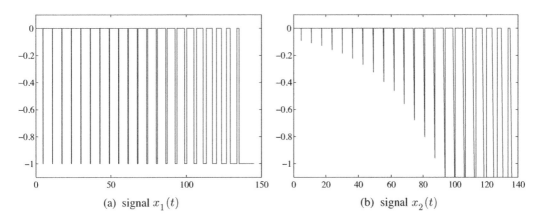

(a) signal $x_1(t)$ (b) signal $x_2(t)$

Figure 5.61   Intermediate signals in the system.

5.5.3   *Model Parameter Display and Model Browser*

A parameter display of a single block can be done by simply double clicking it. To display the parameters or the whole system, the interface of model browser can be used. The model browser can be launched with the menu **View** → **Model Browser**, or by clicking the button ▓ in the toolbar.

For instance, when the Simulink model shown in Fig. 5.1 is studied, the model browser in Fig. 5.62 can be used to browse and modify the parameters in all the blocks. For large-scale models, the **Search** option can be used to find a particular block and to display its parameters.

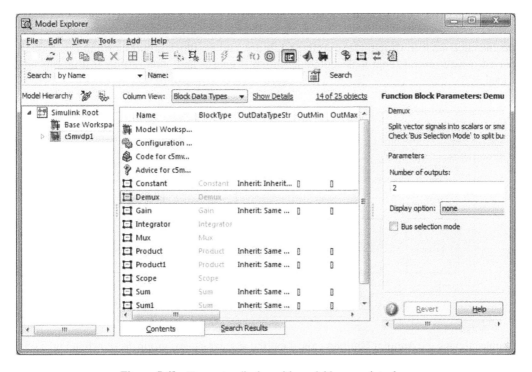

Figure 5.62   Parameter display with model browser interface.

5.5.4   Gauge Display of Signals

As has already been discussed, various of blocks can be used to display the data and curves. In real applications, other forms of signal output, such the waveform shown in various industrial instruments, are also useful. Global Majic Inc. has designed a lot of instrument gauge display blocks using ActiveX techniques. ActiveX is a protocol defined under Microsoft and it is an extension of the Visual Basic toolbox. ActiveX components are defined as executable files with the suffixes of .exe, .dll or .ocx. Once ActiveX components are introduced, they become part of the program development environment and provide new facilities for applied executable programs. ActiveX components retain the properties, events and methods of ordinary VB controls.

The Gauges Blockset, which was named the Dials & Gauges Blockset, can be called directly by Simulink library, or by command `gaugeslibv2` in the MATLAB command window. The interface in Fig. 5.63 will then be displayed. This is the main interface for the blockset. All the components in the blockset can be viewed from the interface.

Figure 5.63   Main interface of the Gauges Blockset.

Double click the icon **Global Majic ActiveX Library**, and the block group shown in Fig. 5.64 is displayed, where several subgroups containing different gauges and other display instruments are provided.

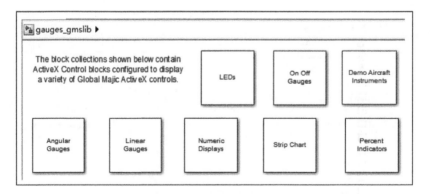

Figure 5.64   Global Majic ActiveX component library.

Double clicking the **Angular Gauges** icon opens the angular gauges group, as shown in Fig. 5.65. In this group, various instrument gauges are provided, such as **Amp Meter**, **Vacuum**, **Stop Watch** and many others. All these gauges are directly derived from the **Generic Angular Gauge**.

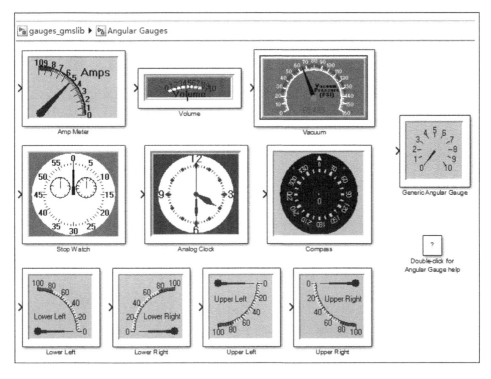

Figure 5.65   Angular gauges library.

Example 5.27   *Consider the PI control shown in Fig. 5.66. The method of moving an ActiveX gauge component is slightly different from the standard blocks in Simulink. You have to click the white area on the icon to drag the block.*

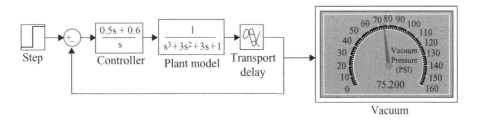

Figure 5.66   PI control system with gauges output (model name: c5mdng1).

It can be seen that the measuring range, preset at 0~160, of the dial is too large, since the range of the signal is within 2. Thus the readability of the meter is very poor. Double click the block, and a dialog box shown in Fig. 5.67 can be displayed, and the range of the meter can be set when the **Scales** *pane is selected, and the dialog box is shown in Fig. 5.68. The range of the signal can be set to 0 to 1.5 in the dialog box. In the* **Ticks** *pane, the property* **DeltaValue** *can be set to 0.1. The Simulink model can then be set as shown in Fig. 5.69, where the range of signal is changed to the new settings. Simulation can be performed and simulation results will be displayed on the meter. Actually the simulation process is extremely fast and it is not possible to observe the output signal in this way, unless an extremely small step size is chosen.*

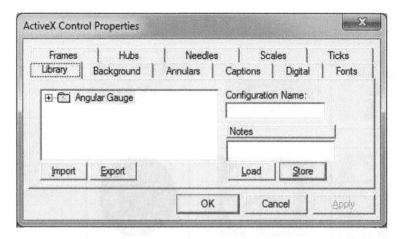

Figure 5.67   Dial property setting dialog box.

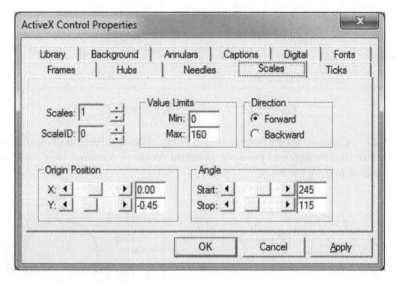

Figure 5.68   Scale setting dialog box.

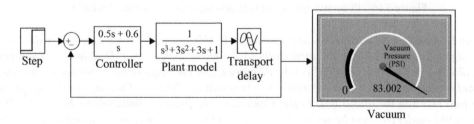

Figure 5.69   Simulation model when setting is changed (model name: c5mdng7).

Alternatively, other meters can be used to display the output signals as well. For instance, in the dialog box shown in Fig. 5.67, the **Library** *pane is selected, and double click the* **Volume** *option, and the Simulink model shown in Fig. 5.70 can be obtained as well. Note that the range of the meter is changed. The previous method can be used to set the range of the signal again.*

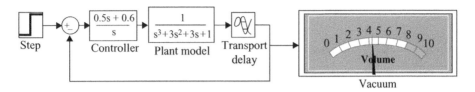

Figure 5.70   Simulink model when the meter is changed (model name: c5mdng8).

5.5.5   Digital Signal Processing Outputs

In the **Additional Sinks** blockset of the group **Simulink Extras**, several signal processing blocks are provided, as shown in Fig. 5.71. In the DSP System Toolbox blockset, more powerful blocks are also provided as shown in Fig. 5.72. Here an example is given to demonstrate the digital signal processing facilities.

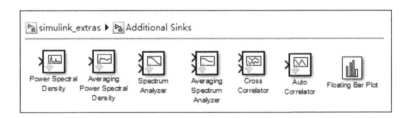

Figure 5.71   Additional sinks library in Simulink.

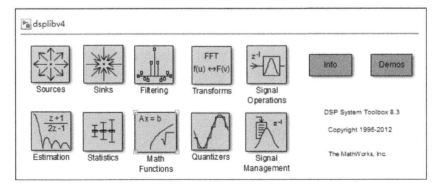

Figure 5.72   DSP System Toolbox blockset.

Example 5.28 *Assume that the signal to be studied is $x(t) = \sin t + 1.2 \cos 2t$. The Simulink model shown in Fig. 5.73(a) can be constructed. In the model, two sinusoidal signal blocks with different parameters are used. For instance, the dialog box of the second sinusoidal input can be specified as Fig. 5.73(b), with an amplitude of 1.2, an initial phase of $\pi/2$ and a frequency of 2 rad/sec. Thus $1.2 \sin(2t + \pi/2) = 1.2 \cos 2t$ is one of the signals in the input. The output signal can be connected to an autocorrelation block.*

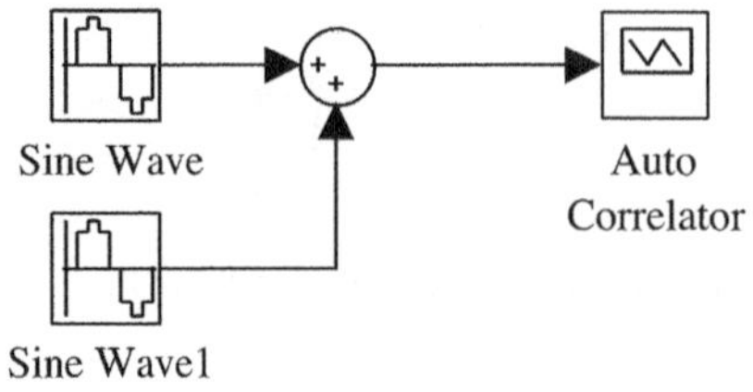

(a) Simulink model (c5mfft)

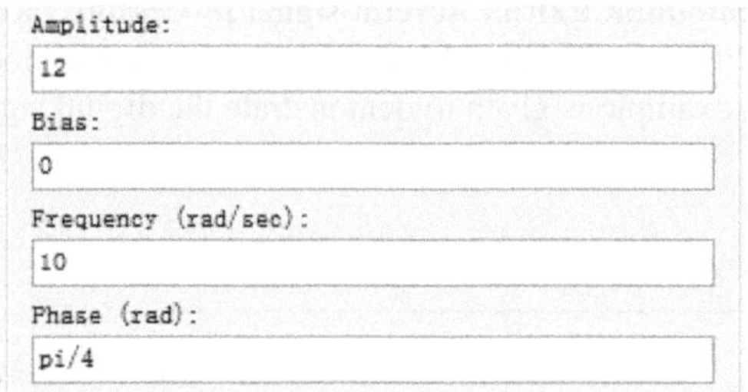

Amplitude:

12

Bias:

0

Frequency (rad/sec):

10

Phase (rad):

pi/4

(b) sinusoidal input dialog box

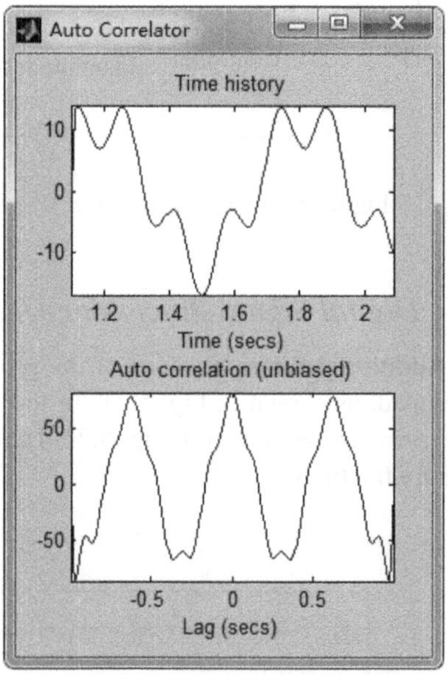

(c) autocorrelation function

Figure 5.73   Autocorrelation analysis of input signals.

If an autocorrelation block is used, the simulation algorithm should be selected as fixed step size ones, and a step size of 0.01 s can be selected. The time response and autocorrelation function of the input signal are shown in Fig. 5.73(c).

5.6   Three-dimensional Animation of Simulation Results

The Virtual Reality Toolbox in earlier versions of MATLAB has been renamed as the Simulink 3D Animation Blockset, and with its blockset, simulation results can be displayed in three-dimensional animation.

5.6.1   Fundamentals of Virtual Reality

Virtual reality (VR) is a term that applies to computer-simulated environments that can simulate physical presence in places in the real world, as well as in imaginary worlds[6]. The imaginary world can be generated by computers and can provide user interaction with audio, visual and sensing techniques, so that users can experience the imaginary world as if they were right inside the environment [7].

The three-dimensional animation technique supported by MATLAB and Simulink allows the user to display simulation results with virtual reality approaches. In this section, the fundamental concepts of virtual reality are briefly introduced first, and the Virtual Reality Modeling Language (VRML) and world model creation methods are illustrated. Finally, simulation results driving three-dimensional animations will be demonstrated with MATLAB and Simulink.

The VRML language is a commonly used language for describing virtual reality models. This language is used as a fundamental way of presenting three-dimensional animation of simulation results.

When the VRML language is used to describe three-dimensional space, the framework of the three axes satisfies the right hand principle, as shown in Fig. 5.74(a). It can be seen that the arrangement of the axes is different from the ones conventionally used in MATLAB graphics. The rotational direction of the given axes also satisfies the right hand principle, as shown in Fig. 5.74(b).

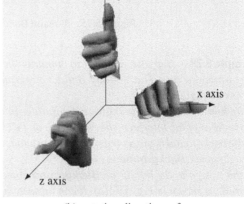

(a) right hand principle of axes   (b) rotation directions of axes

Figure 5.74   Illustration of right hand principle.

5.6.2   V-realm Software and World Modeling

V-realm Builder 2.0 software, provided with MATLAB, is a practical visual tool to create and edit VRML models. It allows the user to create scenes essential for imaginary worlds or virtual reality. The V-realm program is provided with the 3D Animation Blockset and can be installed automatically in the folder sl3d"vrealm"program.

For the modeling facilities in V-realm software to be fully used, certain installation tasks must be set in MATLAB command window, with the commands

```
>> vrinstall -install viewer
   vrinstall -install editor
```

After answering some questions, the software can fully be installed.

When the installation and setting up processes are complete, the executable file vrbuild2.exe in the program folder can be activated directly and the interface in Fig. 5.75 can be used immediately.

It can be seen that a large number of buttons and menu items are provided in the interface. A simple example is given below to show the virtual world modeling program with the software.

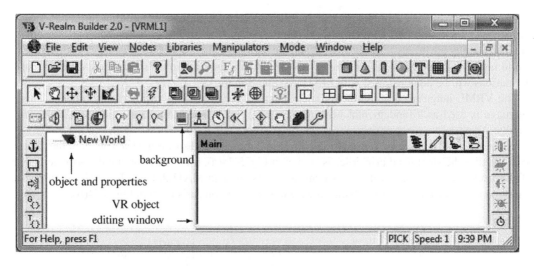

Figure 5.75   V-realm Builder 2.0 interface.

📖 **Example 5.29**   *Suppose we want to simulate the virtual display of a plane flying in circular trajectory around a tree. The V-realm Builder program can be used to establish the VRML file to describe the virtual world file.*

With the **File** → **New** *menu in the V-realm Builder interface, a new editing window will be opened. From the toolbar of the interface, the* ▦ *button can be clicked to add the background of the world. By default, the background is presented as sky and ground, and the color changes gradually. Meanwhile, an objected named* **Background** *is created.*

Click the + *sign to the left of the* **Background** *object, and the properties of the background will be displayed, as shown in Fig. 5.76(a). It can be seen that there are four options in the* **groundColor** *property, allowing the user to specify the color from the near end (default of level 0) to the far*

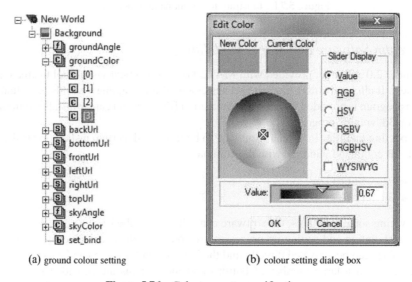

(a) ground colour setting   (b) colour setting dialog box

Figure 5.76   Color property specification.

end (default of level 3). Double click option [3], and the dialog box shown in Fig. 5.76(b) will be displayed. The user can select a color from it, or assign the color in an alternative way.

As can be seen on the first row of the toolbar icons in V-realm Builder 2.0, a great many objects, such as cylinders, cuboids and cones, can be added to the world edit window. Also, a lot of existing objects can also be added, by selecting the **Libraries** → **Import From** → **Object Library** *menu item, and the menu shown in Fig. 5.77(a) will be displayed. The often-used object libraries are shown in Fig. 5.77(b).*

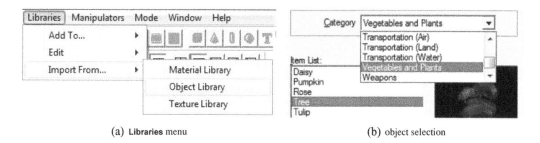

(a) **Libraries** menu   (b) object selection

Figure 5.77   Object selection.

For instance, the **Tree** *object can be selected from the* **Vegetables and Plants** *group, and copied into the world edit window. Similarly, the* **Boeing 737** *object can be chosen from the* **Transportation (Air)** *group and also copied into the edit window. In the framework on the left hand side, two objects are added, all labeled with* **Transform***. We can simply edit the names to* **Tree** *and* **Plane***, respectively. In fact, the automatically generated objects are too large to display properly. We can modify the size by double clicking the* **scale** *option under the* **Tree** *object, and a dialog box shown in Fig. 5.78(a) will be displayed. The scales can be modified to make the object easier to display. Also we can double click the* **center** *option of the object to open another dialog box, as shown in Fig. 5.78(b). The property can also be modified with the dialog box.*

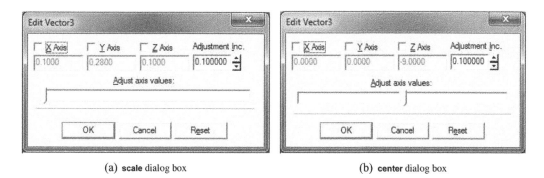

(a) **scale** dialog box   (b) **center** dialog box

Figure 5.78   Object properties dialog boxes.

In the **Tree** *object, the four parameters of the* **center** *option can be set to* $(0, 0, -9, 0.1)$, *and the* **scale** *parameters can be set to* $(0.1, 0.3, 1, 0.1)$. *For the* **Plane** *object, the* **center** *property can be set to* $(-4.7, -0.6, 1, 0.1)$, *and the* **scale** *property can be set to* $(0.15, 0.15, 0.15, 0.1)$. *The world shown in Fig. 5.79 can be established. The world model can be saved as myvrml2.wrl file.*

Figure 5.79   World scene modeling constructed.

5.6.3   Browsing Virtual Reality World with MATLAB

A series of MATLAB functions can be used to manipulate and browse the properties of the objects in the virtual world file *.wrl, as if manipulating an object defined in the MATLAB environment. The whole VRML file can be loaded to extract the scene (i.e. the virtual world file) and the nodes (i.e. the objects). The following statements can be used to implement the relevant tasks.

1) The `vrworld()` function can be used to open a VRML file *.wrl. For instance, the following command can be used to open myvrml2.wrl file and assign the handle of the scene to the variable `myworld`.

```
>> myworld=vrworld('myvrml2.wrl');
```

2) The command `open()` can be used to load the scene, and the `view()` can be used to display the scene in a virtual reality browser

```
>> open(myworld)  % open the scene with the handle myworld
   view(myworld)  % display the scene with a VR viewer
```

3) When the scene is loaded, the function `vrnode()` can be used to extract the handles of the nodes, and the functions `set()` and `get()` can be used to access the properties of the nodes, such that the virtual reality display can be implemented.

5.6.4 Virtual Reality World Driven by Simulink Models

The myvrml2.wrl file discussed earlier was created with the V-realm Builder software, and it is a static virtual world file. A further VRML model can be written to make the static objects move, but it might be too demanding for the average user to do so.

The Three-dimensional Animation Blockset provided in Simulink can be used to drive the VRML virtual world in a simpler and more straightforward way. The blocks provided in the blockset are shown in Fig. 5.80, including the virtual reality display block **VR Sinks**, virtual reality file saving block **VR To Video**. Also, the blocks to external virtual reality input devices such as **Joystick Input**, **Space Mouse Input**, **VR Text Output** and **VR Tracer** are provided. These blocks can be used later to manipulate the virtual reality world.

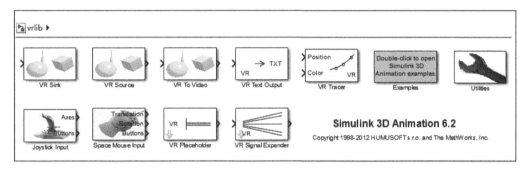

Figure 5.80 Simulink Three-dimensional Animation Blockset.

Example 5.30   *With the myvrml2.wrl file, let us assume that the plane is starting from the initial position, and flying in a circular trajectory centered on the tree object. The Simulink model can be created to drive the virtual reality world and to display the results.*

Assume that the starting position is $(-4.7, -0.6, 1)$ and the center of the trajectory is $(0, 0, -9)$, with a radius of 15. The trajectory of the plane can be described as

$$x = 15\cos(t + 118°), \quad y = -0.6 + 0.1t, \quad z = -9 + 15\sin(t + 118°)$$

where a time variable in one period is assigned as $t \in (0, 360°)$. It should be noted that since radians are used in MATLAB, unit transformation should be performed first. The following statements can be used, and the trajectory of the plane can be calculated as shown in Fig. 5.81.

```
>> t=0:.1:2*pi; t0=118*pi/180; % generate t vector and change units
   x=15*cos(t+t0); y=-0.6+0.1*t; z=-9+15*sin(t+t0);
   plot3(x,z,y), grid, set(gca,'box','off')
   set(gca,'xdir','reverse','ydir',reverse') % inverse the directions of axes
   view(-67.5,52)                            %  view the axis from this angle
```

*With the use of Simulink modeling technique, a model shown in Fig. 5.82 can be established, and the block **VR Sink** can be used to process the three-dimensional animation of the scene.*

*We can copy the **VR Sink** block into the simulation model, and in the default setting, no input port is provided, since its link with the world file has not been established yet. Double clicking the block opens a dialog box shown in Fig. 5.83. We need to enter myvrml2.wrl in the edit box labeled*

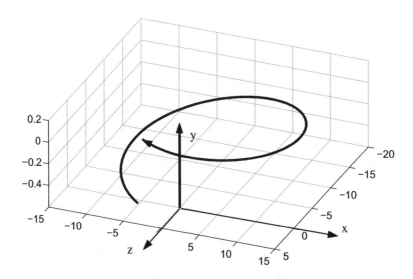

Figure 5.81   Three-dimensional trajectory of the plane.

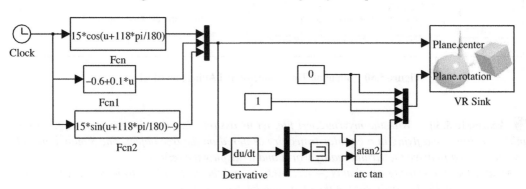

Figure 5.82   Plane simulation model (model name: c5mvrs1).

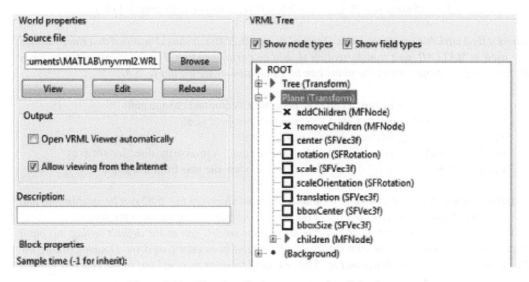

Figure 5.83   Virtual reality input port setting dialog box.

*Source file to establish the link. Then the properties in the **VRML Tree** will be displayed in tree structure. Since the **center** and **rotation** properties of the **Plane** object are to be accessed, the two properties should be checked. Two input ports are then created automatically with the labels **Plane.center** and **Plane.Rotation**. The Simulink model in Fig. 5.82 can finally be established, and during the simulation, the animation viewer can be shown to display the plane in three-dimensional animation.*

5.7   Subsystems and Block Masking Techniques

In the modeling and simulation process, sometimes it can be very difficult to model the whole system with a single block diagram model. The whole model can be divided by the functions of the blocks to several subsystems. Different Subsystem structures are supported in Simulink. Moreover, some of the subsystems can be masked into reusable independent blocks, and the user's own block library or blockset can be constructed. In this section, the topics of the construction and application of subsystems and the masking procedures will be presented. Finally, an application example will be given to demonstrate the procedure of the subsystem based modeling technique for a really complicated system.

5.7.1   Building Subsystems

To establish a subsystem we should first assign the input and output ports of it. The input ports can be connected to the **In** blocks in the **Sources** group. The output ports can be connected to the **Out** blocks in the **Sinks** group. It should be noted that in very earlier versions of Simulink, the two blocks are provided in the **Signals & Systems** group. With the input and output ports, the user can arbitrarily design the internal structure of the subsystem model.

If there is already a Simulink model constructed, the blocks in the expected subsystem can be selected by mouse dragging action. Then the menu **Edit** → **Create Subsystem** can be used to establish a subsystem. If the input and output ports are not specified, Simulink will automatically arrange the signals, that is, the signals entering the selected area are assumed to be input signals and those leaving the area the outputs.

■ **Example 5.31**   *PID controllers are widely used in control engineering. The standard form of an actual PID controller is described as*

$$U(s) = K_p \left(1 + \frac{1}{T_i s} + \frac{s T_d}{1 + s T_d / N} \right) E(s), \tag{5.6}$$

where a first-order term with lag is used to approximate pure derivative action. If $N \geq 10$, the approximation is usually acceptable. With the use of Simulink, a PID controller model can be constructed easily, as shown in Fig. 5.84(a). In the model, there are four parameters: Kp, Ti, Td and N. They should be assigned in the MATLAB workspace, before the simulation process is invoked.

*If all the blocks in the PID controller model are selected, either by **Edit** → **Select All** menu or by mouse dragging action, the **Edit** → **Create Subsystem** menu item can be used to construct a subsystem, as shown in Fig. 5.84(b). Double clicking the subsystem block constructed, the internal structure of original model in Fig. 5.84(a) can be brought back again.*

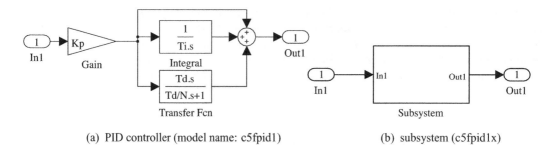

(a) PID controller (model name: c5fpid1)   (b) subsystem (c5fpid1x)

Figure 5.84   Simulink model of PID controller.

5.7.2   Conditional Subsystems

In Simulink, certain subsystems are allowed to be controlled by other signals. These kinds of systems are called conditionally executed subsystems. Three kinds of control structures are supported in Simulink.

- **Enabled subsystem**: An external signal to control such a subsystem is referred to as the control signal, with "enabled" and "disabled" control actions. In Simulink, it is assumed that when the control is positive the subsystem is enabled, otherwise it is disabled. If the subsystem is enabled, it behaves normally, and if it is disabled, in order to keep consistency, the current output signals can be maintained or reset to zero.

- **Triggered subsystem**: If the control signal detects a certain event, the system can be trigged to perform certain actions, and then maintain its output until the next trigger event. The allowed trigger events are usually the change of control signal as a rising edge, falling edge or both edges.

- **Enabled and triggered subsystem**: If the subsystem is enabled, the trigger event can be used to trigger the system. If the disable signal is active, the trigger actions can also be disabled.

The following example can be used to demonstrate the behaviors of the conditional subsystems. The concepts of rising and falling edges in triggers are also illustrated.

Example 5.32   *Consider a new subsystem composed of two nonlinear elements, a saturation block and a dead zone block, in series connection. An enable subsystem can be constructed as follows: First, open a blank model window and copy the block* **Enabled subsystem** *in the* **Subsystems** *group into the blank model. Double click the subsystem block, copy the two nonlinear elements into the window, and assign the parameters of the nonlinear blocks to* $(-1, 1)$ *and* $(-3, 3)$ *respectively. Thus the subsystem shown in Fig. 5.85(a) can be constructed.*

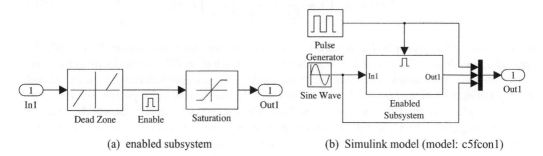

(a) enabled subsystem   (b) Simulink model (model: c5fcon1)

Figure 5.85   Block diagram of enabled system.

In the model window, a sinusoidal input block with an amplitude of 4 is added, and the enable control signal, driven by a pulse signal is added to the window. The model shown in Fig. 5.85(b) can be constructed. Fixed step simulation with a step size of 0.02 s can be performed, and the simulation results shown in Fig. 5.86(a) can be obtained.

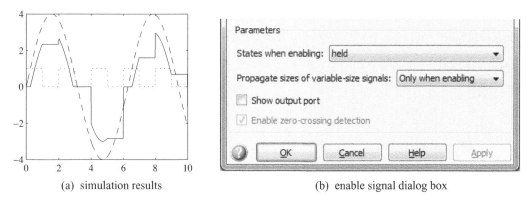

(a) simulation results   (b) enable signal dialog box

Figure 5.86 Enable signal simulation and setting.

*Double click the subsystem icon, and the dialog box shown in Fig. 5.86(b) will be opened. We can set the option **States when enabling** to **reset** and **held**, respectively. In the static nonlinear systems, there is no difference between these options.*

Example 5.33 *This example illustrates the properties of the trigger subsystem and the internal structure of the subsystem model so that the input port is connected directly to the output port. A trigger subsystem block can be copied to the model window. The internal structure of the subsystem model is shown in Fig. 5.87(a). The simulation model containing such a block is shown in Fig. 5.87(b).*

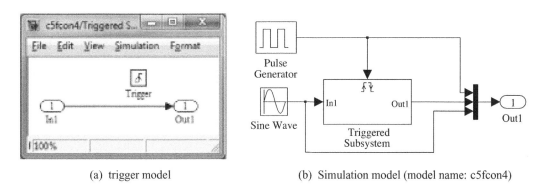

(a) trigger model   (b) Simulation model (model name: c5fcon4)

Figure 5.87 Model trigger subsystem.

*Double click the trigger block in the subsystem model. The dialog box shown in Fig. 5.88(a) is displayed. It can be seen that the **Trigger type** listbox provides the trigger type such as **rising** (meaning the rising edge trigger), **falling** and **either** and **function-call**, that is, callback function, options.*

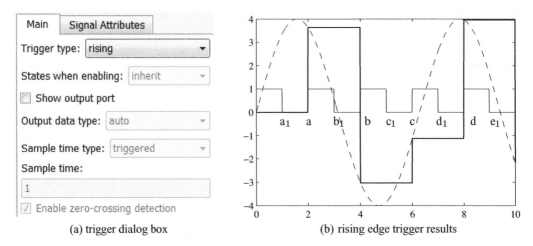

(a) trigger dialog box   (b) rising edge trigger results

Figure 5.88   Setting and simulation of triggers.

Starting the simulation process, the command plot(tout,yout) *can be executed and the results are shown in Fig. 5.88(b). The time instances a, a_1, b, b_1 are used to indicate signal changes in the pulse control to the trigger. The ones with subscript indicate the falling edges and the ones without subscripts are for rising edges. It can be seen that if the trigger is active, the signal of the output changes, and then remains constants until the next trigger signal.*

*If the trigger type in Fig. 5.88(a) dialog box is set to **falling**, the simulation results in Fig. 5.89(a) will be obtained, while if the trigger type is set to **either**, the simulation results in Fig. 5.89(b) will be obtained.*

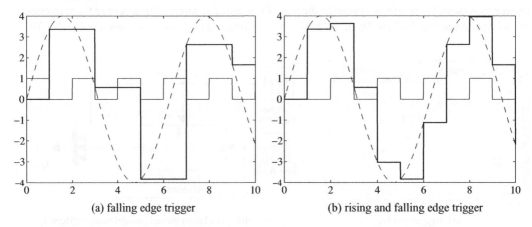

(a) falling edge trigger   (b) rising and falling edge trigger

Figure 5.89   Simulation results with other trigger types.

Simulink also allows a trigger port and an enable port to control the subsystem. If the subsystem is enabled, triggers can activate the subsystem, while if it is disabled, trigger actions will be neglected.

Further flexible control structures, such as conditional loops and switches, are also allowed. These structures are constructed by Stateflow, and these blocks can also be implemented with statements.

5.7.3 Masking Subsystems

To mask a subsystem means to hide the internal structure of the subsystem, and when the block is double clicked, a parameter setting dialog box will be given instead, allowing the users to specify the parameters of the masked block. In fact, most of the blocks provided in Simulink are masked from low-level models. For instance, in a transfer function block, the internal structure is not accessible to users. Only numerator and denominator polynomials can be assigned to the block. Also, consider the PID controller subsystem; it can be masked such that the four parameters can be entered with dialog boxes.

To mask a user-designed model, the first thing to do is to convert the whole block diagram of the Simulink model into a subsystem. Select the subsystem icon, and then, using the **Edit** → **Mask Subsystem** menu, the dialog box of the model mask editor interface shown in Fig. 5.90 can be opened. The user can use this interface to design the mask block properties.

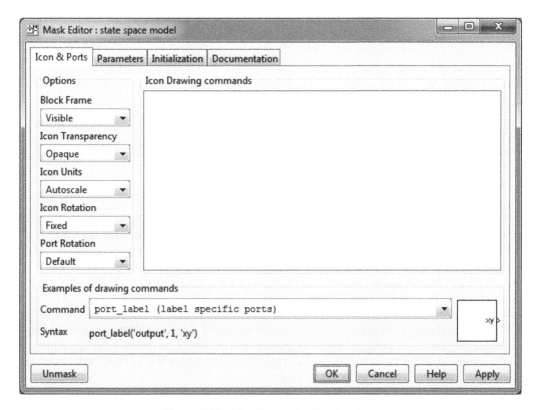

Figure 5.90 Simulink mask editor interface.

- **Drawing commands** pane: In the mask editor interface, three ways of representing icons of the masked block are supported. The MATLAB commands `plot()` can be used to draw curves, `disp()` can be used to write text and `image()` allows you to display images. If a "smiling face" is to be drawn as the icon of the masked blocks, the following MATLAB statements can be used to draw four curves, one for the face, two for the eyes and one for the mouth. The icon drawn is shown in Fig. 5.91(a).

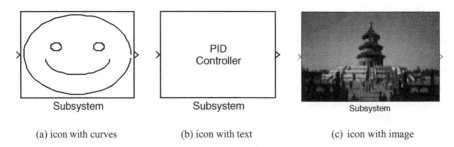

(a) icon with curves   (b) icon with text   (c) icon with image

Figure 5.91   Icon representation.

```
plot(cos(0:.1:2*pi),sin(0:.1:2*pi),-0.4+0.1*cos(0:0.1:2*pi),...
    0.2+0.1*sin(0:0.1:2*pi),...
    0.4+0.1*cos(0:0.1:2*pi),0.2+0.1*sin(0:0.1:2*pi),...
    0.6*cos(0:.1:pi),-0.1-0.4*sin(0:.1:pi))
```

It can be seen from the above commands that the three curves used the array $0:.1:2*\text{pi}$ for drawing curves. We can try simplifying the drawing commands by introducing a vector *t*, such that the simpler code can be filled in the edit box

```
t=0:.1:2*pi; plot(cos(t),sin(t));
line(-0.4+0.1*cos(t),0.2+0.1*sin(t));
line(0.4+0.1*cos(t),0.2+0.1*sin(t));
t=0:.1:pi; line(0.6*cos(t),-0.2-0.4*sin(t));
```

It can be seen that the above command can draw a "smiling face" in the MATLAB graphics window. However, if these commands are placed in the text box of the mask dialog box, a warning message "Warning: Unrecognized function encountered in mask display command." will be displayed. The message indicates that the variable vector *t* cannot be transferred to the text box. So the variable should be embedded in the drawing statement, and cannot be used alone. The user can define the *t* vector in the **Initialization** pane of the dialog box shown in Fig. 5.90. Then in the **Icon & Port** pane, the vector *t* can be used directly.

The command `disp('PID\nController')` can be used in the **Icon & Port** pane, and the icon display in the block is as shown in Fig. 5.91(b), where in the string, \n means a carriage return.

The command `image(imread('tiantan.jpg'))` can be used so that an image will be shown in the icon of the block as shown in Fig. 5.91(c).

• **Initialization** pane: Since some variables cannot be used to draw icons, these variables can be initialized in the pane. For instance, the vector *t* can be specified in the pane, and then, in the **Icon & Port** pane, the variable *t* can be used directly. The command $t = 0:.1:2*\text{pi}$ in this case can be defined in the **Initialization** pane, and the commands below can then be used to draw the icon of the masked block.

```
plot(cos(t),sin(t)); plot(-0.4+0.1*cos(t),0.2+0.1*sin(t));
plot(0.4+0.1*cos(t),0.2+0.1*sin(t));
plot(0.6*cos(t(1:end/2)),-0.2-0.4*sin(t(1:end/2)));
```

- **Block Frame** option: There are two options, **Visible** and **Invisible**. The former is the default option, so most of the Simulink blocks have visible boxes.

- **Icon Transparency** options: Two options, **Opaque** and **Transparent** are available. The former is the default, and the information in the block port label may be hidden. If the name of the ports is to be displayed, the option **Transparent** should be selected.

- **Icon Rotation** option: Two options, **Fixed** and **Rotates**, are supported. If the latter is selected, the icon of the block is flipped if the flipping option is selected. For instance, if the **Rotates** option is selected, the effect will be as shown in Figs 5.92(a) and (b). If the **Fixed** option is selected, the image will not be flipped, as shown in Fig. 5.92(c).

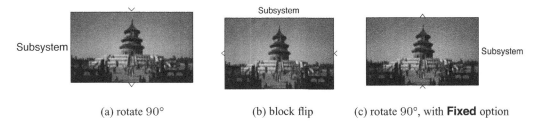

(a) rotate 90°   (b) block flip   (c) rotate 90°, with **Fixed** option

Figure 5.92   Icon rotating and flipping.

- **Drawing coordinates** specification: Three options are supported: with **Pixels**, **Autoscale** and **Normalized**.

An important step in block masking is to establish the internal block variables with those in the masking dialog box. Selecting the **Parameters** pane, as shown in Fig. 5.93, variable names can be specified and they can be linked to the variables in the block directly.

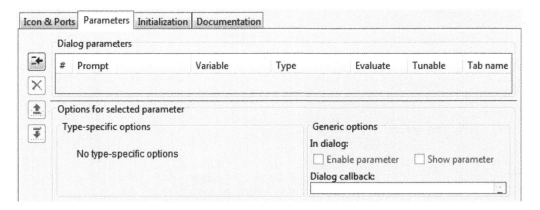

Figure 5.93   Parameters pane.

The buttons ⊒⁺ and ✕ are used to assign and delete variable names. For instance, in the PID controller block, click the ⊒⁺ button four times, and four data entries are prepared. Clicking for the first time gives the dialog box shown in Fig. 5.94. In the **Prompt**, prompt information can be entered. For instance, the message `Proportional Kp` can be specified. In the **Variable** edit box, the variable name can be entered `Kp`. Note that the variable names should be the same as the one in the block diagram.

#	Prompt	Variable	Type	Evaluate	Tunable	Tab name
1	Proportional Kp	Kp	edit ▾	✓	✓	
2	Integral Ti	Ti	edit ▾	✓	✓	
3	Derivative Td	Td	edit ▾	✓	✓	
4	Filter constant N	N	popup ▾	✓	✓	

Dialog parameters

Figure 5.94   Masking dialog box of associated variables.

The following method can be adopted to set up connections with the variables in the masked subsystems. In the edit box **Type**, the default option is **edit**, meaning that the edit box should accept data. For instance, N can be used to allow it be selected from a few values, and the other available options are in **popup**. The **Popup string** string can be entered as $10 \rightarrow 100 \rightarrow 1000$.

The positions of the variables can be arranged, and the ⬆ and ⬇ buttons can be used to move the item up and down respectively. The user may further select the **Documentation** pane to arrange the help information. The subsystem can be masked in this way. Double click the mask box, and the dialog box shown in Fig. 5.95 is displayed, allowing the user to enter the parameters of the PID controller. Note that the value of the filter constant N can be assigned in the listbox, and as it is expressed, the allowed options are 10, 100 or 1000.

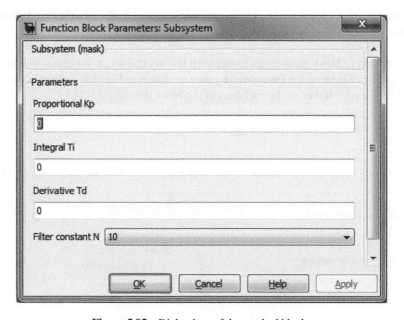

Figure 5.95   Dialog box of the masked block.

Click the right mouse button to display the shortcut menu, select the **Look under mask** item, and the internal structure of the masked model will be opened as shown in Fig. 5.96(a). We can modify the names of the input and output ports. For instance, the input port can be renamed as **error**, and the output port can be renamed as **control**; the new masked block is changed automatically to the form shown in Fig. 5.96(b). Note that to display the names of the ports properly, the option **Icon transparency** in the masking dialog box should be changed to **Transparent**.

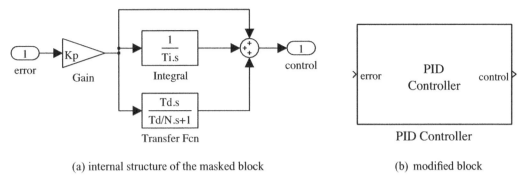

(a) internal structure of the masked block   (b) modified block

Figure 5.96   Masked block with modified ports.

■ **Example 5.34**   *Consider again the state space model studied in Example 5.7. The subsystem of the state space model is shown in Fig. 5.97(a). The subsystem can be masked into a new state space block, as shown in Fig. 5.97(b). Again, the **Icon transparency** property can be set to **Transparent**, and in the **Drawing commands** box, the command* disp('x'' = Ax + Bu\ny = Cx + Du') *should be filled in.*

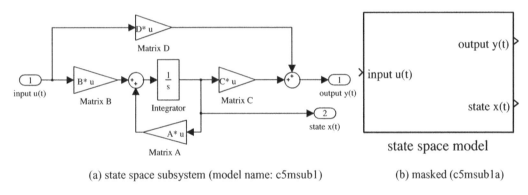

(a) state space subsystem (model name: c5msub1)   (b) masked (c5msub1a)

Figure 5.97   State space model with modified ports.

*The matrices A, B, C, D and x_0 can be assigned to the edit boxes as shown in Fig. 5.98(a), and the parameter dialog box of the masked block will be as shown in Fig. 5.98(b), where x_0 is an empty vector representing initial states. Also, a **checkbox** labeled **States** is designed. If it is unchecked, the state vector output port is to be hidden. In this example, the method of hiding output ports is not demonstrated, and readers may consider how to do so with S-functions, which will be presented in the next chapter.*

■ **Example 5.35**   *As stated earlier, factional-order operator s^γ can be approximated with the Oustaloup filter. The transfer function model is represented as*

$$G_f(s) = K \prod_{k=1}^{N} \frac{s + \omega_k'}{s + \omega_k}, \tag{5.7}$$

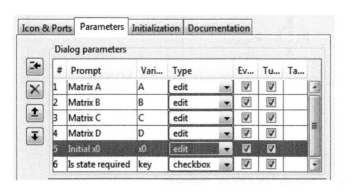

(a) **Parameters** pane   (b) parameter dialog box

Figure 5.98   Masking of the state space model.

where the zeros, poles and gain are calculated from

$$\omega_k' = \omega_b \omega_u^{(2k-1-\gamma)/N}, \ \omega_k = \omega_b \omega_u^{(2k-1+\gamma)/N}, \ K = \omega_h^\gamma, \tag{5.8}$$

where $\omega_u = \sqrt{\omega_h/\omega_b}$, (ω_b, ω_h) specifies the expected range of frequency and N is the selected order of the filter. The following function can be written

```
function G=ousta_fod(gam,N,wb,wh)
k=1:N; wu=sqrt(wh/wb); wkp=wb*wu.^((2*k-1-gam)/N);
wk=wb*wu.^((2*k-1+gam)/N); G=zpk(-wkp,-wk,wh^gam); G=tf(G);
```

*A Simulink subsystem model can be constructed as shown in Fig. 5.99(a), where apart from the Oustaloup filter, a low-pass filter can be appended. The parameter box of the expected system is as shown in Fig. 5.99(b). The masking dialog box for parameter specification is designed in Fig. 5.100. In the **Initialization** pane, the following code can be entered*

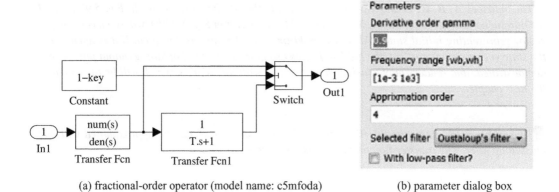

(a) fractional-order operator (model name: c5mfoda)   (b) parameter dialog box

Figure 5.99   Masking of fractional-order differentiators.

```
wb=ww(1); wh=ww(2); G=ousta_fod(gam,n,wb,wh);
num=G.num1; den=G.den1; T=1/wh; str='Fractional n';
if isnumeric(gam), if gam>0, str=[str, 'Der s^' num2str(gam) ];
   else, str=[str, 'Int s^' num2str(gam) '']; end
else, str=[str, 'Der s^gam']; end
```

Dialog parameters

#	Prompt	Variable	Type	Evaluate	Tunable	Tab name
1	Derivative order gamma	gam	edit	[x]	[x]	
2	Frequency range [wb,wh]	ww	edit	[x]	[x]	
3	Apprixmation order	n	edit	[x]	[x]	
4	Selected filter	key1	popup	[x]	[x]	
5	With low-pass filter?	key	checkbox	[x]	[x]	

Figure 5.100 Dialog box of fractional-order differentiator model.

In the ***Icon*** *edit box, the string* `disp(str)` *should be entered. The masked blocks can be used directly in simulation models, see Example 5.23.*

Example 5.36 *From MATLAB version R2010a, a PID controller block is provided, and it is a very sophisticated block, where different kinds of PID controllers can be converted. For instance, continuous, discrete and PID controllers with saturation, PID with integrator reset and anti-windup PID controllers can be converted directly with the PID controller block. Interested readers can read the low-level programming details of the masked blocks.*

1) Click the PID controller block with right mouse button, from the shortcut menu, and the ***View Mask*** *menu item can be selected to open the dialog box. In the* ***Parameters*** *pane, the parameter listbox can be selected as shown in Fig. 5.101. It can be seen that there are more than 100 parameters with this block. In the parameter dialog box, some of the items have checked* ***Evaluate*** *properties. These options are assigned automatically according to the PID parameters.*

Icon & Ports | Parameters | Initialization | Documentation

Dialog parameters

#	Prompt	Variable	Type	Evaluate	Tunable	Tab name
1	Controller:	Controller	popup	[ ]	[ ]	
2	Time-domain:	TimeDomain	popup	[ ]	[ ]	
3	Sample time (-1 for inherite...	SampleTime	edit	[x]	[ ]	
4	Integrator method:	IntegratorMethod	popup	[ ]	[ ]	
5	Filter method:	FilterMethod	popup	[ ]	[ ]	

Figure 5.101 Parameter specifications in the PID controller block.

2) *If the **Initialization** pane in Fig. 5.101 is selected, the command* `pidpack.` `PIDConfig.configPID(gcbh)` *can be used to find the file from the MATLAB search path.*

In the folder `toolbox/simulink/blocks`*, there are two subfolders,* `@pidpack` *and* `+pidpack`*, meaning that* `pidpack` *domain is designed. The related function can be defined in the domain. Some of the objects are related to the callback function. Some of the functions can be used to redraw the Simulink block diagram and the icons of the blocks.*

5.7.4 Constructing Users' Own Block Library

The menu **Edit** → **Edit mask** can be used to modify the masked blocks. The command can be used to open the dialog box shown in Fig. 5.90. The masked parameters can be modified using the masking procedure described earlier. To modify the internal structure of the model, in the shortcut menu, use the **Look under mask** submenu to open the model window. The internal block diagram can be modified in this way.

If the users have already established some masked blocks, a model library can be constructed to save all the blocks. The users may establish their own block library.

The menu **File** → **New** → **Library** in the Simulink model can be used to open a blank window. The model can be saved to a new file, named my_blks.mdl. The menu **File** → **Model properties** can be used to assign the properties of the block library. The model library thus created is locked. The blocks in the library cannot be moved or modified. To modify the blocks, the menu **Edit** → **Unlock library** can be used to unlock the library, and then you can modify the library. When the file is saved, it is locked again.

With such a library, a group of commonly used blocks can be copied to the library window. The model library shown in Fig. 5.102 can be constructed. The next time you want to draw a new model, the Simulink library does not need to be opened. The block `my_blks` or `open_system('my_blks')` can be used to open the user library.

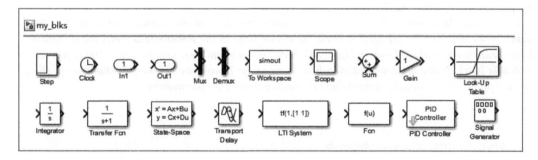

Figure 5.102 User designed new model library (model name: my_blks).

To have the new library appear in the Simulink library, a file slblocks.m should be copied to the folder. Find an existing slblocks.m file, and copy it to the folder where my_blks.mdl is located. The following statements in the file should be modified.

```
blkStruct.Name=sprintf('%s\n%s','Commonly Used Blocks','for Simulink');
blkStruct.OpenFcn = 'my_blks';
```

5.7.5   An Illustrative Example: F-14 Aircraft Simulation

Now consider a complicated example of Simulink modeling. When computer programs were not well established, Dean Frederick proposed a benchmark problem, which was used for testing the functions and modeling accuracy of computer software to be tested. The F-14 aircraft was used as the benchmark problem [8]. Since the introduction of the benchmark problem, many researchers worldwide have used it to test their algorithms and computer software. All these methods and software were extremely complicated by modern standards. Since the emergence of the new generation of computer software, such as Simulink, interactive graphical modeling software can easily be used to handle the complicated models.

The benchmark problem of the F-14 aircraft system is shown in Fig. 5.103. The system has two input signals. The vector representation of the input signal is $\boldsymbol{u} = [n(t), \alpha_c(t)]^T$, where $n(t)$ is the white noise signal with zero mean and a variance of one. The input signal $\alpha_c(t) = K\beta(e^{-\gamma t} - e^{-\beta t})/(\beta - \gamma)$ is the angle of attack command. The constants are $K = \alpha_{c_{max}} e^{\gamma t_m}$, $\alpha_{c_{max}} = 0.0349$, $t_m = 0.025$, $\beta = 426.4352$ and $\gamma = 0.01$. The parameters of the system are given by

$$\tau_a = 0.05, \sigma_{wG} = 3.0, a = 2.5348, b = 64.13,$$
$$V_{\tau_0} = 690.4, \sigma_\alpha = 5.236 \times 10^{-3}, Z_b = -63.9979, M_b = -6.8847,$$
$$U_0 = 689.4, Z_w = -0.6385, M_q = -0.6571, M_w = -5.92 \times 10^{-3},$$
$$\omega_1 = 2.971, \omega_2 = 4.144, \tau_s = 0.10, \tau_\alpha = 0.3959,$$
$$K_Q = 0.8156, K_\alpha = 0.6770, K_f = -3.864, K_F = -1.745.$$

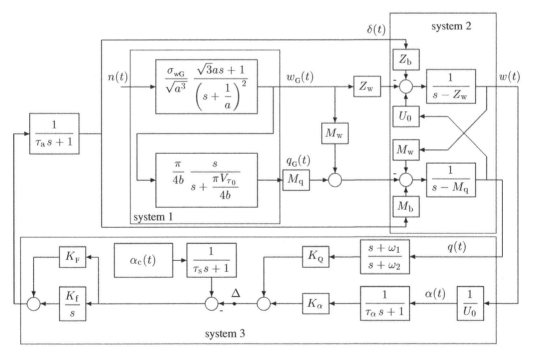

Figure 5.103   Block diagram model of an F-14 aircraft.

The following MATLAB statements can be used to enter the variables, and the variable names are the same as the specified variables

```
tA=0.05; Swg=3.0; a=2.5348; b=64.1300;
Vto=690.4; Sa=0.005236; Zb=-63.9979; Mb=-6.8847;
U0=689.4; Zw=-0.6385; Mq=-0.6571; Mw=-0.00592;
w1=2.971; w2=4.144; ts=0.1; ta=0.3959;
KQ=0.8156; Ka=0.677; Kf=-3.864; KF=-1.7450;
g=0.01; be=426.4352; tm=0.025; K=0.0349*exp(g*tm);
```

where in the final line, the variable names g and be correspond to the variables γ and β. The above statements can be saved into the MATLAB file c5f14dat.m. Before simulation is invoked, this program should be executed first to assign variables in the MATLAB workspace.

There are three output signals in the system, $\mathbf{y}(t) = [N_{Z_p}(t), \alpha(t), q(t)]^T$, and the output signal $N_{Z_p}(t)$ is defined as

$$N_{Z_p}(t) = \frac{1}{32.2}[-\dot{w}(t) + U_0 q(t) + 22.8\dot{q}(t)]. \tag{5.9}$$

It can be seen that the original system is rather complicated, and it is much easier to represent the original system as four subsystems. The first three are depicted in Fig. 5.103, and the fourth is described by (5.9).

In the first subsystem, there is one input signal $n(t)$, and two output signals $w_G(t)$ and $q_G(t)$, so the subsystem model shown in Fig. 5.104(a) is established. Selecting all the blocks in the subsystem, and selecting the **Edit → Create Subsystem** menu item, the icon of the subsystem block is shown in Fig. 5.104(b).

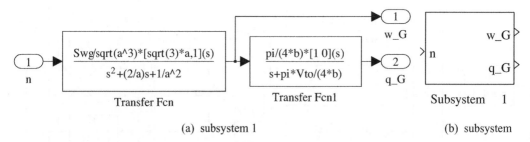

(a) subsystem 1   (b) subsystem

Figure 5.104   Simulink representation of subsystem 1 (overall model: c5f14).

In the second subsystem, there are three inputs: $\delta(t)$; $w_G(t)$ through block Z_w; and $-(M_w w_G(t) + M_q q_G(t))$. The subsystem has two output signals, $w(t)$ and $q(t)$. The subsystem can be constructed as shown in Fig. 5.105(a), and the subsystem block is shown in Fig. 5.105(b).

In the third subsystem, there are two input signals, $w(t)$ and $q(t)$, and one output signal. The subsystem model is shown in Fig. 5.106(a) and the subsystem block icon is shown in Fig. 5.106(b).

Now the fourth subsystem is considered for implementation (5.9) with Simulink. The model constructed in shown in Fig. 5.107(a), and the icon is shown in Fig. 5.107(b).

The overall Simulink model of F-14 aircraft system can be constructed, with the subsystems established earlier, as shown in Fig. 5.108. It can be seen that with the idea of subsystem modeling,

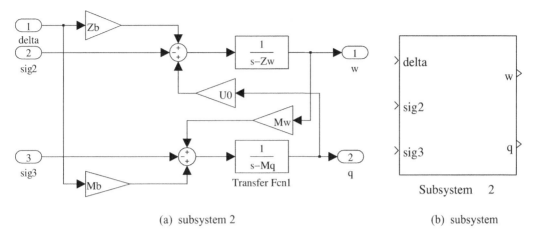

(a) subsystem 2   (b) subsystem

Figure 5.105   Simulink model of subsystem 2.

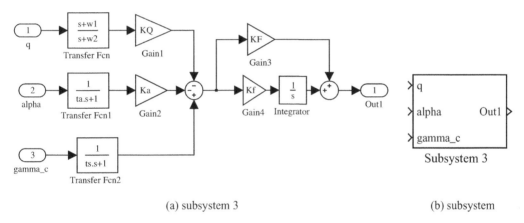

(a) subsystem 3   (b) subsystem

Figure 5.106   Simulink model of subsystem 3.

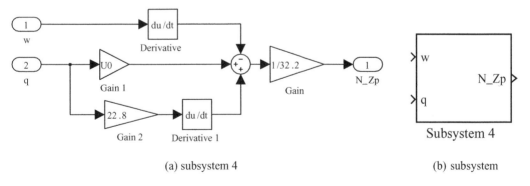

(a) subsystem 4   (b) subsystem

Figure 5.107   Simulink model of subsystem 4.

the methodology is more systematic and the system model is easier to maintain, and the final model can reliably be constructed.

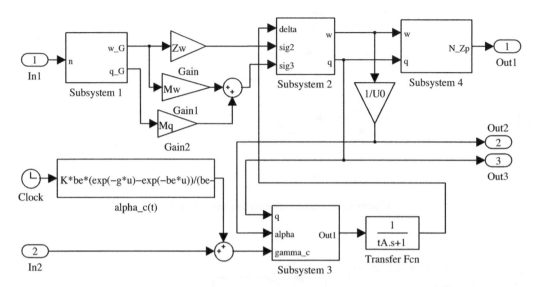

Figure 5.108   F-14 model in Simulink (model name: c5f14).

Exercises

5.1 Create a Simulink model to describe the Lorenz equation with vectorized integrators, and compare the simulation results with numerical results.

5.2 Consider the Lorenz equation. There is no input signal to the equations. If the three state signals $x_i(t)$ are used as the output signals, and the parameters β, σ, ρ and $x_i(0)$ are used as additional parameters, mask the Simulink model and draw the system output in phase space for different additional parameters.

5.3 Consider the two Simulink models given in Fig. 5.109. Analyze the systems and see whether there are any algebraic loops in the systems. Validate the conclusion and analyze whether the existence of the algebraic loop affects the simulation results.

5.4 Consider the linear differential equation

$$y^{(4)} + 3y^{(3)} + 3\dot{y} + 4\dot{y} + 5y = e^{-3t} + e^{-5t}\sin(4t + \pi/3).$$

If the initial values are $y(0) = 1$, $\dot{y}(0) = \ddot{y}(0) = 1/2$, $y^{(3)}(0) = 0.2$, establish a Simulink model for the equation and draw the system response.

5.5 Consider the model in Problem (5.4). If the differential equation is a time varying one

$$y^{(4)} + 3ty^{(3)} + 3t^2\ddot{y} + 4\dot{y} + 5y = e^{-3t} + e^{-5t}\sin(4t + \pi/3),$$

and the initial values are still $y(0) = 1$, $\dot{y}(0) = \ddot{y}(0) = 1/2$, $y^{(3)}(0) = 0.2$, use Simulink to establish a model and observe the simulation results.

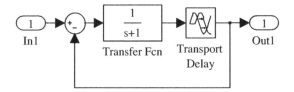

(a) system model 1 (model name: c5exala)

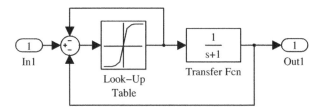

(b) system model 2 (model name: c5exalb)

Figure 5.109 Problem (5.3).

5.6 Represent the following two transfer function matrices in Simulink and observe the step responses of the system under unity negative feedback.

(a) $G_1(s) = \begin{bmatrix} \dfrac{0.5e^{-0.2s}}{3s+1} & \dfrac{0.07e^{-0.3s}}{2.5s+1} & \dfrac{0.04e^{-0.03s}}{2.8s+1} \\[3mm] \dfrac{0.004e^{-0.5s}}{1.5s+1} & \dfrac{-0.003e^{-0.2s}}{s+1} & \dfrac{-0.001e^{-0.4s}}{1.6s+1} \end{bmatrix}$

(b) $G_2(s) = \begin{bmatrix} \dfrac{0.66e^{-2.6s}}{6.7s+1} & \dfrac{-0.0049e^{-s}}{9.06s+1} \\[3mm] \dfrac{1.11e^{-6.5s}}{3.25s+1} & \dfrac{-0.012e^{-1.2s}}{7.09s+1} \\[3mm] \dfrac{-33.68e^{-9.2s}}{8.15s+1} & \dfrac{0.87(11.61s+1)e^{-s}}{(3.89s+1)(18.8s+1)} \end{bmatrix}$

5.7 Assume that an ODE is given by [9]

$$\begin{cases} \dot{u}_1 = u_3 \\ \dot{u}_2 = u_4 \\ 2\dot{u}_3 + \cos(u_1 - u_2)\dot{u}_4 = -g\sin u_1 - \sin(u_1 - u_2)u_4^2 \\ \cos(u_1 - u_2)\dot{u}_3 + \dot{u}_4 = -g\sin u_2 + \sin(u_1 - u_2)u_3^2, \end{cases}$$

where $u_1(0) = 45, u_2(0) = 30, u_3(0) = u_4(0) = 0$, g$= 9.81$. Solve the ODE with Simulink and draw the states versus time curves.

5.8 Assume that the position (x, y) of the Apollo satellite is given by

$$
\begin{cases}
\ddot{x} = 2\dot{y} + x - \dfrac{\mu^*(x + \mu)}{r_1^3} - \dfrac{\mu(x - \mu^*)}{r_2^3} \\[2mm]
\ddot{y} = -2\dot{x} + y - \dfrac{\mu^* y}{r_1^3} - \dfrac{\mu y}{r_2^3},
\end{cases}
$$

where $\mu = 1/82.45$, $\mu^* = 1 - \mu$, $r_1 = \sqrt{(x + \mu)^2 + y^2}$, $r_2 = \sqrt{(x - \mu^*)^2 + y^2}$. Assume that the initial values of the system are given by $x(0) = 1.2$, $\dot{x}(0) = 0$, $y(0) = 0$, $\dot{y}(0) = -1.04935751$. Establish a Simulink model and perform a simulation analysis, and draw the trajectory (x, y) of the Apollo satellite.

5.9 Solve numerically the following implicit differential equation

$$
\begin{cases}
\dot{x}_1 \ddot{x}_2 \sin(x_1 x_2) + 5\ddot{x}_1 \dot{x}_2 \cos(x_1^2) + t^2 x_1 x_2^2 = e^{-x_2^2} \\
\ddot{x}_1 x_2 + \ddot{x}_2 \dot{x}_1 \sin(x_1^2) + \cos(\ddot{x}_2 x_2) = \sin t,
\end{cases}
$$

with $x_1(0) = 1$, $\dot{x}_1(0) = 1$, $x_2(0) = 2$, $\dot{x}_2(0) = 2$, and draw the trajectories.

5.10 Consider the following delay differential equation

$$
y^{(4)}(t) + 4y^{(3)}(t - 0.2) + 6\ddot{y}(t - 0.1) + 6\ddot{y}(t) + 4\dot{y}(t - 0.2) + y(t - 0.5) = e^{-t^2},
$$

where it is known that when $t \leq 0$, the value of $y(t)$ is zero. Use Simulink to represent the above equation, and draw the $y(t)$ curve.

5.11 Consider the following delay different equation

$$
\frac{\mathrm{d}y(t)}{\mathrm{d}t} = \frac{0.2y(t - 30)}{1 + y^{10}(t - 30)} - 0.1y(t).
$$

Assuming that $y(0) = 0.1$, establish a Simulink model and perform the simulation. Draw the $y(t)$ curve.

5.12 For the fractional-order linear differential equation [10]

$$
0.8\mathscr{D}_t^{2.2} y(t) + 0.5\mathscr{D}_t^{0.9} y(t) + y(t) = 1, \quad y(0) = y'(0) = y''(0) = 0,
$$

solve numerically the differential equation. If the fractional order 2.2 is approximated to 2, and 0.9 to 1, the solution to the integer-order ODE can be obtained. Assess the accuracy with integer-order approximation.

5.13 Solve the following nonlinear differential equation of fractional-order, using an approximate method

$$
\frac{3\mathscr{D}^{0.9} y(t)}{3 + 0.2\mathscr{D}^{0.8} y(t) + 0.9\mathscr{D}^{0.2} y(t)} + \left| 2\mathscr{D}^{0.7} y(t) \right|^{1.5} + \frac{4}{3} y(t) = 5\sin(10t),
$$

with zero initial conditions. Validate the simulation results.

5.14 Assume that a nonlinear fractional-order differential equation is given by the Simulink model shown in Fig. 5.110. Write out the mathematical description of the system. Draw the output signal $y(t)$.

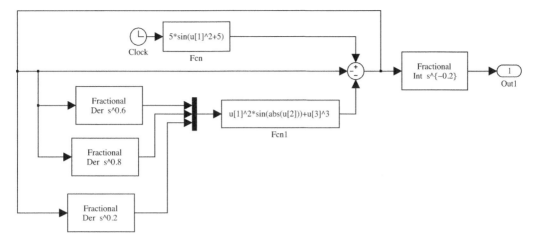

Figure 5.110 Simulink representation of a nonlinear fractional-order model (model: c5mfode4).

5.15 For a plant model with large time delay $G(s) = 10e^{-20s}/2s + 1$, assume that a controller is given by $G_c(s) = 0.6 + 0.008/s$, and evaluate the output signal and show it in a proper gauge.

5.16 Consider the following Lorenz equation

$$\begin{cases} \dot{x}_1(t) = -\beta x_1(t) + x_2(t)x_3(t) \\ \dot{x}_2(t) = -\rho x_2(t) + \rho x_3(t) \\ \dot{x}_3(t) = -x_1(t)x_2(t) + \sigma x_2(t) - x_3(t). \end{cases}$$

There is no input signal in the system. If the three state variables $x_i(t)$ are assigned as the output signals of the block, and the parameter vectors β, σ, ρ and $x_i(0)$ are used as masking parameters, mask the subsystem and draw the trajectories of the Lorenz equations under different parameters.

5.17 Assuming that an error signal $e(t)$ is used as the input signal of a block, construct a masked subsystem to represent ITAE, ISE and ISTE criteria. If the input signal to the block is the error signal $e(t)$, on double clicking the constructed subsystem, a listbox should be presented in a dialog box so that the criteria and the signal in the criteria can be returned in the output of the block. The ISTE criterion is defined as

$$J_{\text{ISTE}} = \int_0^t \tau e^2(\tau) \mathrm{d}\tau.$$

5.18 For the Van der Pol equation given by $\dot{x}_1 = x_2$, $\dot{x}_2 - \mu(x_1^2 - 1)x_2 - x_1$, establish a Simulink model, and mask the model. The parameters μ and the initial values of the integrators x_{10}, x_{20} can be used as masking parameters. In the model to be masked, there is no input signal, but there are two output ports: $x_1(t)$ and $x_2(t)$. If the parameter μ is to be used as an input port, redraw the Simulink model and mask the new model.

References

[1] N Munro. Multivariable control 1: the inverse Nyquist array design method, In: Lecture notes of SERC vacation school on control system design. UMIST, Manchester, 1989

[2] D P Atherton. Nonlinear control engineering – describing function analysis and design. London: Van Nostrand Reinhold, 1975

[3] D Xue, Y Q Chen. MATLAB solutions to advanced applied mathematical problems. Beijing: Tsinghua University Press, 2nd Edition, 2008. In Chinese

[4] D Liberzon, A S Morse. Basic problems in stability and design of switched systems. IEEE Control Systems Magazine, 1999, 19(5):59–70

[5] A Oustaloup, F Levron, F Nanot, *et al.* Frequency band complex non integer differentiator: characterization and synthesis. IEEE Transactions on Circuits and Systems I: Fundamental Theory and Applications, 2000, 47(1):25–40

[6] Wikipedia. Virtual reality. `http://en.wikipedia.org/wiki/Virtual_reality`

[7] C W Wang, W Gao, X R Wang. Theory, realization and application of virtual reality techniques. Beijing: Tsinghua University Press, 1996. In Chinese

[8] D K Frederick, M Rimer. Benchmark problem for CACSD packages. Abstracts of the second IEEE symposium on computer-aided control system design, 1985. Santa Barbara, USA

[9] C B Moler. Numerical computing with MATLAB. The MathWorks Inc, 2004

[10] I Podlubny. Fractional differential equations. San Diago: Academic Press, 1999

6

Advanced Techniques in Simulink Modeling and Applications

In the previous two chapters, the focus was on building fundamental knowledge of Simulink and tactics for the application of Simulink. The main approaches used in those chapters were direct graphical based programming-free methods. In practice, graphical methods may not be sufficient due to their own limitations. For instance, some block diagrams may be too complicated to draw manually. Command-line based modeling and design methods including graphical methods, then have to be used. In this chapter, advanced techniques of command-line modeling and application are presented. Section 6.1 introduces the MATLAB commands to create Simulink models. With the use of such command-line drawing techniques, complicated Simulink models can be created. In Section 6.2, the execution of Simulink models is introduced. Linearization techniques of nonlinear systems are also addressed in this section. In particular, the Padé approximation to pure time delays is further explored. It can be seen that not all the models can be constructed with graphical methods. Some of the complicated models can be created and analyzed using MATLAB commands. Thus, in Section 6.3, advanced techniques will be presented for creating complicated models. S-function programming techniques will be presented and illustrated and their use in simulation of automatic disturbance rejection control (ADRC) systems will be demonstrated as a case study. In Section 6.4, command-line based optimal controller design technique with Simulink models is introduced, and optimal controller design methods for nonlinear plants are also presented.

6.1 Command-line Modeling in Simulink

6.1.1 Simulink Models and File Manipulations

File manipulation of Simulink models can be completed simply with the **File** menu in the model window. Also, we can read and write Simulink files using MATLAB commands. In the MATLAB command window, the function `new_system()` can be used to create a blank modelwindow in the MATLAB workspace. However, the model created in this way will not be displayed automatically, since it is a logical model. To display it, another command, `open_system()` can be used, with the syntax `new_system(model_name,options)`, where `model_name` is specified in a string, and the `options` item can be set to `'Library'` or `'Model'`, where the former creates a blank model library and the latter a blank model. If the `options` argument is not specified, a blank model window

System Simulation Techniques with MATLAB® and Simulink®, First Edition. Dingyü Xue and YangQuan Chen.
© 2014 John Wiley & Sons, Ltd. Published 2014 by John Wiley & Sons, Ltd.

is created. For instance, we can use the `new_system('MyModel')` command to create a logical model named `MyModel`, then use the `open_system('MyModel')` command to open it.

If a model file exists, the command `open_system()` can still be used. This will be illustrated later.

A series of Simulink model processing functions can also be used. For instance, function `find_system()` can be used to obtain all the model names of currently opened model windows. If more than one model windows is open, the names are returned in cells.

When a model is created, the `save_system()` function can be used to save the Simulink model into a model file, with the suffix of .slx (in the old versions, .mdl). The syntax of the function is `save_system(model_name,options)`. If the model name is not specified in the function call, the current filename will be used. If the `options` argument is not given, the current model will be saved as a new file.

It should be noted that *.slx file is binary, thus to have a readable ASCII file, it is suggested to save the file in `mdl` form, with `save_system('model_name.mdl')`.

6.1.2 Simulink Models and Model Files

As shown earlier, the `new_system()` function can be used to create a blank logical Simulink model. The main statements in the automatically generated file will be presented through an example.

Example 6.1 *The following commands can be used to create and open a blank model window.*

```
>> new_system('newmodel');      % open a blank logical window
   open_system('newmodel');      % open a model window
   save_system('newmodel.mdl');  % save the model into an mdl file
```

The model can then be saved to the newmodel.mdl file. The models created by different versions of Simulink are different, and sometimes they may not even be compatible. Manual editing of the model files becomes more and more difficult. We can open the Simulink model file with the edit *command, and the main statements in the file are*

```
Model {
    Name                "newmodel"
    Version             8.0
    MdlSubVersion       0
    GraphicalInterface {
        ......
    }
    SavedCharacterEncoding   "windows-1252"
    SaveDefaultBlockParams   on
```

From the above, it can be seen that the file consists of models and blocks, along with their properties. It is not necessary to modify these manually, or even to understand all the properties in the file. The best way to modify the properties is to use dialog boxes. To modify certain properties, the function set_param() *should be used, and more on this topic will be given later.*

The function `close_system()` can be used to close an open Simulink model. If the window to be closed is not saved, a warning message will be given to ask the user whether to save it or not. If the command `close_system('MyModel', 1)` is given, the window will be forced to close and the unsaved modification will be abandoned.

6.1.3 Drawing Block Diagrams with MATLAB Commands

In earlier chapters, we saw that Simulink models can easily be constructed with the easy-to-use graphical facilities provided by Simulink, and the modeling is quite simple and straightforward. However, in certain applications, the user may need to draw Simulink block diagrams with MATLAB commands. The following procedure can be used to draw Simulink models.

1) Create a new model: As discussed before, the `new_system()` and `open_system()` functions can be used to create and display blank model windows.

```
>> new_system('newmodel'); open_system('newmodel');
```

Once a Simulink model is created, the function `set_param()` can be used to define certain properties, with the syntax

```
set_param(f,property 1,property_value 1,···)
```

where f is the Simulink model name, `property` is a string representing the name of a property and `property_value` is the relevant values of the property. The property names can be listed with the $f_1 =$ `simget(model_name)` function, but of these properties, only a few are commonly used. For instance, in the ODE solver, the default value of the relative error tolerance property, **RelTol** is 10^{-3}, which is sometimes too large. We can set it to smaller values such as 10^{-8}. Other properties, such as the minimum and maximum step sizes, that is, **MinStep** and **MaxStep**, can be modified with the following statements

```
>> f1=gcs;   % get the model handle and modify the properties
   set_param(f1,'MinStep','1e-5','MaxStep','1e-3','RelTol','1e-8');
   save_system(f1) % save the model into a model file
```

Please note that the property values should be set to strings. Sometimes, the function `set_param()` can also be substituted with the following form

```
>> f1.MinStep='1e-5'; f1.MaxStep='1e-3'; f1.RelTol='1e-8';
```

Example 6.2 *Consider the F-14 aircraft simulation model in Example 5.7.5. Before the model can be simulated, the data loading file c5f14dat.m should be executed. This process may be rather complicated. An alternative way is to load data into MATLAB automatically when the model is opened. This can be realized by means of its* `'PreLoadFcn'` *property in the model window. The following command can be used to associate the model with the c5f14dat.m file.*

```
>> set_param('c5f14','PreLoadFcn','c5f14dat');  % change the PreLoadFcn
   save_system('c5f14.mdl')     % save the model to an mdl file
```

*After the execution of the above commands, the model is saved. Next time the model is opened, the c5f14dat.m program will be called automatically to load the necessary data into the MATLAB workspace. Similar work can also be done when the **File** → **Model Properties** menu is selected. In this case, a dialog box shown in Fig. 6.1 will be opened, and we can select from the **Callbacks*** *menu the **PreLoadFcn** option, and write* c5f14dat *into the corresponding edit box. Other types of callback functions can also be specified in this way.*

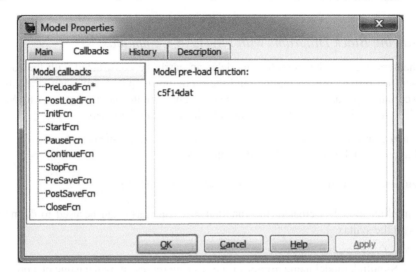

Figure 6.1   Model parameter setting dialog box.

2) Adding blocks: The add_block() function can be used to add a block into the model window with the following syntax

```
add_block(s_name,t_name,property 1,property_value 1,···)
```

where s_name is the prototype of the block to be copied. It should be the same as one defined in the built-in library, or in an opened Simulink model. The 'built-in/Clock' block, for instance, is the name of the built-in **Clock** block. The t_name argument is the target block name to be copied to. For instance, the following statement can be used to copy the standard **Clock** block into the model window named **newmodel**, under the new block name **My Clock**.

```
>> add_block('built-in/Clock','newmodel/My Clock');
```

The commonly used properties are

- **Position** property: the vector $[x_{min}, y_{min}, x_{max}, y_{max}]$ is used to describe the position of the block, that is, the vector represents respectively the coordinates of the lower-left and upper-right corners of the block. Simulink validates the property automatically, and it requires $x_{min} < x_{max}$, otherwise error messages will be given.
- **Name** property: This is the name of the block, and it can be any string, where \n string can be used to represent a carriage return.
- Each type of block may have its own properties. For instance, in the dialog box shown in Fig. 4.24(a), the transfer function block has two properties named respectively **Numerator** and **Denominator**. To change these properties, they can be set to relevant strings.

Furthermore, sometimes the source block name can be specified using the block name in the group. For instance, the source block name 'simulink3/Sources/Clock' can be used to represent the

Clock block in the Simulink group **Sources**. Apart from the blocks, the **Blocksets & Toolboxes** group contains a great variety of useful blocks. For instance, the **Controls Toolbox** group provides the LTI block. It can be copied into the model window with the following statements

```
>> add_block('cstblocks/LTI System','newmodel/tf')
```

where `cstblocks` is the name of the Control System Toolbox group. The source name can be obtained by double clicking the model group icon. The easiest way to find the name of a given existing block is to copy the block into a blank model window, save it as an `mdl` file, and read the ASCII file to find out the names of the block and its parameters.

To delete a block from a model window, the `delete_block()` function can be used, with the syntax `delete_block(t_name)`.

3) Connection of blocks: Two blocks in a Simulink model window can be connected with the `add_line()` function. There are two ways to use the function

```
add_line(model_name,'mod1/outport','mod2/inport')
add_line(model_name, M)
```

In the former statement, we need to specify the starting block name as `mod1`, and the terminating block as `mod2`. Also, specify respectively the output port number and the input port number of the two blocks. The connection between the two blocks can be established automatically. Sometimes the layout automatically generated may not be satisfactory. The latter syntax can be used and the connection is expressed by a matrix M, whose first and second columns specify the coordinates of the intermediate turning points.

To delete a certain connection, the `delete_line()` function can be used and the syntaxes of this function are the same as the `add_line()` function.

4) Modifying block parameters: Block parameters can be assigned or modified with `set_param()` function. The syntax of the function is

```
set_param(model_name,property1, property_value1,···)
```

where the definition of the arguments, `property` and `property_value`, is exactly the same as that given earlier. This function can modify either model properties or block properties. Examples of using this function will be demonstrated later. The function `get_param()` can be used to extract property values from a model or a block.

Example 6.3   *Assume that we want to establish a new model window and there are three blocks in the window: a sinusoidal block, a saturation block and a scope block. In the model, the sinusoidal input block is connected to the saturation block, then the output of it is connected to the scope block. The model can be created with the following MATLAB statements, as shown in Fig. 6.2(a).*

```
>> new_system('c6msys1'); open_system('c6msys1');
   set_param('c6msys1','Location',[100,100,500,400]);
   add_block('built-in/Sine Wave','c6msys1/Input signal');
   add_block('built-in/Saturation','c6msys1/Nonlinear element');
   add_block('built-in/Scope','c6msys1/My Scope');
   set_param('c6msys1/Input signal','Position',[40, 80, 80, 120]);
   set_param('c6msys1/Nonlinear element','Position',[140, 70, 230, 130]);
   set_param('c6msys1/My Scope','Position',[290, 80, 310, 120]);
   add_line('c6msys1','Input signal/1','Nonlinear element/1');
   add_line('c6msys1','Nonlinear element/1','My Scope/1');
```

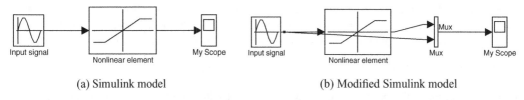

(a) Simulink model   (b) Modified Simulink model

Figure 6.2 Simulink model created and modified by MATLAB commands.

It can be seen that the above statements are easy to understand. The model was created and opened first, and the position of the window was set. Three blocks were copied from Simulink built-in library to the model window, and were assigned new names and positions. The three blocks were then connected as required.

*Now the modification of the established model will be demonstrated. If we want to display both the input signal and the output signal on the same scope, we should disconnect the last line, move the scope block further away, and insert a built-in **Mux** block in between the saturation and scope blocks. The input and output signals should be connected to the new **Mux** block, and the **Mux** block is then connected to the scope block. The following statements can then be given, and the new model is established as shown in Fig. 6.2(b).*

```
>> delete_line('c6msys1','Nonlinear element/1','My Scope/1'); % delete line
   set_param('c6msys1/My Scope','Position',[370,80,390,120]); % move scope
   add_block('built-in/Mux','c6msys1/Mux',...                  % add Mux block
          'Position',[290,80,295,120],'Inputs','2');
   add_line('c6msys1','Nonlinear element/1','Mux/1');          % redraw the 3 lines
   add_line('c6msys1','Input signal/1','Mux/2');
   add_line('c6msys1','Mux/1','My Scope/1');
```

It can be seen from Fig. 6.2(b) that the automatic connection of the block does not look nice. The intermediate turning points for the new connections should be calculated manually, and the following statements can be used to redraw the latter connection, as shown in Fig. 6.3. It can be seen that the connections can be specified by assigning the coordinates of the turning points, but this method is somewhat tedious.

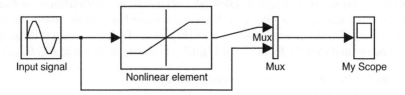

Figure 6.3 Simulink model when connections were changed.

```
>> delete_line('c6msys1','Input signal/1','Mux/2'); % delete line first
   add_line('c6msys1',[100,100; 100,150; 250,150; 250,110; 290,110]);
```

*The following command can be used to move the **Mux** block to a new place*

```
>> set_param('c6msys1/Mux','position',[310,110,315,140]); % move Mux block
```

and the new Simulink model is obtained as shown in Fig. 6.4. It can be seen that although the block is moved to another place, the connections are still valid.

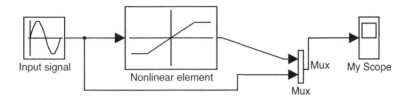

Figure 6.4   The new Simulink model after a block was moved.

*Now we can modify the parameters in the nonlinear saturation block. Assume that we want to change the lower and upper bounds of the saturation block to −0.3 and 0.8 respectively. We need to know first the appropriate property names. Double clicking the saturation block, it can be seen from the dialog box that the property names are respectively **Lower limit** and **Upper limit**. The following commands can then be issued to change the parameters.*

```
>> set_param('c6msys1/Nonlinear element',...
           'Lower limit','-0.3','Upper limit','0.8')
```

Note that the property values to be specified here should be strings. For instance, the lower bound should be set to `'-0.3'`*, rather than* −0.3*. If more than one property needs to be specified, the **MaskValueString** property can be specified, such that*

```
>> set_param('c6msys1/Nonlinear element','MaskValueString','-0.3|0.8')
```

■ **Example 6.4**   *Command-line based techniques are useful for the Simulink modeling of complicated systems, if they follow certain rules. For example, if we wants to establish a Simulink model for the system shown in Fig. 6.5 and want to find the equivalent model from Port A to Port B, where $L = 7$, it might be very complicated to draw the corresponding Simulink model manually. With the use of the following commands the Simulink model shown in Fig. 6.6 can be created automatically.*

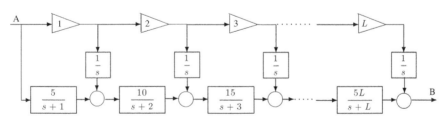

Figure 6.5   Block diagram of a complicated system.

```
>> L=7; mod='ssss'; new_system(mod); open_system(mod);
   for i=1:L, i1=int2str(i); i0=int2str(i-1);
      pos1=[100+(i-1)*85 50 125+(i-1)*85 70];
      pos2=pos1+[10 50 10 50]; pos3=pos1+[-35 100 -30 110];
      pos4=pos1+[15 105 10 100];
      add_block('built-in/Gain',[mod '/gain' i1],'Position',pos1,'Gain',i1)
```

```
add_block('built-in/Integrator',[mod '/int' i1],'Position',pos2,...
    'Orientation','down')
add_block('built-in/Transfer Fcn',[mod '/tf' i1],'Position',pos3,...
    'Numerator',[int2str(5*i)],'Denominator',['[1 ' i1 ']'])
add_block('built-in/Sum',[mod '/sum' i1],...
    'Position',pos4,'IconShape ','round','Inputs','++|')
add_line(mod,['gain' i1 '/1'],['int' i1 '/1'])
add_line(mod,['int' i1 '/1'],['sum' i1 '/1'])
add_line(mod,['tf' i1 '/1'],['sum' i1 '/2'])
if i>1,
    add_line(mod,['gain' i0 '/1'],['gain' i1 '/1'])
    add_line(mod,['sum' i0 '/1'],['tf' i1 '/1'])
end, end
add_block('built-in/Inport',[mod '/in'],'Position',[25 50 45 70])
pos5=pos4+[50 0 45 0];
add_block('built-in/Outport',[mod '/out'],'Position',pos5)
add_line(mod,'in/1','gain1/1'); add_line(mod,'in/1','tf1/1')
add_line(mod,['sum' int2str(L) '/1'],'out/1'); save_system(mod)
```

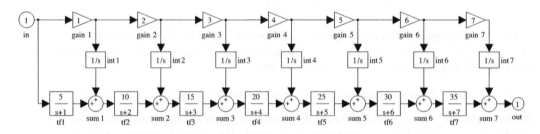

Figure 6.6   The Simulink model generated by commands.

With the use of linearization commands, which will be presented in the next section, the overall equivalent model can be extracted easily.

With the command-line model creation function, it is easier to draw the complicated simulation model for $L = 70$ or even $L = 500$.

6.2   System Simulation and Linearization

6.2.1   Execution of Simulation Process

We have already shown that the simulation process can be invoked by the use of the **Simulation** menu item. Also, we can initiate the simulation task with the sim() function, with the syntax $[t, x, y] = \text{sim}(f1, \text{tspan}, \text{options}, \text{ut})$, where f1 is the Simulink model name and tspan is the time span for simulation, which can be expressed as $[t_0, t_f]$, that is, the start and end time of simulation. If tspan argument is a scalar, it represents just the end time. The argument options specifies the control variable, while ut specifies the external input signal. The returned arguments are t, x and y, where t is the time vector, x returns the internal states and y is the output signal. The syntax of the function is similar to the MATLAB function ode45() function discussed in Chapter 3.

Similar to the numerical solutions of ODEs in MATLAB, simulation options can also be set for Simulink models. In the MATLAB command window, the `gcs` command can be used to extract the handle of the current Simulink model. The functions `simset()` and `simget()` can be used to set or extract properties of a Simulink model. These functions are quite similar to `set()` and `get()` functions in user graphical interface programming. Commonly used properties of the Simulink block include **Solver** (algorithm selection, such as `ode45`, `ode15s` etc.) and **RelTol** (relative error tolerance with a default value of 10^{-3}). These properties can either be set through the interface, or with MATLAB commands.

Example 6.5 *Assume that the Simulink model for the F-14 aircraft is saved in the file c5f14.mdl. The following statements can be used to initiate the simulation process*

```
>> c5f14dat; [t,x,y]=sim('c5f14',[0,10]);
```

The execution of the simulation is quite similar to the case where we click the ▶ button in the Simulink model. The difference is that the button clicking method returns the simulation results to the yout *and* tout *variables, while the command-line method transfers the simulation results to the variables* t, *and* y *in the MATLAB workspace directly.*

Example 6.6 *Now consider again the Simulink model designed for the Van der Pol equation studied in Example 4.3. It has been shown that when* $\mu = 1000$, *the corresponding ODE is a ill-conditioned ODE. The stiff ODE solver should be selected to solve the problem. For instance, the* ode15s *algorithm can be selected to solve the problem. To demonstrate the problem solution process, the Simulink model is redrawn as shown in Fig. 6.7. The following commands can be used to initialize the system.*

```
>> mu=1000; x01=1; x02=2;
```

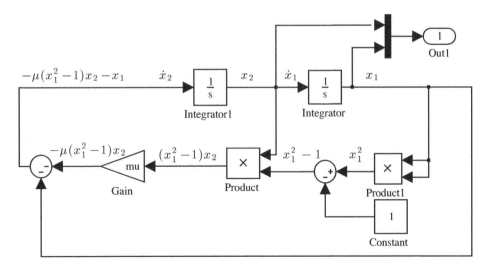

Figure 6.7 Modified Simulink model (model name: c6mvdp).

The function `simget()` *can be used to extract the default simulation control template, and the value of the **Solver** property can be modified with the* `simset()` *function. The simulation results can be obtained and the simulation results are identical to that obtained in Fig. 3.5.*

```
>> f=simget('c6mvdp'); f.Solver='ode15s'; % set algorithm to ode15s
   [t,x,y]=sim('c6mvdp',[0,3000],f); plot(t,y(:,1)),
   figure; plot(t,y(:,2))
```

It can be seen that the command-line execution of simulation and model creation is as powerful as the menu item selection in Simulink model windows. The command-line style is also suitable to be embedded into a higher level program for certain levels of automation in a simulation process with no or reduced user interaction. When command-line execution is used, the variables `tout` *and* `yout` *will no longer be automatically returned to the MATLAB workspace.*

6.2.2 Linearization of Nonlinear Systems

Compared with nonlinear systems, linear systems have certain advantages, since they can be analyzed and designed easily and in a systematic way. In actual applications, nonlinear behaviors exist everywhere. Strictly speaking, all models should be nonlinear. Sometimes we can use linear models to approximate nonlinear ones. The linearization technique extracts the approximate linear behavior from nonlinear systems, thus it is an effective way in dealing with nonlinear systems. The linearized model extracted may fit the behavior of the original nonlinear system within a region of the operating point. On the other hand, if the original system is linear, and the structure of the system is complicated, the linearization technique can also be used to extract the overall linear model from inputs to outputs.

Consider the general form of a typical nonlinear system given by

$$\dot{x}_i(t) = f_i(x_1, x_2, \cdots, x_n, \boldsymbol{u}, t), i = 1, 2, \cdots, n. \tag{6.1}$$

The so-called operating point of the system is defined as the working point at which the state variables settle down. In other words, it is the point at which the first-order derivatives of all the states equal zero. Thus the operating point can be obtained by solving directly the following nonlinear equation:

$$f_i(x_1, x_2, \cdots, x_n, \boldsymbol{u}, t) = 0, \quad i = 1, 2, \cdots, n. \tag{6.2}$$

Numerical methods can be used for solving nonlinear equations. A MATLAB function `findop()` is provided to find the operating point from a given Simulink model, but before that we have to use another function `operspec()` to get the input, state and output information.

The syntaxes of these functions are

`op1 = operspec(modelname); op = findop(modelname,op1)`

where `modelname` is the filename of the Simulink model, while the returned variable `op` is an operating point structured array. In the variable `op1`, all the states, inputs and outputs information is contained, and we can extract the information with `op1.States`, `op1.Inputs` and `op1.Outputs`. The vectors x_0, $\boldsymbol{u}_0$ and $\boldsymbol{y}_0$ can be extracted from the structured array with the following function, $x_0, \boldsymbol{u}_0, \boldsymbol{y}_0] = $ `getopinfo(op1)`. If we want to change the default setting of the initial x_0 or $\boldsymbol{u}_0$, the structured array `op1` should be modified to `op2 = initopspec(op1, `$x_0, \boldsymbol{u}_0$`)`. Then `findop()`

should be called again. The function $[x_0, u_0]$ = `getopinfo(op)` can also be used to find the finalized steady-state states x_0 and inputs u_0.

```
function [x0,u0,y0]=getopinfo(op)
x=op.States; u=op.Inputs; x0=[]; u0=[]; y0=[];
for i=1:length(x); x0=[x0; x(i).x]; end
for i=1:length(u); u0=[u0; u(i).u]; end
if nargout==3, y=op.Outputs; for i=1:length(y); y0=[y0;y(i).y]; end,end
```

Within the neighborhood of the operating point (u_0, x_0), the nonlinear dynamic system can be linearized such that

$$\Delta \dot{x}_i = \sum_{j=1}^{n} \left. \frac{\partial f_i(x, u)}{\partial x_j} \right|_{x_0, u_0} \Delta x_j + \sum_{j=1}^{p} \left. \frac{\partial f_i(x, u)}{\partial u_j} \right|_{x_0, u_0} \Delta u_j, \tag{6.3}$$

and the model can be written as

$$\Delta \dot{x}(t) = A_1 \Delta x(t) + B_1 \Delta u(t). \tag{6.4}$$

Selecting new state and input vectors as $z(t) = \Delta x(t)$ and $v(t) = \Delta u(t)$, the linearized state space model is written as

$$\dot{z}(t) = A_1 z(t) + B_1 v(t), \tag{6.5}$$

where the Jacobian matrix is defined as

$$A_1 = \begin{bmatrix} \partial f_1/\partial x_1 & \cdots & \partial f_1/\partial x_n \\ \vdots & \ddots & \vdots \\ \partial f_n/\partial x_1 & \cdots & \partial f_n/\partial x_n \end{bmatrix}, \quad B_1 = \begin{bmatrix} \partial f_1/\partial u_1 & \cdots & \partial f_1/\partial u_p \\ \vdots & \ddots & \vdots \\ \partial f_n/\partial u_1 & \cdots & \partial f_n/\partial u_p \end{bmatrix}. \tag{6.6}$$

Example 6.7 *Consider the nonlinear system model shown in Fig. 6.8. It can be seen that there are two nonlinear elements. With the use of Simulink, the system model can easily be established, as shown in Fig. 6.9. Note that here we use input and output ports to describe the input and output signals.*

The following statements can be used to evaluate the operating point, and it is found that $x_0 = [0, 0, 0]$ and $u_0 = 0$.

```
>> op1=operspec('c6nlsys'); op=findop('c6nlsys',op1)
```

The operating point is then evaluated and we discover that the input signal (the external signal) has been assumed to be zero. This is not what we are expecting. Assuming that the system is driven by a step input signal, the following statements can be used to evaluate the new operating point, and it is found that $x_0 = [0.1281, 0, 0.0905]$, and $u_0 = 1$.

```
>> [x0,u0,y0]=getopinfo(op1); u0=1;
   op2=initopspec(opspec,x0,u0); op3=findop('c6nlsys',op2)
```

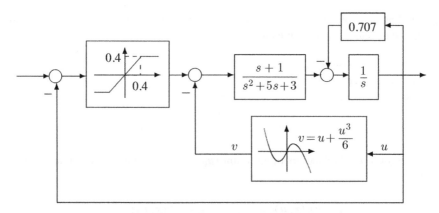

Figure 6.8 Block diagram of a nonlinear system model.

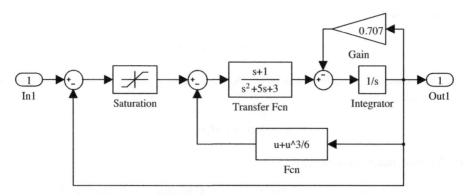

Figure 6.9 Simulink model (model name:c6nlsys).

Having obtained the operating point, the function `linmod2()` can be used to extract the linearized model such that

G = `linearize(model_name,op)`

where `op` is the operating point structured array. The linearized continuous state space model is returned in a state space object G. If the variable `op` is not specified, the default operating point will be used. In fact, if a Simulink model consists only of linear elements, the operating point object is not necessary. The overall linear model can still be extracted from the Simulink model.

Example 6.8 *Consider again the DC motor drive system described in Fig. 4.4. A Simulink model was established and saved in file c4mex2.mdl. The following statements can be used to extract the overall model, and the zero-pole-gain model can be converted*

$$G(s) = \frac{1118525021.9491(s + 6.667)(s + 5.882)}{(s + 179.6)(s + 98.9)(s + 8.307)(s^2 + 0.8936s + 5.819)(s^2 + 68.26s + 2248)}.$$

```
>> G=minreal(zpk(linearize('c4mex2')));
```

🔲 **Example 6.9**   *Consider the complicated system given in Example 6.4. The equivalent model can be obtained with the following statements, and the equivalent model from port A to port B can be obtained as*

$$G(s) = \frac{5040(s+0.2987)(s^2+22.66s+138.5)(s^2-0.8458s+13.56)(s^2+10.89s+71.76)}{s(s+7)(s+6)(s+5)(s+4)(s+3)(s+2)(s+1)}.$$

```
>> G=minreal(zpk(linearize('ssss')))
```

🔲 **Example 6.10**   *Consider again the F-14 fighter problem presented in Example 5.7.5. It can be seen that there are no nonlinear elements in the system. Thus* linearize() *function can be used directly to extract the overall system model, without the necessity of evaluating the operating points.*

```
>> G=linearize('c5f14')
```

The matrices in the state space model are

$$A = \begin{bmatrix} -0.639 & 689.4 & 2.0839 & 0.4746 & 0 & -1280 & 0 & 0 & 0 & 0 \\ -0.006 & -0.657 & 0.0456 & 0.0104 & -0.068 & -137.69 & 0 & 0 & 0 & 0 \\ 0 & 0 & -0.789 & -0.1556 & 0 & 0 & 0 & 0 & 0 & 0 \\ 0 & 0 & 1 & 0 & 0 & 0 & 0 & 0 & 0 & 0 \\ 0 & 0 & 3.2637 & 0.7434 & -8.4553 & 0 & 0 & 0 & 0 & 0 \\ 0 & 1.4232 & 0 & 0 & 0 & -20 & -1.6694 & 2.984 & -17.45 & 1 \\ 0 & 1 & 0 & 0 & 0 & 0 & -4.144 & 0 & 0 & 0 \\ 0.0015 & 0 & 0 & 0 & 0 & 0 & 0 & -2.5259 & 0 & 0 \\ 0 & 0 & 0 & 0 & 0 & 0 & 0 & 0 & -10 & 0 \\ 0 & 3.1515 & 0 & 0 & 0 & 0 & -3.6967 & 6.6075 & -38.64 & 0 \end{bmatrix},$$

$$B^T = \begin{bmatrix} 0 & 0 & 1 & 0 & 0 & 0 & 0 & 0 & 0 & 0 \\ 0 & 0 & 0 & 0 & 0 & 0 & 0 & 0 & 1 & 0 \end{bmatrix},$$

$$C = \begin{bmatrix} 0 & 21.41 & 0 & 0 & 0 & 0 & 0 & 0 & 0 & 0 \\ 0.0014505 & 0 & 0 & 0 & 0 & 0 & 0 & 0 & 0 & 0 \\ 0 & 1 & 0 & 0 & 0 & 0 & 0 & 0 & 0 & 0 \end{bmatrix}, \quad D = \begin{bmatrix} 0 & 0 \\ 0 & 0 \\ 0 & 0 \end{bmatrix}.$$

🔲 **Example 6.11**   *Consider the nonlinear system described in Example 6.7. The operating points can be obtained, and the linearized model can then be obtained as*

$$A = \begin{bmatrix} -0.707 & 1 & 1 \\ -2 & -5 & -3 \\ 0 & 1 & 0 \end{bmatrix}, \quad B = \begin{bmatrix} 0 \\ 1 \\ 0 \end{bmatrix}, \quad C = [1, 0, 0], D = 0.$$

```
>> op1=operspec('c6nlsys'); op=findop('c6nlsys',op1);  % find the operating point
   G=linearize('c6nlsys',op)                            % linearization
```

If step input is used, the operating point is computed again, the new linearized state space model can be obtained as

$$A = \begin{bmatrix} -0.707 & 1 & 1 \\ -2.0082 & -5 & -3 \\ 0 & 1 & 0 \end{bmatrix}, \quad B = \begin{bmatrix} 0 \\ 1 \\ 0 \end{bmatrix}, \quad C = \begin{bmatrix} 1 & 0 & 0 \end{bmatrix}, \quad D = 0.$$

```
>> [x0,u0,y0]=getopinfo(op1); op2=initopspec(op1,x0,1);
   op=findop('c6nlsys',op2); G=linearize('c6nlsys',op)
```

*It can be seen that different operating points may yield different linearized models. When step input is used, the linearized model has zero **B** vector. Thus the linearized model is useless for this example.*

6.2.3   *Padé Approximation to Pure Time Delays*

Using Padé approximation techniques, more accurate linearization models can be obtained for systems with pure time delays. The nth order Padé approximation to pure delay term can be written as

$$P_{n,\tau}(s) = \frac{1 - \tau s/2 + p_1(\tau s)^2 - p_2(\tau s)^3 + \cdots + (-1)^n p_{n-1}(\tau s)^n}{1 + \tau s/2 + p_1(\tau s)^2 + p_2(\tau s)^3 + \cdots + p_{n-1}(\tau s)^n}, \tag{6.7}$$

where for $n \le 8$, the coefficients of Padé approximation can be obtained, as shown in Table 6.1.

Table 6.1 Padé approximation coefficients.

Order n	p_1	p_2	p_3	p_4	p_5	p_6	p_7
1							
2	1/12						
3	1/10	1/120					
4	3/28	1/84	1/1680				
5	1/9	1/72	1/1008	1/30240			
6	5/44	1/66	1/792	1/15840	1/665280		
7	3/26	5/312	1/686	1/11440	1/308880	1/17297280	
7	7/60	1/60	1/624	1/9360	1/205920	1/7207200	1/518918400

The MATLAB function `pade()` provided in the Control System Toolbox can be used to approximate the pure time delay term by a rational transfer function. The syntax of the function is $[n,d] = \mathtt{pade}(\tau,k)$, where τ is the delay constant and k is the expected order of Padé approximation. The rational approximation to $P_{k,\tau}(s)$ can be obtained in the variable vectors n and d.

■ **Example 6.12**   *Consider a unity negative feedback system, whose components in the forward path are composed of a transfer function followed by a pure time delay term. The Simulink model shown in Fig. 6.10(a) can be constructed. Double click the time delay block, and fill in the edit box* *of **Pade order (for linearization)** to 2 and 4 respectively, as shown in Fig. 6.10(b), and the following models will be obtained*

```
>> G1=zpk(linearize('c6fdly'))
```

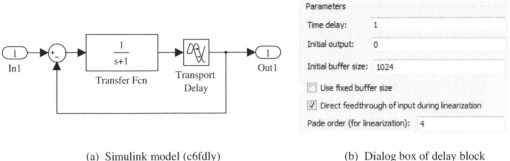

(a) Simulink model (c6fdly)   (b) Dialog box of delay block

Figure 6.10   Simulink model of a system with a time delay.

If the orders of Padé approximation are selected as 2 and 4, respectively, the linearized models can be obtained

$$G_2(s) = \frac{(s^2 - 6s + 12)}{(s + 6.749)(s^2 + 1.251s + 3.556)},$$

$$G_4(s) = \frac{(s^2 - 11.58s + 36.56)(s^2 - 8.415s + 45.95)}{(s + 14.88)(s^2 + 1.21s + 3.564)(s^2 + 5.908s + 63.36)}.$$

It can be seen from (6.7) that, the orders of numerator and denominator in Padé approximation are the same, and there are negative coefficients in the numerator. The closed-loop system thus obtained may become unstable. To solve the problem, more work should be done to better find Padé approximations.

The pure time delay term $\mathrm{e}^{-\tau s}$ can be expressed as a Taylor series

$$\mathrm{e}^{-\tau s} = c_0 + c_1 s + c_2 s^2 + \cdots = 1 - \frac{1}{1!}\tau s + \frac{1}{2!}\tau^2 s^2 - \frac{1}{3!}\tau^3 s^3 + \cdots. \tag{6.8}$$

Assume that an r/mth-order Padé approximation can be expressed by the following rational transfer function model

$$G_m^r(s) = \frac{\beta_{r+1}s^r + \beta_r s^{r-1} + \cdots + \beta_1}{\alpha_{m+1}s^m + \alpha_m s^{m-1} + \cdots + \alpha_1}, \tag{6.9}$$

where $\alpha_1 = 1$, $\beta_1 = c_1$, then the α_i ($i = 2, \cdots, m + 1$) and β_i ($i = 2, \cdots, k + 1$) coefficients can be obtained from the following equations

$$\boldsymbol{Wx} = \boldsymbol{w}, \quad \boldsymbol{v} = \boldsymbol{Vy}, \tag{6.10}$$

where

$$x = [\alpha_2, \alpha_3, \cdots, \alpha_{m+1}]^T, \ w = [-c_{r+2}, -c_{r+3}, \cdots, -c_{m+r+1}]^T$$
$$v = [\beta_2 - c_2, \beta_3 - c_3, \cdots, \beta_{r+1} - c_{r+1}]^T, \quad y = [\alpha_2, \alpha_3, \cdots, \alpha_{r+1}]^T,$$

(6.11)

and

$$W = \begin{bmatrix} c_{r+1} & c_r & \cdots & 0 & \cdots & 0 \\ c_{r+2} & c_{r+1} & \cdots & c_1 & \cdots & 0 \\ \vdots & \vdots & \cdots & \vdots & \ddots & \vdots \\ c_{r+m} & c_{r+m-1} & \cdots & c_{m-1} & \cdots & c_{r+1} \end{bmatrix}, \quad V = \begin{bmatrix} c_1 & 0 & 0 & \cdots & 0 \\ c_2 & c_1 & 0 & \cdots & 0 \\ \vdots & \vdots & \vdots & \ddots & \vdots \\ c_r & c_{r-1} & c_{r-2} & \cdots & c_1 \end{bmatrix}.$$

(6.12)

A MATLAB function `paderm()` can be written and it can be used to find the Padé approximation, with independently chosen orders of numerator and denominator.

```
function [nP,dP]=paderm(tau,r,m)
c(1)=1; for i=2:r+m+1, c(i)=-c(i-1)*tau/(i-1); end
w=-c(r+2:m+r+1)'; vv=[c(r+1:-1:1)'; zeros(m-1-r,1)];
W=rot90(hankel(c(m+r:-1:r+1),vv)); V=rot90(hankel(c(r:-1:1)));
x=[1 (W\w)']; y=[1 x(2:r+1)*V'+c(2:r+1)];
dP=x(m+1:-1:1)/x(m+1); nP=y(r+1:-1:1)/x(m+1);
```

The syntax of the function is $[n,d] = $ `paderm`(τ,r,m), where r and m are the expected orders of numerator and denominator respectively. The numerator and denominator vectors of the new Padé approximation can be returned in vectors n and d.

■ **Example 6.13** *Consider a simple example of a pure time delay model given by* $G(s) = \mathrm{e}^{-s}$. *The following statements can be used to find Padé approximations of different orders*

$$G_1(s) = \frac{-6s + 24}{s^3 + 6s^2 + 18s + 24}, \quad G_2(s) = \frac{6}{s^3 + 3s^2 + 6s + 6}, \quad G_3(s) = \frac{-s^3 + 12s^2 - 60s + 120}{s^3 + 12s^2 + 60s + 120}.$$

```
>> tau=1; [nP1,dP1]=paderm(tau,1,3); G1=tf(nP1,dP1)
   [nP2,dP2]=paderm(tau,0,3); G2=tf(nP2,dP2)
   [nP3,dP3]=pade(tau,3); G3=tf(nP3,dP3)
```

Although there are certain advantages in the new Padé approximation, it cannot be embedded into the existing linearization function, and the quality of linearization cannot be improved.

6.3 S-function Programming and Applications

In many real applications, some complicated systems cannot be easily constructed using building-block approaches provided in Simulink, while it might be more easily modeled by MATLAB functions. The MATLAB function blocks presented in Chapter 5 can only be used to express static nonlinear functions given by $y = f(u)$, not dynamical systems. That is to say, systems expressed by complicated differential or difference equations cannot be modeled in this way. S-function modeling (S for system) technique should be introduced to solve these problems. S-functions can be written in MATLAB, as well as in C, C++, Fortran and Ada languages. The programmed S-function blocks can be used as easily as those standard ones provided in Simulink libraries.

S-functions have their own program structures. Examples will be given to show the application and programming techniques. If the S-function is not modeled with MATLAB, the compilation process should be performed and the S-function file should be made into executable files (.mexw32 files or .dll files in earlier versions of MATLAB). The executable S-functions are usually faster and can be used directly in real-time simulation.

S-functions are suitable for describing state space equations. The state space equations can be continuous, discrete-time or both (referred to as hybrid state space equations). The general form of the hybrid state space equation can mathematically be expressed by

$$
\begin{cases}
y = g(t, x, u), & \text{output equation} \\
\dot{x}_c = f_c(t, x_c, u), & \text{continuous state equation} \\
x_d(t+1) = f_d(t, x_d, u), & \text{discrete state equation,}
\end{cases}
\tag{6.13}
$$

where the state vector x of the system is composed of the subvectors of continuous states x_c and discrete-time states x_d, that is, $x = [x_c^T, x_d^T]^T$.

It has been pointed out that S-functions can be written in MATLAB as well as in other languages. The S-function written in MATLAB can only be used in the MATLAB/Simulink environment. However, S-functions written in other languages can be converted into stand-alone programs and can be used in real-time simulation and control.

6.3.1 Writing S-functions in MATLAB

An S-function can be declared with the statement

```
function [sys,x0,str,ts] = funname(t,x,u,flag,p1,p2,···)
```

where `funname` is the filename of the S-function, and t, x and u are respectively time, state vector and input vector of the system. The use of the argument `flag` is presented in Table 6.2. Also, an arbitrary number of additional parameters p_1, p_2, $\cdots$ can be used in S-functions. These parameters can be declared later in the dialog box of the S-function. This will be demonstrated later through examples.

Table 6.2 The `flag` argument explanation.

Flag	Task of the function	Syntax of the function
0	initialization	`[sys,x0,str,ts]=mdlInitializeSizes(t,x,u,p1,···)`
1	continuous states update	`sys=mdlDerivatives(t,x,u,p1,···)`
2	discrete states update	`sys=mdlUpdate(t,x,u,p1,···)`
3	compute the output	`sys=mdlOutputs(t,x,u,p1,···)`
4	set next time instance	`sys=mdlGetTimeOfNextVarHit(t,x,u,p1,···)`
9	terminate simulation	`mdlTerminate(t,x,u,p1,···)`

S-functions can be written using the following procedure:

1) Setting of initial parameters: The statement `sizes = simsizes` can be used first to accept the default parameter template `sizes`, which is a structured array. The essential and frequently used members in the template are:

- `NumContStates`: the number of continuous states in the system.
- `NumDiscStates`: the number of discrete-time states.
- `NumInputs` and `NumOutputs`: the number of inputs and outputs respectively.
- `DirFeedthrough`: which indicates whether state variable is explicitly used in the output equation. This member is useful in the sorting of blocks.
- `NumSampleTimes`: the number of sampling intervals. The S-function supports the description of systems with multiple sampling intervals.

After the members are set, the template `sizes` can be returned with the statement `sys = simsizes(sizes)`. Also, the initial state vector x_0, string variable `str` and the sampling information variable `ts` should also be returned, where `ts` is composed of a matrix of two columns. The first column stores the sampling intervals while the other stores the offsets. For continuous systems and the systems with only one sampling interval, the variable can be expressed by $[t_1, t_2]$, where t_1 is the sampling interval, and the offset is normally 0. If $t_1 = -1$, it means that the sampling interval of the block inherits the sampling interval of its input signal.

2) Updates of the states: The continuous states can be updated by the function called `mdlDerivatives`, while the discrete ones are updated by the `mdlUpdate` function. The updated state variables are returned in the `sys` argument. For hybrid systems, the two functions should be called together.

3) Output signal evaluation: Function `mdlOutputs` can be used to evaluate the output signal of the block. The signal is returned in the argument `sys`.

A MATLAB template file `sfuntmpl.m` is provided, and you can write your own S-functions based on this template file. The following examples provide sufficient information in the programming of S-functions.

An S-function can be executed in a Simulink model in the following way. At the beginning of the whole simulation process, the `flag` variable is set to 0 automatically, and the initialization function is called. Then in each simulation step, the `flag` value is set to 3 first, to compute the output signal of the block. Then it is set to 1 and 2, to allow the updates of the continuous and discrete-time state variables, respectively. In the next simulation step, the value of the `flag` is set to $3 \rightarrow 1 \rightarrow 2$ sequentially, to repeat the above procedure until the end of the simulation process.

Example 6.14 *Consider the state space model with state output, illustrated in Example 5.7. The state space equation can be written as*

$$\begin{cases} \dot{x}(t) = Ax(t) + Bu(t) \\ y(t) = Cx(t) + Du(t) \end{cases}, Y(t) = \begin{bmatrix} y(t) \\ x(t) \end{bmatrix},$$

where $Y(t)$ is the output of the block. The S-function can be written to express the system. The additional parameters in this case are A, B, C and D matrices for the state space equation. The number of continuous states equals the number of rows of matrix A. The number of inputs equals the number of columns of matrix D, and the number of outputs should be the number of rows of *matrix **D**, plus the number of states. There is no discrete-time state in the system. The S-function for such a system can be written as*

```
function [sys,x0,str,ts]=c6exsf1(t,x,u,flag,A,B,C,D) % additional arguments
switch flag,
case 0, [sys,x0,str,ts]=mdlInitializeSizes(A,D);      % initialization
case 1, sys = mdlDerivatives(t,x,u,A,B);             % updating states
case 3, sys = mdlOutputs(t,x,u,C,D);                 % compute output
case {2, 4, 9}, sys = [];                            % unused
otherwise, error(['Unhandled flag = ',num2str(flag)]); % handling errors
end
%   initialization function mdlInitializeSizes
function [sys,x0,str,ts] = mdlInitializeSizes(A,D)
sizes = simsizes;   % extract default size template
sizes.NumContStates = size(A,1); % set number of continuous states
sizes.NumDiscStates = 0;            % no discrete-time states, case 2 is bypassed
sizes.NumOutputs=size(A,1)+size(D,1); % number of outputs
sizes.NumInputs = size(D,2);        % number of inputs
sizes.DirFeedthrough = 1;           % since state is explicitly used in output
sizes.NumSampleTimes = 1;           % number of sampling intervals
sys = simsizes(sizes);              % renew the template
x0 = zeros(size(A,1),1);            % zero initial state vector
str = [];                           % empty string for the block
ts = [-1 0]; % sampling interval, -1 for inhiritation, no offset
%   continuous state updates mdlDerivatives
function sys = mdlDerivatives(t,x,u,A,B)
sys = A*x + B*u;    % the state space equation
%   output signal evaluation mdlOutputs
function sys = mdlOutputs(t,x,u,C,D)
sys = [C*x+D*u; x]; % compute the augmented output signal Y(t)
```

Thus a Simulink model with the S-function block can be constructed as shown in Fig. 6.11(a). Double click the S-function block to open the dialog box shown in Fig. 6.11(b).

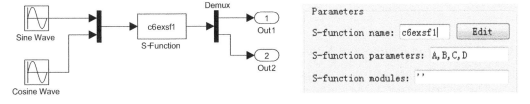

(a) Simulink model (model file: c6msf2)   (b) parameter setting dialog box

Figure 6.11   Simulink model with S-function block.

*In the edit box labeled **S-function name**, enter the string* c6exsf1. *Thus the connection of the block with the* c6exsf1.m *file can be established. Additional parameters **A**, **B**, **C** and **D** should be filled into the edit box labeled **S-function parameters**. The following statements can be entered* *into the MATLAB workspace before simulation can be performed. After simulation, results identical to those in Example 5.7 can be obtained.*

```
>> A=[2.25, -5, -1.25, -0.5;  2.25, -4.25, -1.25, -0.25;
      0.25, -0.5, -1.25,-1;  1.25, -1.75, -0.25, -0.75];
   B=[4,6; 2,4; 2,2; 0,2]; C=[0,0,0,1; 0,2,0,2]; D=zeros(2,2);
```

Two kinds of block construction methods using MATLAB, that is, the M-function method in Chapter 5 and the S-function method, have been presented. It can be seen that the M-function method is suitable for static nonlinearities, while the S-function method is usually used to model dynamic systems. It was also pointed out in Chapter 5 that the M-function method cannot accept additional parameters, which limits the use of M-function blocks. For static nonlinearities with additional parameters, S-function method should be used instead. In this case, only the code for `flag=0` and 3 are required. This will be demonstrated through the following example.

Example 6.15   *To establish a staircase signal generator block, and the key time instances are $t_1, t_2, \cdots, t_N$, and the corresponding output levels are $r_1, r_2, \cdots, r_N$. It might be quite complicated to model it using existing low-level Simulink blocks, and the M-function block do not allow the use of additional parameters t_i and y_i. Thus the S-function is the only choice for modeling such a signal generator with one block.*

From this application, the block obviously has no input port, and has only one output port. There are no state signals since it is static. Two additional vectors, $\texttt{tTime} = [t_1, t_2, \cdots, t_N]$ and $\texttt{yStep} = [r_1, r_2, \cdots, r_N]$ can be used to describe the staircase required. The following MATLAB function can be written to describe such a block.

```
function [sys,x0,str,ts]=multi_step(t,x,u,flag,tTime,yStep)
switch flag,
case 0                                              % initialization
   sizes = simsizes;                                % read the template
   sizes.NumContStates=0; sizes.NumDiscStates=0;    % no state required
   sizes.NumOutputs=1; sizes.NumInputs=0;           % number of I/O
   sizes.DirFeedthrough=0; sizes.NumSampleTimes=1;
   sys=simsizes(sizes); x0=[]; str=[]; ts=[0 0];
case 3, i=find(tTime<=t); sys=yStep(i(end));        % evaluate output
case {1,2,4,9},  sys = [];                          % unused flags
otherwise, error(['Unhandled flag=',num2str(flag)]); % error handling
end
```

6.3.2   Application Example of S-functions: Simulation of ADRC Systems

The automatic disturbance rejection controller (ADRC) was proposed by the late Professor Jingqing Han and his collaborators [1], and it is a good example to demonstrate the design of S-functions and their applications. This approach may also provide a novel solution scheme for certain control applications.

Example 6.16   *Let us now consider a tracker-differentiator block, whose discrete-time mathematical model is written as*

$$\begin{cases} x_1(k+1) = x_1(k) + Tx_2(k) \\ x_2(k+1) = x_2(k) + T\mathrm{fst}(x_1(k), x_2(k), u(k), r, h), \end{cases} \tag{6.14}$$

where T is the sampling interval, u(k) is the input signal in its discrete-time form. Additional parameters r, h and T are used, where r determines the speed of signal tracking, and r is used to specify the filter effect. The nonlinear function fst(·) is defined as follows.

$$\delta = rh, \delta_0 = \delta h, y_0 = x_1 - u + hx_2, \quad a_0 = \sqrt{\delta^2 + 8r|y_0|}, \tag{6.15}$$

$$a = \begin{cases} x_2 + y_0/h, & |y_0| \leq \delta_0 \\ x_2 + 0.5(a_0 - \delta)\text{sign}(y_0), & |y_0| > \delta_0, \end{cases} \tag{6.16}$$

$$\text{fst} = \begin{cases} -ra/\delta, & |a| \leq \delta \\ -r\text{sign}(a), & |a| > \delta. \end{cases} \tag{6.17}$$

It can be seen that the nonlinear function cannot be created with existing low-level Simulink blocks. With the use of an S-function, the problem can be solved easily. In this system, there is one input signal, u(k), and two discrete-time states $x_1(k)$ and $x_2(k)$. We expect to have two output signals, such that $y_i(k) = x_i(k)$. The following S-function can be created to implement the above algorithm:

```
function [sys,x0,str,ts]=han_td(t,x,u,flag,r,h,T)
switch flag,
case 0, [sys,x0,str,ts] = mdlInitializeSizes(T);     % Initialization
case 2, sys = mdlUpdates(x,u,r,h,T);                 % update states
case 3, sys = x;                                     % evaluate outputs
case {1, 4, 9}, sys = [];                            % unused flags
otherwise, error(['Unhandled flag = ',num2str(flag)]); % error handling
end;
function [sys,x0,str,ts] = mdlInitializeSizes(T)     % initialization function
sizes = simsizes;              % get the parameter template
sizes.NumContStates = 0;       % no continuous state
sizes.NumDiscStates = 2;       % two discrete states
sizes.NumOutputs = 2;          % two output, follows the two states
sizes.NumInputs = 1;           % one input signal
sizes.DirFeedthrough = 0;      % state is not explicitly used in computing output
sizes.NumSampleTimes = 1;      % one sampling interval
sys = simsizes(sizes);         % set the new template
x0 = [0; 0];                   % set initial states to zeros
str = []; ts = [-1 0];         % other augments
% update the discrete-time state variables
function sys = mdlUpdates(x,u,r,h,T)    % discrete state equation
sys=[x(1)+T*x(2); x(2)+T*fst2(x,u,r,h)];
% subfunction fst2
function f=fst2(x,u,r,h)
delta=r*h; delta0=delta*h; y0=x(1)-u+h*x(2);
a0=sqrt(delta*delta+8*r*abs(y0));
if abs(y0)<=delta0, a=x(2)+y0/h;
else, a=x(2)+0.5*(a0-delta)*sign(y0); end
if abs(a)<=delta, f=-r*a/delta; else, f=-r*sign(a); end
```

A Simulink model established as shown in Fig. 6.12(a) is saved as ex_han.mdl. Double clicking the **S-function** *block, a dialog box shown in Fig. 6.12(b) will be displayed. We can specify* han_td *in the edit box labeled **S-function name**. A connection is then established to the S-function file* han_td.m. *In the edit box labeled **S-function parameters**, the three additional parameters r, h and T can be specified. For instance, we can assign them to r = 30, h = 0.01 and T = 0.01 in the MATLAB workspace, or set them dynamically with the* set *command*

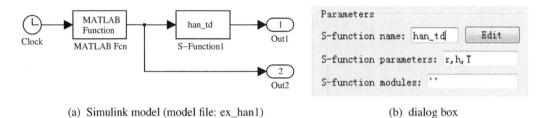

(a) Simulink model (model file: ex_han1)   (b) dialog box

Figure 6.12   Simulation of tracker-differentiation block.

```
>> set('ex_han','PreLoadFcn','r=30; h=0.01; T=0.01;');  % PreloadFcn setting
   save_system('ex_han.mdl');                            % save the system
```

For input signal generation purposes, an M-function can be used and the corresponding MATLAB function can be written as

```
function y=han_fun(x)
if x<=2*pi, y=sin(x);                   % sinusoidal wave in the first period
elseif x<=2.5*pi, y=2*(x-2*pi)/pi; %  followed by triangular waves
elseif x<=3.5*pi, y=1-2*(x-2.5*pi)/pi;
elseif x<=4*pi, y=-2+2*(x-3*pi)/pi; end
```

where in the first period, the input is assumed to be a sinusoidal wave, while in the second period a triangular wave is generated. The terminating time is assumed to be 4π. Simulation results can be obtained as shown in Fig. 6.13, with the plot(tout,yout) *command. And it can be seen that the tracking and differentiation signals from the block are satisfactory.*

In the tracking–differentiator, the parameter r determines the tracking speed. The larger the value of r, the faster the tracking. Meanwhile, the tracking error may also increase when r is large and the input signal is subject to random disturbances. A suitable h should be used to suppress the errors.

Example 6.17   *Professor Han also proposed the extended state observer (ESO) for use in ADRC systems. The mathematical expression of the ESO is*

$$\begin{cases} z_1(k+1) = z_1(k) + T[z_2(k) - \beta_{01}e(k)] \\ z_2(k+1) = z_2(k) + T[z_3(k) - \beta_{02}\mathrm{fal}(e(k), 1/2, \delta) + bu(k)] \\ z_3(k+1) = z_3(k) - T\beta_{03}\mathrm{fal}(e(k), 1/4, \delta), \end{cases} \quad (6.18)$$

where $e(k) = z_1(k) - y(k)$, and

$$\mathrm{fal}(e, a, \delta) = \begin{cases} e\delta^{a-1}, & |e| \leq \delta \\ |e|^a\mathrm{sign}(e), & |e| > \delta. \end{cases} \quad (6.19)$$

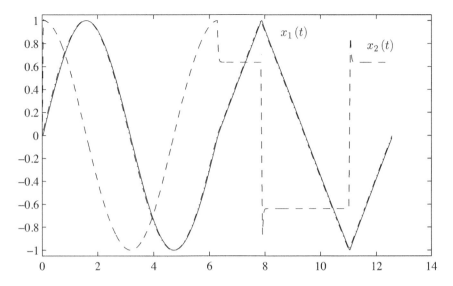

Figure 6.13   Tracking and differentiation of the input signal.

Similar to ordinary observers in control systems, the ESO block takes $u(k)$ and $y(k)$ as its input signals. From the mathematical model, the S-function can be written

```
function [sys,x0,str,ts]=han_eso(t,x,u,flag,a2,d,bet,b,T)
switch flag,
case 0, [sys,x0,str,ts] = mdlInitializeSizes;
case 2, sys = mdlUpdates(x,u,d,bet,b,T);
case 3, sys = x;
case {1, 4, 9}, sys = [];
otherwise, error(['Unhandled flag = ',num2str(flag)]);
end;
%    initialization function
function [sys,x0,str,ts] = mdlInitializeSizes
sizes = simsizes;
sizes.NumContStates = 0;   % no continuous states
sizes.NumDiscStates = 3;   % 3 discrete states
sizes.NumOutputs = 3;      % 3 output signals, that is, the states
sizes.NumInputs = 2;       % two inputs
sizes.DirFeedthrough = 0;  % states do not appear explicitly in output
sizes.NumSampleTimes = 1;
sys = simsizes(sizes);
x0 = [0; 0; 0]; str = []; ts = [-1 0]; % others
%    updates the discrete-time states
function sys = mdlUpdates(x,u,d,bet,b,T)
e=x(1)-u(2);
sys=[x(1)+T*(x(2)-bet(1)*e);
     x(2)+T*(x(3)-bet(2)*fal(e,0.5,d)+b*u(1));
     x(3)-T*bet(3)*fal(e,0.25,d)];
```

```
% subfunction fal
function f=fal(e,a,d)
if abs(e)<d, f=e*d∧(a-1); else, f=(abs(e))∧a*sign(e); end
```

An ADRC controller can be constructed based on the ideas of the tracking–differentiator and the ESO. The mathematical form of the ADRC controller is given by

$$\begin{cases} e_1 = v_1(k) - z_1(k), & e_2 = v_2(k) - z_2(k) \\ u_0 = \beta_1 \text{fal}(e_1, a_1, \delta_1) + \beta_2 \text{fal}(e_2, a_2, \delta_1) \\ u(k) = u_0 - z_3(k)/b, \end{cases} \tag{6.20}$$

where function fal(·) is the same as in (6.19). It can be seen that there are no states in the system. In the proposed block, the signals $v_i(k)$ and $z_i(k)$ are from the tracker-differentiator and the ESO block, respectively. They form the input of the block, such that $U(k) = [v_1(k), v_2(k), z_1(k), z_2(k), z_3(k)]$ and $u(k)$ is the output signal of the block. Since in the controller, there are a few additional parameters, the S-function should be written as

```
function [sys,x0,str,ts]=han_ctrl(t,x,u,flag,aa,bet1,b,d)
switch flag,
case 0, [sys,x0,str,ts] = mdlInitializeSizes(t,u,x);
case 3, sys = mdlOutputs(t,x,u,aa,bet1,b,d);
case {1,2,4,9}, sys = [];
otherwise, error(['Unhandled flag = ',num2str(flag)]);
end;
% Initialization
function [sys,x0,str,ts] = mdlInitializeSizes(t,u,x)
sizes = simsizes;
sizes.NumContStates = 0; sizes.NumDiscStates = 0;
sizes.NumOutputs = 1; sizes.NumInputs = 5;
sizes.DirFeedthrough = 1; sizes.NumSampleTimes = 1;
sys = simsizes(sizes); x0 = []; str = []; ts = [-1 0];
% output computation
function sys = mdlOutputs(t,x,u,aa,bet1,b,d)
e1=u(1)-u(3); e2=u(2)-u(4);
u0=bet1(1)*fal(e1,aa(1),d)+bet1(2)*fal(e2,aa(2),d); sys=u0-u(5)/b;
% common function
function f=fal(e,a,d)
if abs(e)<d, f=e*d∧(a-1); else, f=(abs(e))∧a*sign(e); end
```

■ **Example 6.18** *Assume that a time-varying plant model is given by*

$$\begin{cases} \dot{x}_1(t) = x_2(t) \\ \dot{x}_2(t) = \text{sign}(\sin t) + u(t). \end{cases}$$

A Simulink model can be constructed to implement the ADRC controller, as shown in Fig. 6.14, and the model can be saved in ex_han2.mdl *file. In this model, additional parameters d, $\boldsymbol{\beta}$, b and*

T can be set to the ESO block, and a, β_1, b, d can be set to the ADRC block. They can be stored in the PreloadFcn *with the following statements*

```
>> set_param('ex_han2','PreLoadFcn',['r=10; h=0.01; T=0.01; ',...
           'bet=[100,65,80]; bet1=[100,10]; aa=[0.75,1.25]; d=0; b=1;']);
   save_system('ex_han2.mdl');
```

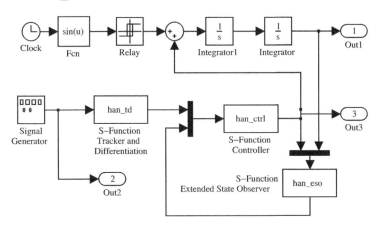

Figure 6.14   Simulink model of ADRC control (model name: ex_han2).

*Now, assume that the external input signal is set to square wave by using the **Signal Generator** block with a frequency of 0.1 Hz. The output of the plant model can be obtained as shown in Fig. 6.15(a), and the signal fed into the plant model is shown in Fig. 6.15(b). It can be seen that with the ADRC control strategy, the time-varying plant can be controlled satisfactorily.*

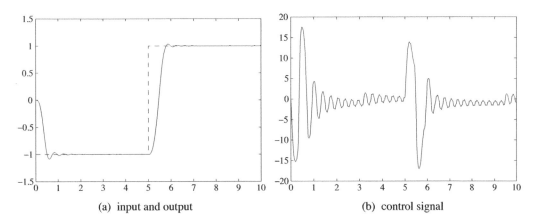

(a) input and output   (b) control signal

Figure 6.15   Control of the time-varying plant model.

One of the advantages of ADRC controllers is that they do not rely too much on the plant model, that is, the control strategy has certain robustness. If the plant model is changed to

$$\begin{cases} \dot{x}_1(t) = x_2(t) \\ \dot{x}_2(t) = 15\,\mathrm{Sat}(\cos t) + u(t), \end{cases}$$

*where Sat(·) is the saturation function, the Simulink model can then be established as shown in
Fig. 6.16. The output of the system is obtained as shown in Fig. 6.17. It can be seen that although
there are significant differences in the plant models, the control results are almost the same.*

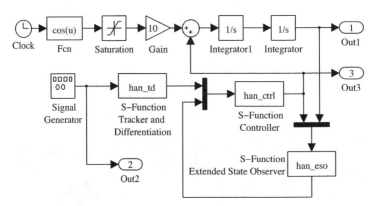

Figure 6.16   Simulink model with a different plant (model name: ex_han5).

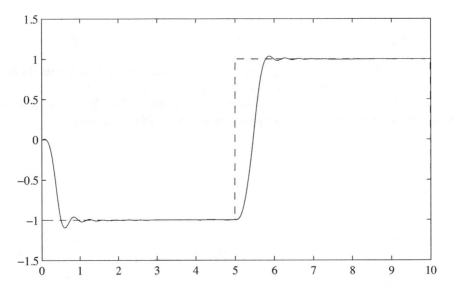

Figure 6.17   Output of the new plant model under the same ADRC controller.

It can be seen from the above example that the advantages of MATLAB and Simulink are
incorporated in the implementation of the whole system. With S-function implementation, the
modeling and simulation can be completed easily and straightforwardly. The ADRC controller
framework can be used to serve as an example in other simulation and design problems.

6.3.3   Level-2 S-function Programming

In the previous subsection, Level-1 S-functions were used. For modeling S-function blocks, the
Level-2 S-function approach can also be used. Object oriented programming methodology is adopted
in Level-2 S-function programming. It is suitable for representing blocks with multiple inputs and

outputs, and complex matrix output signal ports are also supported. Level-2 S-functions can also be started by following the template file msfuntmpl.m. In Level-2 S-functions, the `switch ··· case` structure is no longer used, so the user has to adapt to new program structures. Examples below will be given to demonstrate Level-2 S-function programming.

Example 6.19 *Consider again the continuous state space equation problem demonstrated in Example 6.14. If Level-2 structure is expected, the original S-function can be modified into the following form.*

```
function c6exsf1_l2(block)
setup(block);     % state space equation: main function
% initialization
function setup(block)
block.NumDialogPrms = 4;                           % number of additional parameters
block.NumInputPorts = 1; block.NumOutputPorts = 2;     % I/O port numbers
B=block.DialogPrm(2).Data; C=block.DialogPrm(3).Data;  % extract parameters
block.SetPreCompInpPortInfoToDynamic;                   % set ports to default
block.SetPreCompOutPortInfoToDynamic;
block.InputPort(1).Dimensions = size(B,2);              % number of inputs
block.InputPort(1).DirectFeedthrough = true;
block.OutputPort(1).Dimensions = size(C,1);        % number of output in port 1
block.OutputPort(2).Dimensions = size(B,1);        % number of output in port 2
block.SampleTimes = [0 0];
block.NumContStates = size(B,1);                    % number of continuous states
block.RegBlockMethod('SetInputPortSamplingMode',...
                     @SetInputPortSamplingMode);
block.RegBlockMethod('InitializeConditions ',    @InitConditions);
block.RegBlockMethod('Outputs',                  @Output);
block.RegBlockMethod('Derivatives',              @Derivatives);
% compute the output signals: port one for output, port 2 for states
function Output(block)
x=block.ContStates.Data;
C=block.DialogPrm(3).Data; D=block.DialogPrm(4).Data;
block.OutputPort(1).Data = C*x+D*block.InputPort(1).Data;
block.OutputPort(2).Data = x;
% sampling mode setting
function SetInputPortSamplingMode(block, port, sp)
block.InputPort(port).SamplingMode = sp;
block.OutputPort(1).SamplingMode = sp;
block.OutputPort(2).SamplingMode = sp;
% initialization function, and the initial state space vector is specified
function InitConditions(block)
A=block.DialogPrm(1).Data; block.ContStates.Data = zeros(length(A),1);
% continuous state update
function Derivatives(block)
x=block.ContStates.Data;
A=block.DialogPrm(1).Data; B=block.DialogPrm(2).Data;
block.Derivatives.Data=A*x+B*block.InputPort(1).Data;
```

There are certain similarities and differences in the S-functions designed in the two levels. For ordinary state space models, Level-1 modeling is much easier to program and the statements are quite straightforward. However, for special problems, such as when matrices are used as inputs, Level-2 functions should be written.

■ **Example 6.20**   *Consider the tracking–differentiator problem in Example 6.16. If Level-2 S-function style is expected, the following MATLAB function should be written*

```
function han_td_12(block)
setup(block); % Main program
% initialization program
function setup(block)
block.NumDialogPrms  = 3; block.NumInputPorts = 1;
block.NumOutputPorts = 1; block.NumContStates = 0;
block.SetPreCompInpPortInfoToDynamic;
block.SetPreCompOutPortInfoToDynamic;
block.InputPort(1).Dimensions = 1;
block.InputPort(1).DirectFeedthrough = false;
block.OutputPort(1).Dimensions = 2;
T = block.DialogPrm(3).Data; block.SampleTimes = [T 0]; % sampling interval
% define response functions
block.RegBlockMethod('PostPropagationSetup',    @DoPostPropSetup);
block.RegBlockMethod('InitializeConditions ',    @InitConditions);
block.RegBlockMethod('Outputs',                 @Output);
block.RegBlockMethod('Update',                  @Update);
% define discrete state variables
function DoPostPropSetup(block)
block.NumDworks = 1; block.Dwork(1).Name = 'x0';
block.Dwork(1).Dimensions = 2; block.Dwork(1).DatatypeID = 0;
block.Dwork(1).Complexity = 'Real';
block.Dwork(1).UsedAsDiscState = true;
% initial state variable setting
function InitConditions(block)
block.Dwork(1).Data = [0; 0];
% compute the output signal
function Output(block)
block.OutputPort(1).Data = block.Dwork(1).Data;
% state variable updates
function Update(block)
r = block.DialogPrm(1).Data; h = block.DialogPrm(2).Data;
T = block.DialogPrm(3).Data; u = block.InputPort(1).Data;
x = block.Dwork(1).Data;
block.Dwork(1).Data=[x(1)+T*x(2); x(2)+T*fst2(x,u,r,h)];
% common function, the same as the one in Level-1
function f=fst2(x,u,r,h)
delta=r*h; delta0=delta*h; y=x(1)-u+h*x(2);
a0=sqrt(delta*delta+8*r*abs(y));
if abs(y)<=delta0, a=x(2)+y/h;
else, a=x(2)+0.5*(a0-delta)*sign(y); end
if abs(a)<=delta, f=-r*a/delta; else, f=-r*sign(a); end
```

It should be noted that the discrete state variable is stored in the property `block.Dwork`, *and it should be defined in the* `DoPostPropSetup()` *method.*

6.3.4   Writing S-functions in C

Apart from MATLAB, S-functions can also be written in languages such as C, C++, Fortran and Ada. For instance, the S-function builder can be used to design the template in the C language. We can double click the **S-Function Builder** block in the **User-defined Functions** group in Simulink. The S-function builder is then opened in Fig. 6.18, and a C version of the S-function template can be generated automatically.

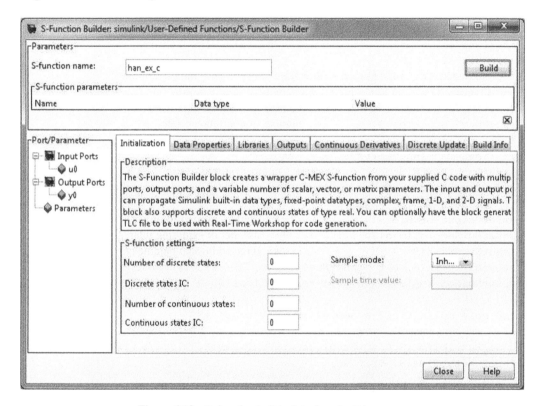

Figure 6.18   S-function builder interface for C language.

A large number of S-function examples are provided in Simulink. Open the **S-function demos** group, you will see that a large variety of examples can be used and referenced. Reading some of the examples will help in understanding the S-function programming fundamentals and tactics.

Example 6.21   *Consider again the tracking–differentiator example in (6.14). Using the C template provided in* `simulink\src\sfuntmpl_basic.c`, *the following C routine can be written, where a lot of lengthy comments and blank lines have been removed.*

```
#define S_FUNCTION_NAME  sfun_han /*  define the name of S-function */
#define S_FUNCTION_LEVEL 2          /*  Level-2 S-function */
#include "simstruc.h"
#include "math.h"      /*  mathematical computation library */
```

```c
double sign(double x)  /*  user-defined sign function */
{
    double f;
    f=1; if (x<=0) f=-1; return(f);
}
double fst(real_T *x,const real_T *u,const real_T *p1,const real_T *p2)
{   /*  fst function, similar to its MATLAB version */
    double delta, delta0, a0, a, y, r, h;
    r=p1[0]; h=p2[0]; delta=r*h; delta0=delta*h; y=x[0]-u[0]+h*x[1];
    a0=sqrt(delta*delta+8.0*r*fabs(y));  /*  absolute value function fabs */
    if (fabs(y) <= delta0) a=x[1]+y/h;
    else a=x[1]+0.5*(a0-delta)*sign(y);
    if (fabs(a) <= delta) return (-r*a/delta); else return(-r*sign(a));
}
static void mdlInitializeSizes(SimStruct *S)  /*  initialization */
{
    ssSetNumSFcnParams(S, 3);    /*  number of additional parameters */
    ssSetNumContStates(S, 0);    /*  number of continuous states */
    ssSetNumDiscStates(S, 2);    /*  number of discrete states */
    if (!ssSetNumInputPorts(S, 1)) return;
    ssSetInputPortWidth(S, 0, 1);  /*  number of inputs */
    ssSetInputPortDirectFeedThrough(S, 0, 0);  /*  input is not reflected */
    if (!ssSetNumOutputPorts(S, 1)) return;
    ssSetOutputPortWidth(S, 0, 2);    /*  output port definition */
    ssSetNumSampleTimes(S, 1);        /*  number of sampling intervals */
    ssSetNumRWork(S, 3);              /*  three additional parameters */
    ssSetNumIWork(S, 0); ssSetNumPWork(S, 0); ssSetNumModes(S, 0);
    ssSetNumNonsampledZCs(S, 0); ssSetOptions(S, 0);
}
static void mdlInitializeSampleTimes(SimStruct *S)
{   /*  sampling interval setting */
    ssSetSampleTime(S,0,*mxGetPr(ssGetSFcnParam(S,2)));  /*  extract T */
    ssSetOffsetTime(S, 0, 0.0);
}
static void mdlOutputs(SimStruct *S, int_T tid)
{   /*  output equation of the block */
    const real_T *x = ssGetRealDiscStates(S);  /*  use pointer */
    real_T       *y = ssGetOutputPortSignal(S,0);
    y[0] = x[0]; y[1] = x[1];    /*  write output equation */
}
static void mdlUpdate(SimStruct *S, int_T tid)
{   /*  discrete state updates */
    real_T       *x = ssGetRealDiscStates(S);
    const real_T *u = (const real_T*) ssGetInputPortSignal(S,0);
    const real_T *r = mxGetPr(ssGetSFcnParam(S,0));  /*  additional pars */
    const real_T *h = mxGetPr(ssGetSFcnParam(S,1));
    const real_T *T = mxGetPr(ssGetSFcnParam(S,2));
    real_T tempX[2] = {0.0, 0.0};  /*  temporary array assignment */
```

```
    tempX[0] = x[0] + T[0]*x[1];
    tempX[1] = x[1] + T[0]*fst(x,u,r,h);
    x[0] = tempX[0]; x[1] = tempX[1];
}
```

In the above C routine, the MATLAB command `mex` can be used to compile it and finally an executable file will be generated, with the suffix of mexw32. Before the first time the `mex` command is used, the environment variables should be specified with the command `mex -setup`.

Example 6.22 *For the above C version of S-function, the command* `mex sfun_fun.c` *should be used to compile the C code, and finally generate an executable file* `sfun_han.mexw32`. *The use of the function is exactly the same as the ones written in MATLAB.*

Similarly, the extended state observers and the ADRC controller can be expressed in S-functions in `sfun_eso.c` *and* `sfun_ctr.c`, *respectively. The listings of the functions are not given here. With the three new S-functions written and compiled in C, the Simulink model can be constructed as shown in Fig. 6.19. In fact, the block diagram thus constructed is exactly the same as the one we obtained earlier. We may compare the two Simulink models and it can be seen that the times elapsed for the two Simulink models are respectively 0.43 s and 0.10 s, which means that the C version is much faster, although the programming of it is much more difficult. Moreover, the C version of the system can be used in real time. Thus in a pure numerical simulation, the MATLAB version of S-functions – in particular, Level-1 S-function – is advised.*

```
>> tic, [t,y]=sim('ex_han2'); toc % S-function in MATLAB
   tic, [t,y]=sim('ex_han3'); toc % S-function in C
```

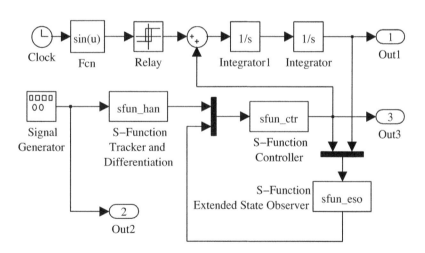

Figure 6.19   Simulink model of ADRC system with S-functions in C (model name: ex_han3).

6.3.5   *Masking an S-function Block*

It can be seen from the previous examples that the S-functions always come with additional parameters. If the additional parameters of the S-function can be linked to the edit boxes, its use can become easier and more straightforward. So, it is necessary to use masking facilities to encapsulate the S-function block into a masked block, and we can further design an appropriate dialog box for it.

Example 6.23   *Consider again the tracking–differentiator studied earlier. The S-function block was previously established as shown in Fig. 6.12(b). Right click its icon to get the shortcut menu, from which the **Edit → Mask S-function** item can be used to mask the S-function block. A dialog box shown in Fig. 6.20(a) is displayed, and the prompt for the block can be designed accordingly.*

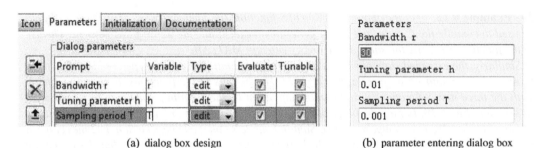

| (a) dialog box design | (b) parameter entering dialog box |

Figure 6.20   Masking the tracking–differentiator block (model name: han_td_m).

After the block is masked, double clicking the block again, a parameter entry dialog box shown in Fig. 6.20(b) will be displayed. The user can specify additional parameters easily, since each parameter is prompted.

Example 6.24   *Consider the staircase signal generator discussed in Example 6.15. There are two additional parameters defined and used in the block:* tStep *and* yStep. *The masked model can be created as shown in Fig. 6.21(a). Two vectors* tvec *and* uvec *are the parameters used in the dialog box. The icon drawing commands in Fig. 6.21(b) should be specified, and in the **Icon & Ports** column, the command* plot(x,y) *can be used to draw the model icon. Once the masking process is completed, the parameter dialog box shown in Fig. 6.21(c) can be obtained, if the block is double clicked. When the staircase data are entered into the block, the icon will be updated.*

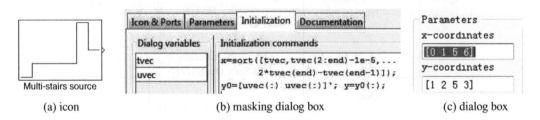

| (a) icon | (b) masking dialog box | (c) dialog box |

Figure 6.21   Masking of a staircase signal generator (model name: c6mstairs).

6.4   Examples of Optimization in Simulation: Optimal Controller Design Applications

"Optimal control" is so-called because it refers to achieving certain objectives under particular conditions or constraints; the aim is to define the control action such that a given criterion is minimized or maximized. Here, the criterion is the same as the objective function in optimization problems. In optimal control, the commonly used objectives can be assigned to integrals of error signals. Similar to optimization techniques, optimal control problems are also classified as being constrained or unconstrained. Calculus of variations is a useful tool for solving unconstrained optimal control problems [2, 3]. For some small-scale problems, analytical solutions may be found. Some constrained optimal control problems can be solved using Pontryagin's maximal principle. Since in the early days computers and powerful software were not widely available, indirect ways of solving optimal control problems were explored. Among these, linear quadratic optimal control problems received much attention in the control community where two weighting matrices Q and R were introduced and some beautiful mathematical formulae for the problem were derived. However, there is no widely accepted method for assigning these two weighting matrices. This makes the optimal criterion superficial, if not meaningless and even misleading.

A Simulink Design Optimization blockset is available in MATLAB. This blockset derives from the earlier Nonlinear Control Design (NCD) blockset and can be used in the design of optimal controllers for general nonlinear systems. Other optimization related problems, such as parameter fitting or estimation, are also solvable with this blockset. However, for optimal control problems, the information such as overshoot and settling time, as well as the shape of the response curve must be provided by the NCD user. It might be too difficult for an average user to provide this information so that the "optimal" controller can be designed. Also, the controller so designed may be too subjective.

With the popularity of powerful computer tools such as MATLAB, many optimal control problems can be converted into numerical optimization problems, and this type of problem can be immediately solved with the use of MATLAB. The optimal controller so designed may not have a beautiful mathematical derivation, but it is much more practical for engineers. We will use some examples to demonstrate optimal controller design problems and their solutions with the use of MATLAB and Simulink.

6.4.1 Optimal Criterion Selection for Servo Control Systems

For servo control problems, we should first address which kind of optimal criteria are meaningful and most practical. Consider a typical servo control system shown in Fig. 6.22. The objective of control is to make the output signal $y(t)$ follow the input signal $r(t)$ as closely as possible, that is, to make the tracking error $e(t) = r(t) - \hat{y}(t)$ as small as possible. Since the error signal $e(t)$ is a dynamic signal, one meaningful way to measure it is to use integral based criteria, for instance

$$J_{\text{ISE}} = \int_0^\infty e^2(t)\mathrm{d}t, \quad J_{\text{ITAE}} = \int_0^\infty t|e(t)|\mathrm{d}t. \tag{6.21}$$

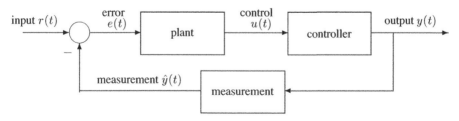

Figure 6.22 Block diagram of a typical servo control system.

Since the J_{ISE} criterion treats the error signal at any time instance equally, oscillation of the error signal is inevitable. To avoid unnecessary oscillation, a time weighted criterion may be more suitable, and the J_{ITAE} criterion may be a feasible solution to the weighting problem. The weighting function increases with time, thus it can force the error signal converge to zero as fast as possible. Thus the ITAE criterion is more meaningful than the traditional ISE one [4].

6.4.2   Objective Function Creation and Optimal Controller Design

For linear control systems, the ISE criterion can be systematically obtained by taking the $\mathcal{H}_2$ norm of the system, or by solving relevant Lyapunov equations. Thus this kind of problem can easily be solved, even without powerful computer tools like MATLAB. That is why ISE criterion based problems are widely available in the literature. The ITAE criterion can only be assessed by simulation methods. With Simulink, simulation results of a control system can be easily obtained. We will demonstrate the optimal controller design procedures through examples.

Example 6.25   *Consider a plant model $G(s) = e^{-2s}/(s+1)^5$. If we want to design an optimal PID controller, a Simulink model such as that shown in Fig. 6.23 should be created first. It can be seen that the system is composed of two parts: the bottom part describing the closed-loop PID control structure, and the upper part which generates the ITAE signal. The ITAE signal is fed into the first output port. If the termination time of the simulation is selected long enough, the last value of the output signal can be regarded as an approximation to the ITAE criterion. For our optimization problem, the decision variables are $x = [K_p, K_i, K_d]$, which are the parameters of the PID controller. Note that the Simulink model uses the PID controller block provided in MATLAB version 2010a onward. If an old version of MATLAB is used, the user should redraw the control model with a low-level PID controller and the saturation manually.*

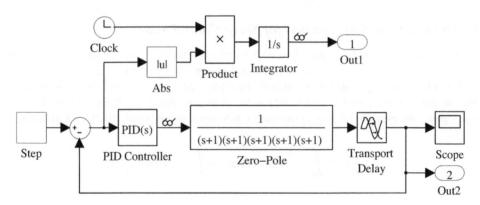

Figure 6.23   Simulink model with ITAE criterion (model name: c6moptim2).

Having constructed the Simulink model shown in Fig. 6.23, the objective function can be expressed by the following MATLAB function.

```
function y=c6foptim2(x)
assignin('base','Kp',x(1)); assignin('base','Ki',x(2));
assignin('base','Kd',x(3)); [t1,x1,y1]=sim('c6moptim2',[0,30]);
y=y1(end,1);
```

In the above, the function `assignin()` *is used to distribute the values of the decision variable* **x** *assigned into the corresponding variable in MATLAB so that the Simulink system can use those values directly. If we select the simulation termination time as 30 s, the* `sim()` *function can be used directly to invoke the simulation process.*

The following MATLAB statements can be used to start the optimization process. While the optimization is running, you can double click the **Scope** *block to visualize the optimization process. After optimization, the optimal PID controller parameters can be found as* $K_p = 0.7445$, $K_i = 0.1807$ *and* $K_d = 1.2229$. *The output signal is shown in Fig. 6.24(a) and the ITAE curve is obtained as shown in Fig. 6.24(b).*

```
>> x=fminunc(@c6foptim2,rand(3,1)), [t,xa,y]=sim('c6moptim2',[0,30]);
   plot(t,y(:,2)), figure,  plot(t,y(:,1))
```

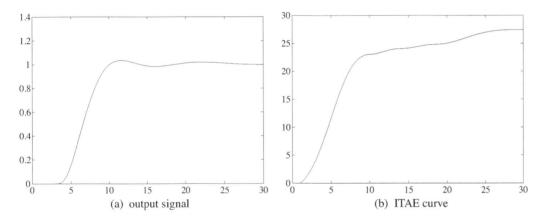

| (a) output signal | (b) ITAE curve |

Figure 6.24 Optimal control responses.

It can be seen from the ITAE curve that the curve settles down after $t > 20s$. *Thus the termination time selected is appropriate. In fact, it is not necessary to select a termination time very strictly. We can check the obtained ITAE curve and find the time instant* t_1 *when it is settled. Studies through examples show that if the termination time is selected in the interval* $(t_1, 2t_1)$, *the controller parameters and control results are almost the same [4]. However, if the termination time selected is too small, the steady-state control result may not be satisfied, while if it is too large, the transient time response might deteriorate.*

It can be seen also from the result that the initial control signal exceeds 100 at the initial time. In real applications, this may cause damage to the hardware. Thus a saturation element is usually appended to the PID controller to protect the system. If the saturation element is implemented, this may make the analysis and design more complicated.

From the previous optimal control design procedure, the algorithm used here is does not affected by whether or not there is any nonlinearity. So a saturation element can simply be inserted into the Simulink model, and the parameters of the saturation element set such that $|u(t)| \leq 10$. *Thus a new Simulink model can be drawn, as shown in Fig. 6.25(a). It should be noted that since the new version of Simulink provides a PID controller block with embedded saturation properties, we can double click the PID controller block and fill in the dialog box shown in Fig. 6.25(b) for the parameters in the saturation element.*

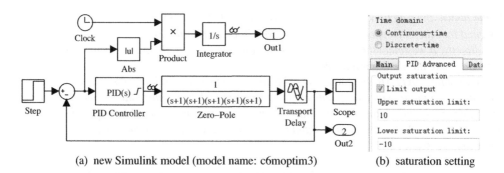

(a) new Simulink model (model name: c6moptim3)   (b) saturation setting

Figure 6.25   Simulink model for PID controller with saturation.

Similar to the above design procedure, we can rewrite the new objective function as

```
function y=c6foptim3(x)
assignin('base','Kp',x(1)); assignin('base','Ki',x(2));
assignin('base','Kd',x(3)); [t1,x1,y1]=sim('c6moptim3',[0,30]);
y=y1(end,1);
```

The following statements can be used to perform optimization again and the new controller parameters can be found as $K_p = 0.7146$, $K_i = 0.1534$ and $K_d = 1.1643$. The output signal of the system is shown in Fig. 6.26(a), and the ITAE signal is shown in Fig. 6.26(b). It can be seen that although the control signal is constrained within the range $(-10, 10)$, the behavior under the new controller is virtually the same as that obtained earlier.

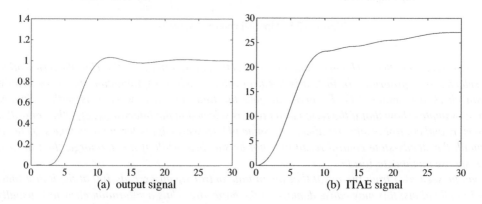

(a) output signal   (b) ITAE signal

Figure 6.26   Optimal control results for controller with saturation.

```
>> x=fminunc(@c6foptim3,x), [t,xa,y]=sim('c6moptim3',[0,30]);
   plot(t,y(:,2)), figure, plot(t,y(:,1))
```

Since the design method is rather systematic, graphical user interfaces can be designed for this kind of problem. For instance, the optimal controller design program (OCD) and the optimum PID controller design program (PID Optimizer), developed by the authors, are examples of these useful interfaces [5]. These interfaces can be downloaded from the service website for this book[1]. If you are interested, you can use these interfaces directly to solve your own controller design and optimization problems.

Example 6.26 *Consider again the DC motor control problem discussed in Example 4.4. In the control system, cascaded PI controllers are used. The inner loop PI controller is $K_{p1} + K_{i1}/s$, while the one for the outer loop is $K_{p2} + K_{i2}/s$. We can set the parameters of the two PI controllers as decision variables in the optimization process. The Simulink model can be established as shown in Fig. 6.27.*

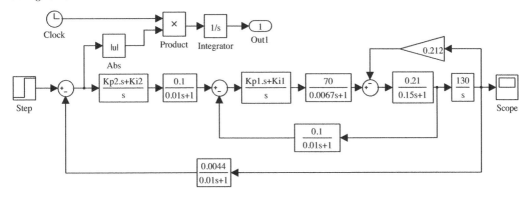

Figure 6.27 Simulink model of DC motor control system (model name: c6mmot1).

Selecting the termination time as 0.6 s, the objective function can be written as

```
function y=c6fmot1(x)
assignin('base','Kp1',x(1)); assignin('base','Ki1',x(2));
assignin('base','Kp2',x(3)); assignin('base','Ki2',x(2));
[t1,x1,y1]=sim('c6mmot1',[0,0.6]); y=y1(end,1);
```

The following optimization commands can be issued and the optimal controllers can be found, where the inner loop controller is $G_1(s) = 10.8489 + 0.9591/s$, while the outer loop controller is $G_2(s) = 37.9118 + 12.1855/s$. The optimal closed-loop system response can be obtained as shown in Fig. 6.28. It can be seen that the control behavior is satisfactory.

```
>> x=fminunc(@c6fmot1,rand(4,1)),
   [t,x1,y]=sim('c6mmot1',[0,0.6]); plot(t,y)
```

6.4.3 Global Optimization Approach

As pointed out earlier, conventional optimization approaches are executed by finding the optimum point from an initially assigned search point. If the initial search point is well selected, the optimization solution may be successful. However, if the initial point is not well chosen, a local minimum, rather than the global minimum point may be found. Global optimization methods should be introduced to solve optimization problems. Normally, the evolution based optimization algorithms tend to be more likely to find the global optimum solutions to the optimization problems. This class

---

[1]URL: http://mechatronics.ucmerced.edu/simubook2013wiley

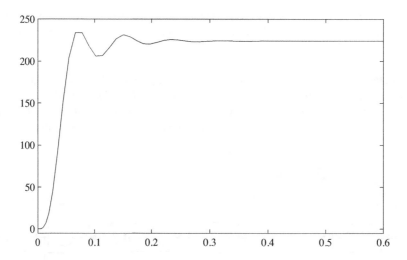

Figure 6.28   Optimal closed-loop step response of the DC motor control system.

of algorithms includes the genetic algorithm (GA) [6], the particle swarm optimization algorithm (PSO) [7] and some other third-party toolboxes [8, 9]. It may not be necessary for the average user to understand low-level details of the algorithms.

MATLAB provides a Global Optimization Toolbox (the earlier version is called the Genetic Algorithm and Direct Search Toolbox The main function is `ga()`, with the syntax

$$x = \texttt{ga}(\texttt{fun}, n, x_0, A, B, A_{\text{eq}}, B_{\text{eq}}, x_{\text{m}}, x_{\text{M}}, \texttt{funcons})$$

where n is the number of decision variables. The rest of the arguments are similar to the ones in the `fmincon()` function defined in Chapter 3.

An alternative MATLAB toolbox is the freely downloadable Genetic Algorithm Optimization Toolbox (GAOT) [10] and the syntax of the function is $x = \texttt{gaopt}(v, \texttt{fun})$, where `fun` is the user written objective function, $v = [x_{\text{m}}, x_{\text{M}}]$, with x_{m} and x_{M} the lower and upper bounds of the decision variables respectively.

Example 6.27   *Consider the following open-loop unstable plant model*

$$G(s) = \frac{s + 2}{s^4 + 8s^3 + 4s^2 - s + 0.4}.$$

The design objective is to maintain the control signal to satisfy $|u(t)| \leq 5$ and minimize the ITAE criterion. To design an optimal PID controller, we build the Simulink model shown in Fig. 6.29(a), where in the saturation block the condition $|u(t)| \leq 5$ has to be specified. An M-function can be written to describe the objective of control

```
function [x,y]=c6foptim5(x,opts)
assignin('base','Kp',x(1)); assignin('base','Ki',x(2));
assignin('base','Kd',x(3)); [t1,x1,y1]=sim('c6moptim4',[0,10]);
y=-y1(end,1);
```

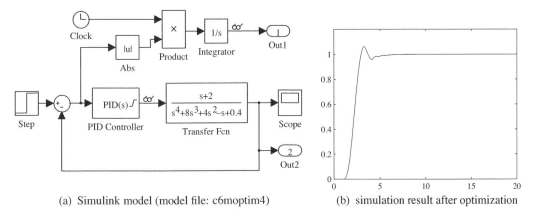

(a) Simulink model (model file: c6moptim4)   (b) simulation result after optimization

Figure 6.29   Global optimal PID controller design for an unstable plant.

It should be noted that if conventional optimization approaches are used, a proper selection of the initial search point should be given first. For systems with an unstable plant model, it might be very difficult to assign a proper initial search point. Thus the evolution methods have their advantages, since parallel searching methodology can be used. For instance, with the GAOT toolbox, the optimal PID controller can be found with the following MATLAB commands such that $K_p = 77.0612$, $K_i = 0.1632$ *and* $K_d = 87.1713$. *Under this controller, the closed-loop step response can be obtained as shown in Fig. 6.29(b). It can be seen that the behavior of the controller is satisfactory.*

```
>> x=gaopt([zeros(3,1) 100*ones(3,1)],'c6foptim5'),
   [t,xa,y]=sim('c6moptim4',[0,20]); plot(t,y(:,2))
```

Exercises

6.1   Create a Simulink model to describe the nonlinear system shown in Fig. 6.30, and find the operational point and the linearized model.

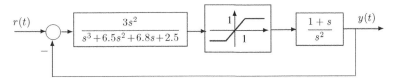

Figure 6.30   Problem (6.1).

6.2   It is known that the mathematical model of an inverted pendulum on a cart is given by

$$\begin{cases} \ddot{y} = \dfrac{f/m + l\theta^2 \sin\theta - g\sin\theta\cos\theta}{M/m + \sin^2\theta} \\ \ddot{\theta} = \dfrac{-f\cos\theta/m + (M+m)g\sin\theta/m - l\theta^2\sin\theta\cos\theta}{l(M/m + \sin^2\theta)}, \end{cases}$$

where θ is the angle between the bar and the vertical (in radians), y is the displacement of the cart (meters), f is the force (newtons), M and m are the masses of the cart and bar respectively (kg). Also, l is half of the bar length (meters) and g is the gravity acceleration constant (9.81 m/s^2). Establish the system model in Simulink. If we have $m = 0.21$ kg, $M = 0.455$ kg and $l = 0.61/2$ m, and assuming that f is the input signal of the system, try to find the linearized model around the equilibrium point of $y = \theta = 0$.

6.3 Consider the block diagram of a control system shown in Fig. 6.31. Find the equivalent transfer function and state space model from the input port $r(t)$ to the output port $y(t)$.

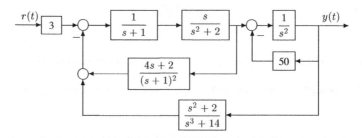

Figure 6.31 Problem (6.3).

6.4 Assume that the open-loop transfer function of a system is given by $G(s) = 1/(s^3 + a_1 s^2 + a_2 s + a_3)$, and the system is in a unity negative feedback structure, as shown in Fig. 6.32. The input signal is a unit step function. Selecting ITAE and ISE criteria, find the parameters a_1, a_2 and a_3 that minimize the criteria, respectively.

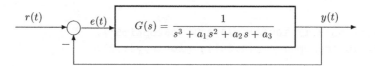

Figure 6.32 Problem (6.4).

Similarly, for high-order system

$$G(s) = \frac{1}{s^n + a_1 s^{n-1} + \cdots + a_{n-1} s + a_n},$$

find the coefficients for $n = 1, 2, 4, 5, 6$ that minimize the two criteria.

6.5 It has been shown that the tracking–differentiator can be used to track and differentiate the original input signal. If the original signal is corrupted by high-frequency noise, assess the behavior of the tracking–differentiator, and experiment with how to select appropriate r and h parameters.

6.6 Rewrite, in Level-2 style, the S-functions of the extended state observer and ADRC controller, and validate them.

6.7 Write S-functions in C for the extended state observer and the ADRC controller. So far, we have a set of six S-function blocks, three with MATLAB and another three with C. Mask these blocks and then establish a group to hold all the six blocks, such that they become reusable.

6.8 Assume that the piecewise linear nonlinearity is defined as $y(x) = k_i x + b_i$, where in the ith segment, $e_i \leq x < e_{i+1}$. If the boundaries $e_1, e_2, \cdots, e_{N+1}$ and the slopes and offsets

$k_1, b_1, \cdots, k_N, b_N$ are specified, model the nonlinearity with an S-function, and then mask the block.

6.9 Assume that a programmable logic device (PLD) has six inputs named A, B, W_1, W_2, W_3, W_4, where W_i are encoding signals. The combination of these inputs leads to the output signal Y, following the truth table shown in Table 6.3 [11]. Write an S-function to model this PLD block.

Table 6.3 Logic relation table of Problem (6.9).

W_1	W_2	W_3	W_4	Y	W_1	W_2	W_3	W_4	Y
0	0	0	0	0	1	0	0	0	$A\overline{B}$
0	0	0	1	AB	1	0	0	1	A
0	0	1	0	$\overline{A+B}$	1	0	1	0	$\overline{B}$
0	0	1	1	$AB+\overline{AB}=A\odot B$	1	0	1	1	$A+\overline{B}$
0	1	0	0	$\overline{A}B$	1	1	0	0	$\overline{A}B+A\overline{B}=A\oplus B$
0	1	0	1	B	1	1	0	1	$A+B$
0	1	1	0	$\overline{A}$	1	1	1	0	$\overline{A}+\overline{B}=\overline{AB}$
0	1	1	1	$\overline{A}+B$	1	1	1	1	1

6.10 For the unstable second-order plant $1/s(s-1)$, find out whether the ADRC controller is still able to maintain a good control behavior. If the plant model is changed to a third-order plant $1/(s^3+3s^2+2s+4)$, check whether the ADRC controller still works.

6.11 For the system demonstrated in Example 5.34, consider further the masking of the state space model. If in the masking process a checkbox is used to select whether the state variable output is needed, write an S-function in C to redefine the block.

6.12 For a plant model with a large time delay $G(s) = 10e^{-20s}/(2s+1)$, check whether ADRC controller can be used directly. If it cannot be used, try to design a better controller.

6.13 Check whether the inverted pendulum can be controlled with the ADRC framework.

6.14 The staircase signal generator was discussed and implemented in Example 6.15. You can use such a signal generator to perform a simulation on the ADRC examples and check the control effect.

6.15 For a plant model given by $G(s) = 100e^{-10s}/[s(s+10)(s+20)(s+30)]$, find the optimum PD controller with ITAE criterion. If the delay constant is changed to 8 s in the plant model, check whether the PD controller still works.

6.16 Determine the optimal PID controller for the complicated plant [12] under ISE and ITAE criteria

$$G(s) = \frac{1 + \dfrac{3e^{-s}}{s+1}}{s+1}.$$

6.17 For a time-varying plant described by the following differential equation

$$\ddot{y}(t) + e^{-0.2t}\dot{y}(t) + e^{-5t}\sin(2t+6)y(t) = u(t),$$

try to design an optimal PI controller that minimizes the ITAE criterion and assess the closed-loop behavior of the system. Analyze through examples the effect of the termination time on the optimization. If other criteria such as ISE and IAE are used, what results can be achieved.

6.18 Assume that a plant model is given by $G(s) = 1/(s+1)^6$. Try to find an optimal reduced model $G_r(s) = e^{-\tau s}/(Ts+1)$ with Simulink, and compare the reduced model with the original model in terms of least squares of the differences in open-loop step responses.

6.19 Design globally optimal PID controllers for the following plants using the genetic algorithm

(a) non-minimal phase model $G(s) = \dfrac{-s+5}{s^3 + 4s^2 + 5s + 6}$;

(b) unstable non-minimal phase model $G(s) = \dfrac{-0.2s + 5}{s^4 + 3s^3 + 5s^2 - 6s + 9}$;

(c) unstable discrete-time model $H(z) = \dfrac{4z - 2}{z^4 + 2.9z^3 + 2.4z^2 + 1.4z + 0.4}$.

References

[1] J Q Han, L L Yuan. Discrete representations of tracking-differentiators. System Science and Mathematics, 1999, 19(3):268–273. In Chinese

[2] S F Cai. Automatic control principles. Beijing: China Machine Press, 1980. In Chinese

[3] X K Xie. Fundamentals of modern control systems. Shenyang: Liaoning People's Publisher, 1980. In Chinese

[4] D Xue. Computer aided design of control systems – MATLAB languages and applications (2nd Edition). Beijing: Tsinghua University Press, 2006. In Chinese

[5] D Xue. Computer aided control systems design with MATLAB (3rd Edition). Beijing: Tsinghua University Press, 2012. In Chinese

[6] D E Goldberg. Genetic algorithms in search, optimzation and machine learning. Reading, MA: Addison-Wesley, 1989

[7] B Birge. PSOt, a particle swarm optimization toolbox for MATLAB. Proceedings of the 2003 IEEE Swarm Intelligence Symposium. Indianapolis, 2003, 182–186

[8] D Xue, Y Q Chen. Solving applied mathematical problems with MATLAB. Boca Raton: CRC Press, 2008

[9] The MathWorks Inc. Global Optimization Toolbox User's Guide, 2010

[10] C R Houck, J A Joines, M G Kay. A genetic algorithm for function optimization: a MATLAB implementation. 1995

[11] R X Peng. Fundamentals of digital electronics. Wuhan: Wuhan University of Technology Press, 2001. In Chinese

[12] C Brosilow, B Joseph. Techniques of model-based control. Englewood Cliffs: Prentice Hall, 2002

7

Modeling and Simulation of Engineering Systems

For engineering systems, the modeling approaches presented in the previous chapters can of course be used. For a normal system, if the mathematical models of the engineering systems are established, Simulink models can easily be constructed since low-level Simulink blocks can be used to model systems with arbitrary complexity into simulation models. For complicated engineering systems, the low-level modeling techniques sometimes become very complicated, if not impossible. With solid knowledge and theoretical background in the field, the user may be able to generate the model needed. If some part of the whole simulation system is not well handled, this may cause inaccuracy or even significant errors in the results. Meanwhile, since the low-level models are too complicated in structure, the model constructed may be too complicated to diagnose or to maintain. Thus for many practical engineering systems, low-level modeling is not a good choice.

The multi-domain physical modeling strategy advocated and implemented in Simulink provides a new methodology for modeling and simulation of engineering and non-engineering systems. Many well-established and specialized blocksets in various disciplines have been developed. Some of the blocksets have been developed in the Simscape framework, and a Simscape language is also provided. The main idea of the multi-domain physical modeling is that a series of blocks or components are masked, which allows the user to assemble Simulink models with these building blocks just as if they were assembling the actual hardware systems. The physical model can be constructed in Simulink in this way, and the mathematical model can be automatically generated with the physical model, such that the facilities in Simulink can be used to simulate and analyze the constructed system.

Since the model is established using building block methods, it is very easy to examine and validate the model. Also, since the well-established blocks can be used directly to assemble the system models, solid background knowledge is no longer necessary. For instance, an electrical engineer can easily establish a mechanical model without reading too much material on mechanical engineering, and the model constructed is much more reliable than one they could establish themselves using low-level modeling methods. In this way, interdisciplinary modeling is possible. More importantly, the model can be established in the same Simulink framework. From the simulation point of view, this modeling strategy has its own advantages.

In Section 7.1, the concept of multi-domain physical modeling is presented, and an introduction to the Simscape blockset is also given. In Section 7.2, electrical system modeling with SimPowerSystems and other blocksets is addressed. In Sections 7.3~7.5, modeling and simulation of electronic

System Simulation Techniques with MATLAB® and Simulink®, First Edition. Dingyü Xue and YangQuan Chen.
© 2014 John Wiley & Sons, Ltd. Published 2014 by John Wiley & Sons, Ltd.

systems, motor drive systems and mechanical systems are presented and a lot of examples are provided.

7.1 Physical System Modeling with Simscape

7.1.1 Limitations of Conventional Modeling Methodology

All the modeling and simulation problems presented earlier were made on the assumption that the system models were known. These models can be established with traditional modeling procedures, where physical laws can be used to write mathematical equations. For instance, Kirchhoff's law can be used to write circuit equations, while Newton's laws can be used to establish system models for mechanical systems. With the mathematical equations, Simulink models can be established, and the models can then be used in simulation processes.

In the modeling of practical systems, background knowledge is required. If the user only has limited background knowledge, the model established may not be good and the reliability of the model not high. Also, models based on physical laws may neglect some of the less important factors, but in actual situations, when these factors can not be neglected, modeling errors will occur, and this will sometimes produce erroneous results.

■ **Example 7.1** *Consider the simple RLC circuit shown in Fig. 7.1. Three electrical current equations can be written for the three loops. Since under Laplace transform, the capacitor and inductor can be represented as an integrator and a differentiator, and the three equations can be written as [1]*

$$\begin{cases} (2s+2)I_1(s) - (2s+1)I_2(s) - I_3(s) = V(s) \\ -(2s+1)I_1(s) + (9s+1)I_2(s) - 4s\,I_3(s) = 0 \\ -I_1(s) - 4s\,I_2(s) + (4s+1+1/s)I_3(s) = 0. \end{cases} \tag{7.1}$$

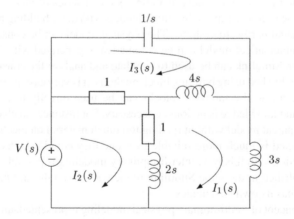

Figure 7.1 A simple circuit.

If the input signal is an AC voltage with a magnitude of 220 V and a frequency of 50 Hz, then the mathematical expression is $v(t) = 220\sin 100\pi t$. Since the original system is linear, the current signals are all sinusoidal waves with a frequency of 50 Hz, while the magnitude and initial phase *are different from voltage v(t). However, these conclusions should be validated by other methods. Of course, in this example, only three equations are involved, so it is very easy to find the solutions with MATLAB. In practice, some circuits may have hundreds of loops, if not more, and hundreds of equations need to be written, and the solution of them could be extremely difficult, if not impossible. Meanwhile, if one loop is neglected due to carelessness, the simulation result may be wrong. Better modeling and simulation methodology is needed.*

Example 7.2 *Consider the spring–damper system shown in Fig. 7.2, where x(t) is the displacement of the sliding block, and f(t) is the external force. In the system, the resistance of the damper is proportional to the speed of the sliding block. From Newton's second law, the mathematical model can be established*

$$M\ddot{x}(t) + f_{\mathrm{v}}\dot{x}(t) + Kx(t) = f(t). \tag{7.2}$$

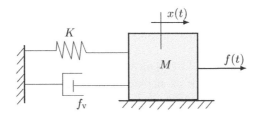

Figure 7.2   Spring–damper system.

Simple systems can easily be derived from Newton's laws, but for systems with many springs and dampers, the modeling could be rather complicated. Again a more powerful modeling and simulation tool is needed for handling complicated mechanical systems. Simulink based modeling of such systems will be presented later in the chapter.

New modeling methodology needs to be adopted for complicated engineering systems. With the new multi-domain physical modeling approach provided with Simulink, the Simscape blockset and other related professional blocksets can be used. These blocksets allow the user to establish physical simulation models in the way that the hardware systems are assembled, component by component, with building blocks. The Simulink model will then be generated automatically, and the whole system can be simulated using the Simulink platform. This kind of simulation cannot currently be performed with other computer languages and software.

7.1.2   Introduction to Simscape

Simscape is a new object oriented multi-domain physical modeling program to work with Simulink. We can issue the `simscape` command in the MATLAB command window, or invoke Simscape from the Simulink model library. The contents of the Simscape blockset is shown in Fig. 7.3.

The Simscape blockset contains the **Foundation Library** composed of basis blocks in electrical, mechanical, magnetic, thermal, hydraulic and pneumatic engineering. More professional blocksets, including SimElectronics, SimDriveline, SimMechanics, SimHydraulics and SimPowerSystems are also provided. The objective of these blocksets is to provide a series of component blocks, and allow

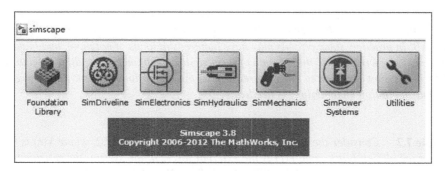

Figure 7.3   Simscape blockset.

users to construct building block simulation models for systems, just like assembling hardware systems. The mathematical model can then be generated automatically with the building blocks of the simulation model. Simscape and its blocksets are aids to Simulink for multi-domain physical modeling. In system modeling, the programmer is no longer required to have a complete background knowledge to the related fields. Users may be able to construct simulation models even in an area they are not very familiar with.

The Simscape language was released by MathWorks allowing users to define new components with syntaxes similar to MATLAB itself, in an object oriented way. This enriches the capability of Simulink for multi-domain physical modeling and simulation.

7.1.3   Overview of Simscape Foundation Library

On double clicking the icon **Foundation Library** shown in Fig. 7.3, the Simscape Foundation Library shown in Fig. 7.4 will be displayed. In this library, several groups are provided, such as **Electrical**, **Mechanical**, **Hydraulic**, **Pneumatic**, **Magnetic** and **Thermal** groups. Also a group called **Physical Signals** can be used to convert different signal types.

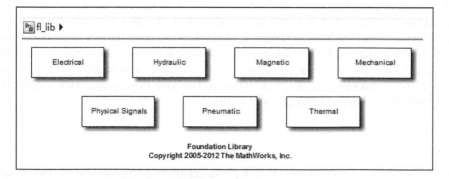

Figure 7.4   Simscape Foundation Library.

Double clicking the **Electrical** group from the library, will show three subgroups, **Electrical Elements**, **Electrical Sources** and **Electrical Sensors**; the **Electrical Elements** subgroup is shown in Fig. 7.5. It can be seen that the commonly used electrical components such as **Resistor, Inductor, Capacitor, Mutual Inductor, Ideal Transformer** and **Variable Resistor** are all provided as blocks.

Electronic devices such as **Diode**, **Op-amp** and **Gyrator** are also provided as blocks. Moreover, the blocks **Rotational Electromechanical Converter** and **Translational Electromechanical Converter** are also given in the subgroup. Each group of components in the library also provides a reference point block. In electrical blocks, the reference point is described by an **Electrical Reference** block.

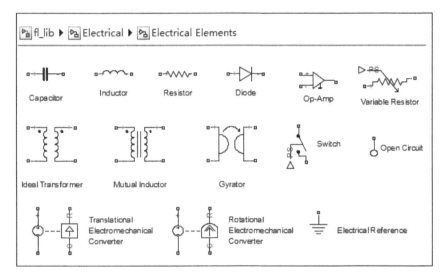

Figure 7.5   Electric element group in the Simscape Foundation Library.

Simscape Foundation Library also includes the following important groups and blocks

- **Mechanical** group: In this group, five subgroups are provided, as shown in Fig. 7.6. The subgroups are **Mechanisms**, **Translational Elements**, **Rotational Elements**, **Mechanical Sources** and **Mechanical Sensors**.

Figure 7.6   Mechanical element group in the Simscape Foundation Library.

- **Translational Elements** group is shown in Fig. 7.7, where **Mass**, **Translational Friction**, **Translational Damper**, **Translational Spring** and **Translational Hard Stop** blocks are provided. As in other element groups, also provided in the subgroup are **Mechanical Translational**

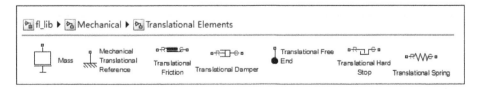

Figure 7.7   **Translational Elements** group.

Reference. It can be seen that the system given in Example 7.2 can be composed directly with the blocks in this subgroup.
- **Rotational Elements** group includes blocks such as **Inertia, Rotational Spring, Rotational Friction, Rotational Damper** and **Rotational Hard Stop**.
- **Mechanism** group includes **Gear Box, Lever, Wheel and Axle**.
- **Mechanical Sources** group includes blocks such as **Ideal Force Source, Ideal Torque Source, Ideal Angular Velocity Source** and **Ideal Translational Velocity Source**.
- **Mechanical Sources** and **Mechanical Sensors** blocks include **Ideal Force Sensor, Ideal Rotational Motion Sensor, Ideal Torque Sensor** and **Ideal Translational Motion Sensor**.

- **Magnetic** group has three subgroups: **Magnetic Elements, Magnetic Sources** and **Magnetic Sensors**. In the **Magnetic Elements** subgroup, the blocks **Reluctance, Electromagnetic Converter, Variable Reluctance** and **Reluctance Force Actuator** blocks are provided. The **Magnetic Sources** subgroup has **Flux Source, MMF Source, Controlled Flux Source** and **Controlled MMF Source**. Included in **Magnetic Sensors** are **Flux Sensor** and **MMF Sensor**.

- **Thermal** group contains the subgroups **Heat Elements, Thermal Sources** and **Thermal Sensors**. The following blocks are provided, **Conductive Heat Transfer, Convective Heat Transfer, Radiative Heat Transfer, Thermal Mass, Ideal Heat Flow Source, Ideal Temperature Source, Ideal Heat Flow Sensor** and **Ideal Temperature Sensor**.

- **Hydraulic** group includes the subgroups **Hydraulic Elements, Hydraulic Sources, Hydraulic Sensors** and **Hydraulic Utilities**. In **Hydraulic Elements**, are **Constant Area Hydraulic Orifice, Variable Area Hydraulic Orifice, Hydraulic Resistive Tube, Linear Hydraulic Resistance, Translational Hydro-mechanical Converter, Rotational Hydro-mechanical Converter, Variable Hydraulic Chamber, Constant Volume Hydraulic Chamber** and **Hydraulic Piston Chamber**. The **Hydraulic Sources** and **Hydraulic Sensors** subgroups provide **Hydraulic Pressure Source, Hydraulic Flow Rate Source, Hydraulic Flow Rate Sensor** and **Hydraulic Pressure Sensor**.

- **Pneumatic** group includes the blocks **Constant Area Pneumatic Orifice, Variable Area Pneumatic Orifice, Adiabatic Cup, Pneumatic Resistive Tube, Pneumatic–Mechanical Converter, Pneumatic Chamber, Rotational Piston Chamber, Pneumatic Pressure Source, Pneumatic Flow Rate Source, Pneumatic Mass & Heat Flow Sensor** and **Pneumatic Pressure & Temperature Sensor**.

7.1.4   Conversions of Two Types of Signals

With the physical modeling tools, two types of signals are supported. One is the typical Simulink signal, the other is the physical signal, labeled PS. These two types of signals coexist in physical models. However, since their definitions are different, they cannot be connected to each other. Relevant conversions should be made before they can be connected to each other.

By double clicking the **Utilities** icon in the main window of Simscape, the auxiliary group can be displayed as shown in Fig. 7.8(a), where conversion blocks such as **PS-Simulink Converter** and **Simulink-PS Converter** are provided. Also, the simulation parameter setting block **Solver Configuration, Two-Way Connection** and **Connection Port** are provided. The **Solver Configuration** block must be used in Simulink modeling.

A blank model window can, of course, be created with the **File → New → Model** menu, but the essential blocks in the Simscape Foundation Library should then be added manually, and it might be complicated to do so. Alternatively, the command `ssc_new` could be used instead, and the Simscape

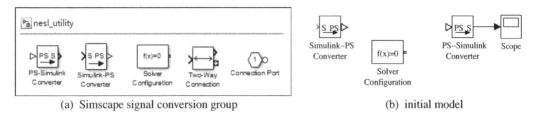

(a) Simscape signal conversion group   (b) initial model

Figure 7.8   Simscape initial model window.

library will be opened automatically. Meanwhile, an initial model window shown in Fig. 7.8(b) will also be created. Users can start modeling tasks with this model.

Example 7.3   *Consider again the circuit studied in Example 7.1. If the voltage across the capacitor, and the current I_1 are needed, a voltmeter and an ammeter should also be added, as shown in Fig. 7.9(a). To establish a Simulink model for the circuit, the command* ssc_new *should be used first to open a new model with the essential blocks, then the circuit can be modeled with relevant blocks as shown in Fig. 7.9(b). The parameters of the blocks should be assigned according to the original circuit.*

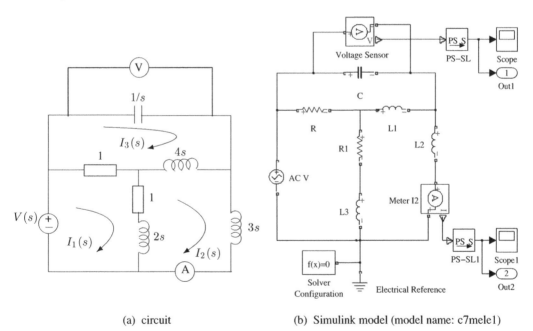

(a) circuit   (b) Simulink model (model name: c7mele1)

Figure 7.9   Circuit and its Simulink model.

Double clicking the capacitor block, the dialog box shown in Fig. 7.10(a) will be opened. Besides the capacitor parameter, other parameters can also be specified, such as the **Series resistance**, **Parallel conductance** *and* **Initial voltage**. *Clicking the hyperlink named* **View source for Capacitor**, *a model edit window shown in Fig. 7.10(b) will be opened. In this edit window, the model is written in Simscape language to describe the capacitor model. The main parameters in the Simscape program*

will be given, and they are exactly the same as they appeared in the dialog box of Fig. 7.10(a). The
values and units are also given in the code.

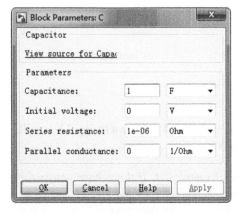

```
component capacitor<foundation.electrical.branch
  parameters
    c  = { 1e-6, 'F' };   % Capacitance
    v0 = { 0,   'V' };    % Initial voltage
    r  = { 1e-6, 'Ohm' }; % Series resistance
    g  = { 0, '1/Ohm' };  % Parallel conductance
  end
  variables
    vc = { 0, 'V' }; % Internal variable
  end
  equations
    v == i*r + vc;
    i == c*vc.der + g*vc;
  end
end
```

(a) capacitor parameter dialog box   (b) Simscape language model

Figure 7.10   Capacitor model and Simscape language file.

After simulation, the voltage u(t) across the capacitor can be obtained. Also, with symbolic com-
putation from (7.1), the analytical solution of the signal can also be obtained, as shown in Fig. 7.11.

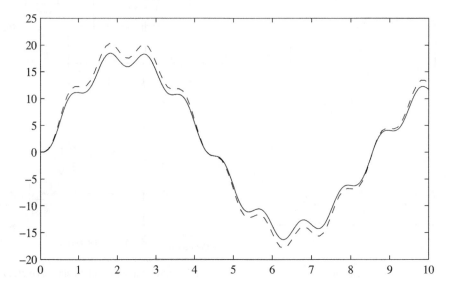

Figure 7.11   Voltage signal via numerical and analytical solutions.

```
>> syms t s;
   A=[2*s+2,-(2*s+1),-1; -(2*s+1),9*s+1,-4*s; -1,-4*s,4*s+1+1/s];
   xx=inv(A)*[laplace(220*sin(2*pi*t));0;0]; U=xx(3)/s; I2=xx(2);
   u=ilaplace(U); t2=0:0.1:10; y2=double(subs(u,t,t2));
   [t1,x,y]=sim('c7mele1',[0,10]); plot(t2,y2,t1,y(:,1),'--')
```

In fact, since the symbolic computation capability in new versions of MATLAB is very weak, the `double()` *command is quite time-consuming. MATLAB R2008a is used to perform analytical solution and the result is transferred back to MATLAB R2011a to draw the plots. In this problem, only three equations are involved, since the circuit is very simple. For complicated circuits, analytical solutions are not possible. A numerical simulation becomes the only feasible way to solve such problems.*

From the capacitor described by Simscape language, it can be seen that the equivalent circuit is shown in Fig. 7.12(a), where r is the **series resistance** and g is the **parallel conductance**. In ideal cases, $r = g = 0$. In some cases, the parameters cannot be neglected. A more exact model can then be used to describe the capacitors with equivalent circuits. The equivalent circuit for an inductor can be expressed as in Fig. 7.12(b), where the series resistance r and parallel equivalent conductance g are used.

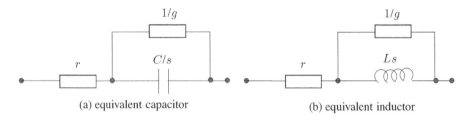

(a) equivalent capacitor   (b) equivalent inductor

Figure 7.12   Equivalent circuits of capacitor and inductor.

From the above example, it can be seen that there are differences in the two curves. The capacitor used in simulation is closer to the practical capacitor element, while in the analytical solution, the model itself is an approximate model, since ideal capacitors and inductors are used.

7.1.5   Brief Description of the Simscape Language

The Simscape language is a new object oriented computer language used in physical modeling. New components can be modeled with this language. The suffix of a Simscape program is .ssc, and the file should be placed in a folder starting with a + sign under a MATLAB search path. The capacitor model in Simscape is given below

```
component capacitor < foundation.electrical.branch % leading sentence
    parameters    % values and units in variables as in the dialog box
        c  =  1e-6, 'F'  ;   % capacitance
        v0 =  0, 'V' ;        % initial voltage
        r  =  1e-6, 'Ohm'  ; % series resistance
        g  =  0, '1/Ohm' ;   % parallel conductance
    end
    variables
        vc=0,'V'; % voltage and unit across the capacitor
    end
    function setup % initial values setup
        if C <= 0
            pm_error('simscape:GreaterThanZero','Capacitance')
        end
```

```
      ... % other detection statements omitted here
      vc = v0; % internal variable for the initial value assignment
   end
   equations
      v == i*r + vc; i == C*vc.der + g*vc;
   end
end
```

The models described by Simscape can be written using a number of modules.

- **Leading statement module**: The block is headed by the keyword `component`, and the group that this block belongs to is also described in this first sentence. The first sentence of a Simscape model can also be headed with the keyword `domain`, where a domain is defined. A domain is similar to a model group.

- **Dialog parameter setting module**: After the leading statement, the `parameters` module can be used to define the dialog box for the block. The value and unit of the parameters are described in this module. For instance, in the capacitor block, four parameters are defined, thus the dialog box appears as shown in Fig. 7.10(a). Within the dialog box, four parameters and their units are prompted.

- **Variable declaration module**: The `variables` keyword has to lead a sentence that defines the key variables in the block. In the above code, the voltage across the capacitor is denoted by v_c.

- **Initial setting module and parameter setting module**: This module is led by `function setup`. The value of C is examined to see whether it is positive or not. If it is not, an error message will be given. Apart from the validation of C, other parameters are also validated, but they are not listed in the above code. Unfortunately, the error messages are not given when the data error occurs in the parameter input, but only when the simulation is invoked. Initial values of v_c can also be assigned in this module.

- **Mathematical model description**: This module is led by the keyword `equations`, and the capacitor block is defined as

$$v = ir + v_c, i = C\frac{dv_c}{dt} + gv_c. \tag{7.3}$$

7.1.6 Modeling and Simulation of Complicated Electrical Network

As indicated earlier, with the use of Simscape's Foundation Library electrical circuits can easily be modeled and simulated. A MATLAB command can also be used to define complicated Simulink models. Here the modeling of a complicated circuit is demonstrated as an example.

▣ **Example 7.4** *Consider the resistance network shown in Fig. 7.13. If the equivalent resistance from port A to port B is needed, the relationship between voltage and current can be established, and linearization can be performed. The network is quite complicated, so it might be difficult to draw the circuit using low-level components manually. Therefore the following Simulink statements can be used instead to create the Simulink model.*

There are two difficulties in using Simscape to draw models with commands: one is how to find the block and parameter names of a resistance block, and the other is how to describe its ports. The simplest way of finding these is to open the Simscape Foundation Library with the `fl_lib`

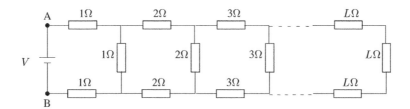

Figure 7.13 Circuit of an electrical network.

command, then open a blank model and copy the resistance block into the model and save it. From the *.mdl file generated, we can find that the block name is

```
'fl_lib/Electrical/Electrical Elements/Resistor',
```

and the parameter name is R. *The port name of the resistance on the left and right side are* LConn1 *and* RConn1 *respectively. For example, if L = 7, the following statements can be used to draw the resistance network*

```
>> L=7; M='ssss1'; new_system(M); open_system(M);
   for i=1:L, i1=int2str(i); i0=int2str(i-1);
     pos1=[100+(i-1)*70 50 125+(i-1)*70 70];
     pos2=pos1+[25 50 25 50]; pos3=pos1+[0 100 0 110];
     scr='fl_lib/Electrical/Electrical Elements/Resistor';
     add_block(scr,[M '/ra' i1],'Position',pos1,'R',i1)
     add_block(scr,[M '/rb' i1],'Position',pos2,...
               'Orientation','down','R',i1)
     add_block(scr,[M '/rc' i1],'Position',pos3,'R',i1)
     add_line(M,['ra' i1 '/RConn1'],['rb' i1 '/LConn1'])
     add_line(M,['rb' i1 '/RConn1'],['rc' i1 '/RConn1'])
     if i>1,
         add_line(M,['ra' i0 '/RConn1'],['ra' i1 '/LConn1'])
         add_line(M,['rc' i0 '/RConn1'],['rc' i1 '/LConn1'])
   end, end
```

When the input and output ports are added manually to the system, the Simulink model shown in Fig. 7.14 is established. Note that for systems with Simscape blocks, stiff equation solvers such as ode15s *are suggested to be used, and a relatively small error tolerance can be set as* 10^{-8}*, or even* eps. *In this case, the equivalent resistance in ports A and B can be obtained* $R_1 = 2.8488\Omega$*, with the following statements, and it is exactly the same as theoretical value.*

```
>> G=linearize('c7fmr1'); R1=1/minreal(G)
```

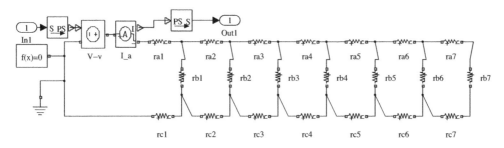

Figure 7.14 Resistance network with Simulink (model name: c7fmr1).

7.2 Description of SimPowerSystems

There are several ways of modeling electrical systems. The models can be established with the Foundation Library of Simscape, or the SimPowerSystems block can also be used. The advantages of the two blocksets can be used together to model and simulate electrical systems.

SimPowerSystems is the electrical system simulation blockset in Simulink; the original name was Power Systems Blockset, mainly developed by HydroQuébec, TECSIM International and Math-Works. This blockset is powerful in the simulation of electrical circuits, power electronic systems, electrical machine and motor drive systems. Circuits or other simulation models can be drawn and converted automatically to state space equations. Strictly speaking, SimPowerSystems is not a part of Simscape, and the scheme for modeling is different. However, the use of the blocks is quite similar to that of Simscape blocks.

In the MATLAB command window, the command `powerlib` can be used to start SimPower-Systems, and the window in Fig. 7.15 will be displayed. Of course, SimPowerSystems can also be invoked from the Simulink model library. In the blockset, there are several groups.

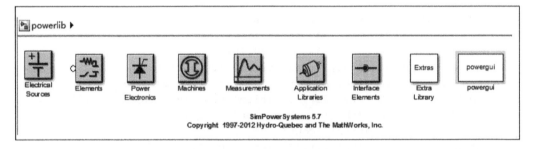

Figure 7.15   SimPowerSystems blockset.

- **Graphical user interface block**: The **powergui** block must be placed in the model window, otherwise simulation cannot be performed.

- **Electrical Sources** group: The blocks in the group are shown in Fig. 7.16. Different AC and DC source blocks, such as **DC Voltage Source**, **AC Voltage Source** and **AC Current Source**, and controlled source blocks, such as **Controlled Voltage Source** and **Controlled Current Source**, are

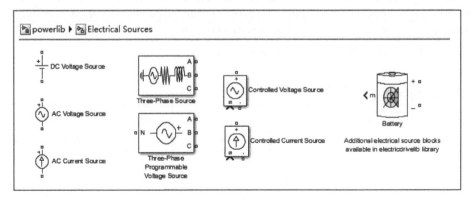

Figure 7.16   Electrical Sources group.

provided in the group. Three-phase source blocks such as **Three-phase Source** and **Three-phase Programmable Voltage Source** and **Battery** blocks are also provided in the group.

- **Measurements** group: The blocks in this group are displayed as shown in Fig. 7.17, including various meters such as **Current Measurement**, **Voltage Measurement** and **Impedance Measurement**, **Three-phase V–I Measurement** and **Multimeter** blocks are provided in the group. Also, subgroups such as **Continuous Measurements** are also provided in the group.

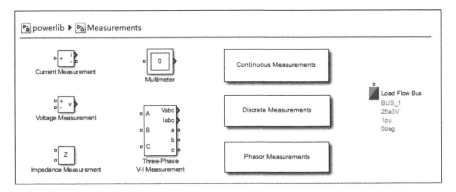

Figure 7.17   Measurements group.

- **Elements** group: The blocks in this group are shown in Fig. 7.18. The commonly used electrical elements such as **Series RLC Load**, **Parallel Series RLC Load**, **Series RLC Branch**, **Parallel Series RLC Branch** and their three-phase versions, and **Mutual Inductance** are provided in this group. Other blocks such as **Transformers**, **Lines** and **Circuit Breaker** are also provided.

  Double clicking the block **Series RLC Branch**, the dialog box shown in Fig. 7.19(a) is displayed, and suitable parameters can be filled in to construct different block types. The RLC elements here are ideal elements. Single resistance, capacitor and inductor blocks can be constructed with the specifications shown in Table 7.1. Or the **Branch type** listbox in Fig. 7.19(b) can be used to select different types of combinations.

Table 7.1   Single resistance, capacitance and inductance parameters.

Elements	Series RLC Branch			Parallel RLC Branch		
Type	Resistance	Inductance	Capacitance	Resistance	Inductance	Capacitance
Single resistance	R	0	inf	R	inf	0
Single inductor	0	L	inf	inf	L	0
Single capacitor	0	0	C	inf	inf	C

Example 7.5   *Consider again the circuit studied in Example 7.3. With SimPowerSystems, the simulation model can be established as shown in Fig. 7.20 (a), and it can be seen that the model is slightly simpler than the one constructed with Simscape foundation library. Since ideal elements are used here, there might be errors. Note that because the* **I Meter** *block is connected in reverse the actual signal measured is* $-I_1$ *signal.*

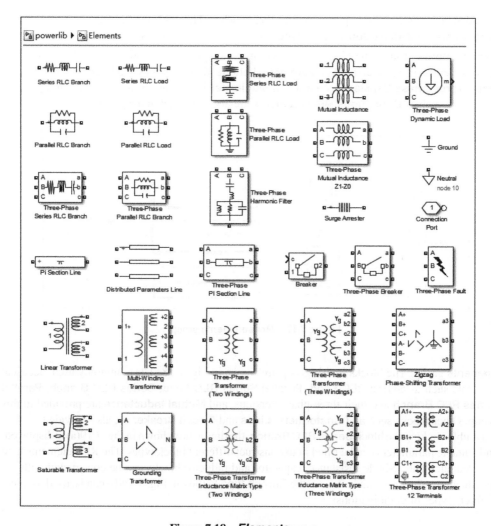

Figure 7.18   *Elements* group.

(a) **Series RLC Branch** dialog box   (b) **Branch type** listbox

Figure 7.19   *Series RLC Branch* dialog box and parameters setting.

With the `power_analyze()` *function, the linear system model can be extracted. Since input and output ports are used, a linearized model can be obtained. The voltage block can then be replaced by a controlled voltage block, as shown in Fig. 7.20(b). The following commands can be used to extract the linearized model:*

$$G(s) = \begin{bmatrix} \dfrac{0.33333(s + 0.08096)(s + 1.544)}{(s + 1.12)(s + 0.0667)(s^2 + 0.06379s + 0.558)} \\[4mm] \dfrac{-0.33333(s + 1)(s^2 + 0.25s + 0.125)}{(s + 1.12)(s + 0.0667)(s^2 + 0.06379s + 0.558)} \end{bmatrix}.$$

```
>> G=linearize('c7mele3'); minreal(zpk(G))
```

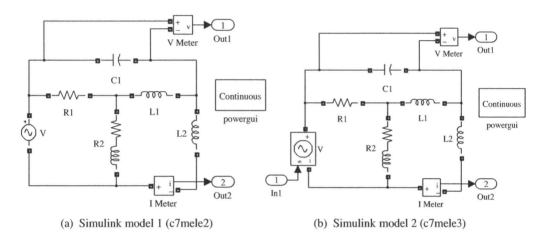

(a) Simulink model 1 (c7mele2)   (b) Simulink model 2 (c7mele3)

Figure 7.20   Electrical circuit models.

- **Power Electronics** group: The blocks in this group are shown in Fig. 7.21, and include the blocks **Diode**, **Thyristor**, **Gto**, **Mosfet** and **IGBT**. It can be seen that each of the blocks has an **m** output port, from which all the internal signals from the block can be extracted. This block can be connected directly to the Simulink output blocks.

  The **Universal Bridge** block in this group is a very useful block, with three ports on the left hand side and two ports on the right. If the left ports are used as input ports, this block can be used as a rectifier. If the right ports are used as the input ports, the block is then an inverter.

- **Machines** group: In this group various electrical machine blocks are provided, as shown in Fig. 7.22, and various **DC Machines**, **Asynchronous Machines** and **Synchronous Machines** are provided.

- **Application Library**: The subgroups are shown in Fig. 7.23(a), and include **Electric Drives Library**, **Flexible AC Transmission Systems (FACTS) Library** and **Distributed Resources Library**.

- **Extras Library** group: The subgroups in the group are shown in Fig. 7.23(b), including **Measurements**, **Discrete Measurements**, **Control**, **Discrete Control** and **Phasor Measurements**.

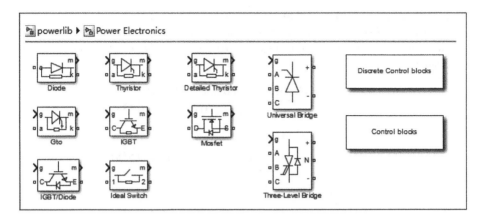

Figure 7.21   **Power Electronics** group.

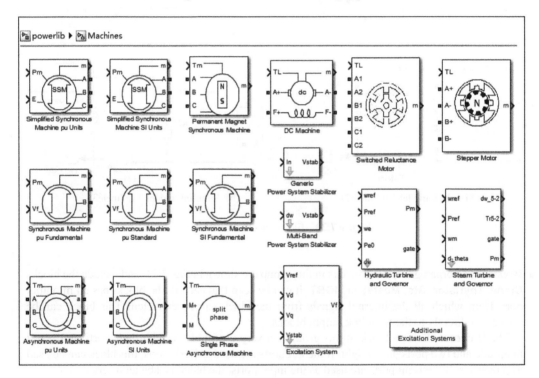

Figure 7.22   **Machines** group.

7.3   Modeling and Simulation of Electronic Systems

Before the release of SimElectronics, the modeling of electronic devices was very difficult. For instance, there was no suitable description of the commonly used transistors, thus only indirect modeling was possible. With the SimElectronics blockset, the problems can be solved easily. In this section, an introduction is given to SimElectronics, and then examples are given to demonstrate analog circuits, digital circuits and operational amplifiers. Finally, the interface with the Spice language is explored.

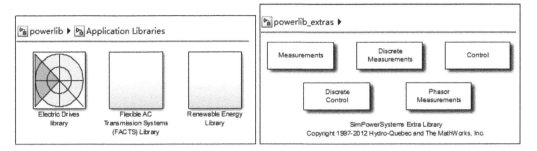

Figure 7.23   **Application Library** group and **Extras Library** group.

7.3.1   *Introduction to the SimElectronics Blockset*

SimElectronics is a component of the Simscape blockset. It can be invoked from Simscape, or by the command `elec_lib`, as shown in Fig. 7.24. It can be seen that there are several groups in the blockset. The commonly used ones are:

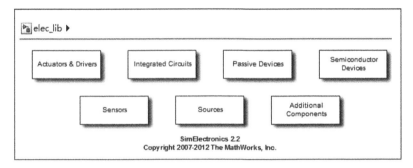

Figure 7.24   SimElectronics blockset.

• **Semiconductor Devices** group: The blocks in this group are shown in Fig. 7.25, including **Diode**, **PNP Bipolar Transistor**, **NPN Bipolar Transistor**, **JFET**, **IGBT** and **MOSFET**. With the blocks, analog circuits can be constructed easily.

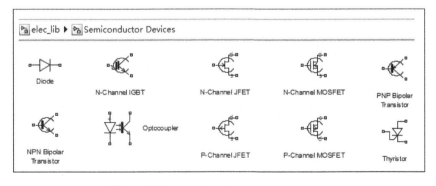

Figure 7.25   **Semiconductor Devices** group.

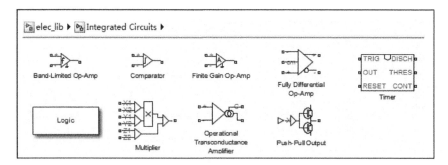

Figure 7.26   Integrated Circuits group.

- **Integrated Circuits** group: the blocks shown in Fig. 7.26 are provided, including **Comparator**, **Finite Gain Op-amp** and **Band-limited Op-amp**. Also the blocks **Logic** and **Timer** are provided.
  A subgroup **Logic** is also provided in this group, with the blocks shown in Fig. 7.27. The subgroup provides blocks for **CMOS AND, CMOS OR, CMOS NOT, CMOS NAND, CMOS NOR, CMOS XOR, CMOS Buffer** and **S-R Latch**. Transient behavior is also considered in the blocks. Ideal logic blocks in the Simulink **Math** group can also be used to model and simulate digital systems.

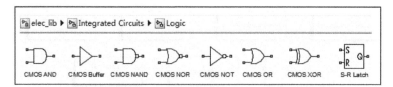

Figure 7.27   Logic subgroup.

- **Passive Devices** group: The blocks in this group is shown in Fig. 7.28, including **Fuse**, **Crystal**, **Thermal Resistor**, **Variable Capacitor**, **Variable Inductor** and **Relay**.

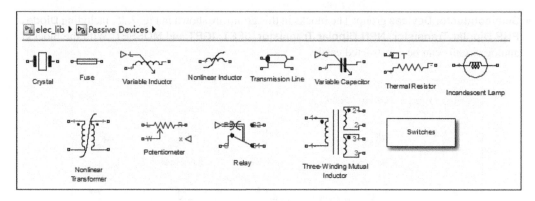

Figure 7.28   Passive Devices group.

- **Sensors** and **Sources**
- **Additional Components** group mainly provides support for the Spice language.

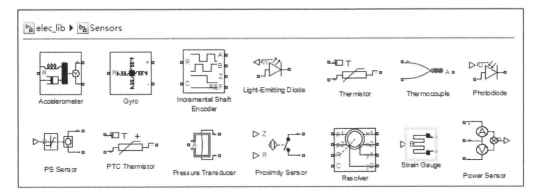

Figure 7.29 Electronic **Sensors** group.

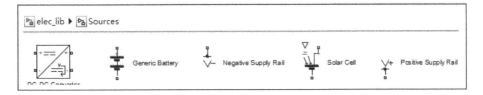

Figure 7.30 Electronic **Sources** group.

7.3.2   Modeling of Analogue Electronic Circuits

In earlier versions of MATLAB and Simulink, commonly used electronic devices such as transistors cannot be modeled directly. The interface to third-party tools, such as the Spice language, can be used to describe such devices. In MATLAB R2008b, a brand new SimElectronics blockset was released, and this kind of electronic system can now be modeled directly with the blockset. Also the Spice language models can be embedded in the Simulink block diagrams. Examples are given below to show the modeling and simulation of analog electronic circuits.

Example 7.6   *Assume that an electronic circuit with a transistor is shown in Fig. 7.31(a). Since there is a transistor in the circuit,* **NPN Bipolar Transistor** *block in SimElectronics can be used in the Simulink model shown in Fig. 7.31(b).*

   Double click the transistor block, and the parameter dialog box shown in Fig. 7.32(a) appears, where various parameters of the transistor are given. These parameters can be modified according to actual circuit elements, otherwise the default parameters will be used. Simulating this system, the output, that is the voltage between node 1 and ground, is shown in Fig. 7.32(b).

   To find the response of the circuit for other signals, the input in Fig. 7.33(a) can be replaced by an input source. For instance, a triangular waveform can be used as an input signal. The block **Repeating Sequence** *can be used to define periodic signals of any waveform, as shown in Fig. 7.33(b).*

Example 7.7   *Operational amplifiers are an important component in continuous controllers; they can be used to construct commonly used structures such as derivative controllers and integral*

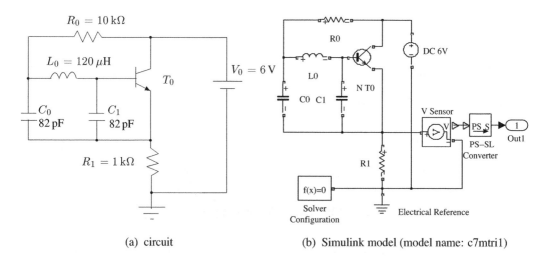

(a) circuit   (b) Simulink model (model name: c7mtri1)

Figure 7.31 Simulink model with a transistor.

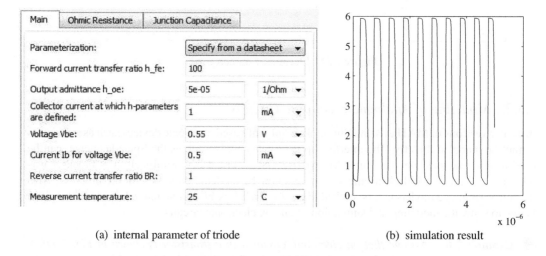

(a) internal parameter of triode   (b) simulation result

Figure 7.32 Transistor and its simulation parameters.

controllers. In control theory, the gain of an operational amplifier is assumed to be infinite. However, such an amplifier does not exist in reality; the gain is in practice finite. Also, the output voltage signal is kept within a certain range, referred to as the clamp voltage.

Consider now the operational amplifier circuit shown in Fig. 7.34(a). A Simulink model shown in Fig. 7.34(b) can be constructed, based on the operational amplifier block. From the model constructed it can be seen that the modeling is straightforward and simple.

Double click the operational amplifier block, and the parameter dialog box opens, as shown in Fig. 7.35(a). The parameters of the amplifier can be assigned in the dialog box. Assuming that the clamp voltage is ±15V, and the input amplitude of the sinusoidal signal v_i is set to 0.3V, the output voltage v_o can be obtained by simulation, as shown in Fig. 7.35(b). Saturation occurred in the output

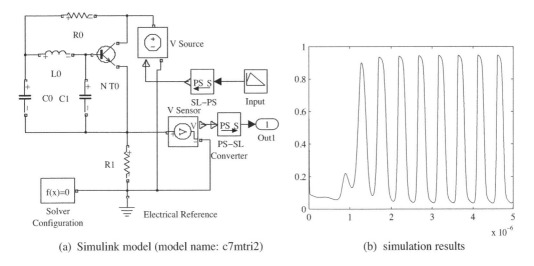

(a) Simulink model (model name: c7mtri2)   (b) simulation results

Figure 7.33   Simulink model and simulation results for a transistor.

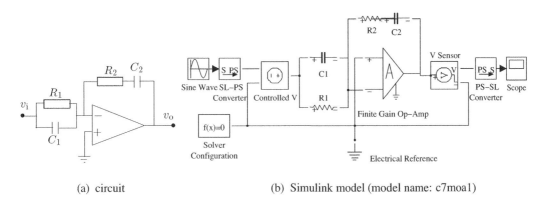

(a) circuit   (b) Simulink model (model name: c7moa1)

Figure 7.34   Circuit and simulation model of an operational amplifier.

of the system, so that the operational amplifier is no longer linear. The nonlinearity in this case cannot be neglected. If the input amplitude is further increased to 0.5V, the output signal is shown in Fig. 7.35(c).

To get the linearized model, the input and output blocks should be replaced with input and output ports, as shown in Fig. 7.36. The following MATLAB commands can be used and the linearized model is then obtained as

$$G(s) = \frac{-999.5452(s + 0.496)(s + 45.45)}{(s + 0.03571)(s + 858.2)}.$$

```
>> G=zpk(linearize('c7moa2'))
```

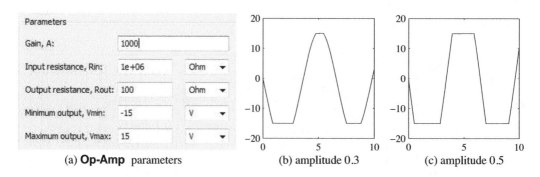

(a) **Op-Amp** parameters   (b) amplitude 0.3   (c) amplitude 0.5

Figure 7.35 Parameters of an operational amplifier and responses.

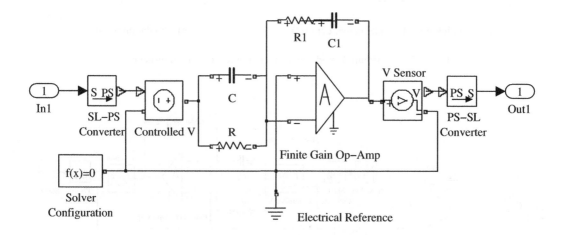

Figure 7.36 Operational amplifier model with input and output ports (model name: c7moa2).

7.3.3 Modeling of Digital Electronic Circuits

A large number of digital electronic devices are modeled in the SimElectronics blockset. Also, ideal logical blocks are provided in the **Logic and Bit Operations** group in Simulink, as shown in Fig. 7.37(a), and the **AND** block can be used to describe other gate blocks, as shown in Fig. 7.37(b). The **Combinational Logic** block can also be used to describe truth tables. With these blocks and other relevant blocks, digital electronic circuits can be constructed.

Also, in the **Simulink Extra** group, as shown in Fig. 7.38(a), the **Flip-flops** subgroup is provided as shown in Fig. 7.38(b), where a lot of flip-flops are provided. These blocks can be used in the modeling and simulation with flip-flops.

In this section, examples are used to demonstrate the modeling and simulation problems of digital electronic circuits.

Example 7.8 *Consider a logic operation* $Z = \overline{A \cdot \overline{A \cdot B}} + B \cdot \overline{A \cdot B}$, *where* $\overline{A \cdot B}$ *is the negative AND gate, while* $\overline{A + B}$ *is the negative OR gate.*

Logic Operations

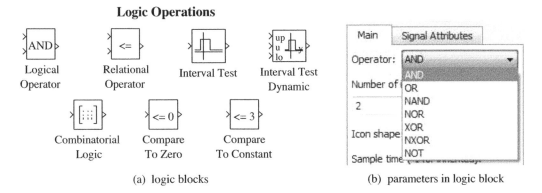

(a) logic blocks   (b) parameters in logic block

Figure 7.37 Simulink **Logic and Bit Operations** group and parameter setting.

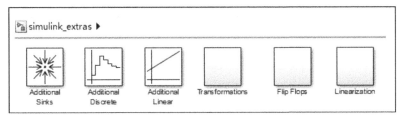

(a) **Simulink Extras** blockset

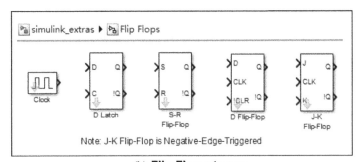

(b) **Flip-Flop** subgroup

Figure 7.38 **Simulink Extras** group and flip-flop blocks.

With the ideal logic operation blocks, the logic operation expression can simply be modeled as shown in Fig. 7.39(a). Since the model thus created is ideal, no transient dynamic responses are considered. To use practical logic gates to model the system, the blocks in SimElectronics can be used and the new model is shown in Fig. 7.39(b).

Example 7.9   *Consider the binary-coded decimal (BCD) decoder problem, where there are four binary inputs A_1, A_2, A_3 and A_4, and a decoder is needed to display the results of the four input signal with a seven-segment LED device. The decoder can be expressed easily by the truth table shown in Table 7.2 [2].*

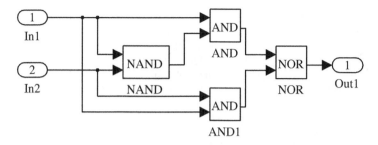

(a) ideal model (model name: c7mdig1)

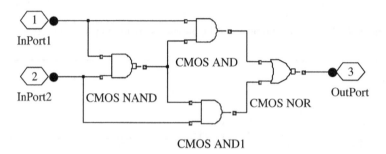

(b) CMOS model (model name: c7mdig2)

Figure 7.39 Simulink model of logic operation expression.

Table 7.2 Truth table of a seven-segment LED decoder.

Input ports				Output ports						
A_1	A_2	A_3	A_4	Y_1	Y_2	Y_3	Y_4	Y_5	Y_6	Y_7
0	0	0	0	1	1	1	1	1	1	0
0	0	0	1	0	1	1	0	0	0	0
0	0	1	0	1	1	0	1	1	0	1
0	0	1	1	1	1	1	1	0	0	1
0	1	0	0	0	1	1	0	0	1	1
0	1	0	1	1	0	1	1	0	1	1
0	1	1	0	0	0	1	1	1	1	1
0	1	1	1	1	1	1	0	0	0	0
1	0	0	0	1	1	1	1	1	1	1
1	0	0	1	1	1	1	1	0	1	1
1	0	1	0	0	0	0	1	1	0	1
1	0	1	1	0	0	1	1	0	0	1
1	1	0	0	0	1	0	0	0	1	1
1	1	0	1	1	0	0	1	0	1	1
1	1	1	0	0	0	0	1	1	1	1
1	1	1	1	0	0	0	0	0	0	0

*The output ports of the truth table can be expressed by a matrix. The matrix can be assigned to the **Combinational Logic** block*

```
>> truTab=[1 1 1 1 1 0; 0 1 1 0 0 0 0; 1 1 0 1 1 0 1; 1 1 1 1 0 0 1;
           0 1 1 0 0 1 1; 1 0 1 1 0 1 1; 0 0 1 1 1 1 1; 1 1 1 0 0 0 0;
           1 1 1 1 1 1 1; 1 1 1 1 0 1 1; 0 0 0 1 1 0 1; 0 0 1 1 0 0 1;
           0 1 0 0 0 1 1; 1 0 0 1 0 1 1; 0 0 0 1 1 1 1; 0 0 0 0 0 0 0];
```

*Thus, a Simulink model can be established as shown in Fig. 7.40. The decoder is represented by the **Combinational Logic** block. Since the data types required for the input and output signals are Boolean, while the data types of the input from the signal generator and output to the LED blocks are both* `double`, ***Data Type Conversion*** *blocks are needed to complete the connections.*

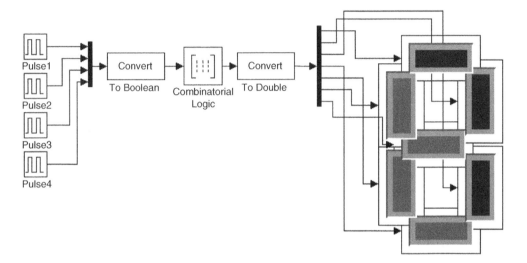

Figure 7.40   Decoder with LCD display (model name: c7mled).

Example 7.10   *In control system identification and communication systems, pseudo random binary sequence (PRBS, or M-sequence) is often used. It is known that such signals can be generated with the digital circuit shown in Fig. 7.41, with flip-flops as the main components.*

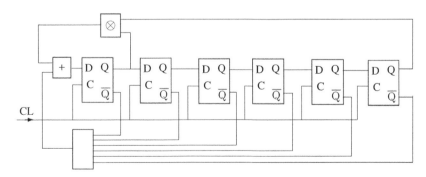

Figure 7.41   PRBS signal generator circuit.

*The flip-flop block in the **Simulink Extra** group can be used directly to create the Simulink model shown in Fig. 7.42, where six **S Flip-flop** blocks are used and driven by a clock pulse signal, with a period of 2 s. Each flip-flop should be controlled by an enable signal. The enable signal should be connected to the **!CLR** port in each flip-flop. The **Logical Operator** block can also be used in the system.*

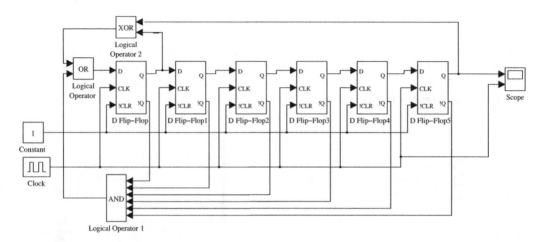

Figure 7.42   Simulation model of a flip-flop based PRBS signal generator (model file: c7mflip).

Simulation results are obtained as shown in Fig. 7.43, and the period of the PRBS signal is 63, that is, 126 s. It can be seen that this digital system works well in generating PRBS signals.

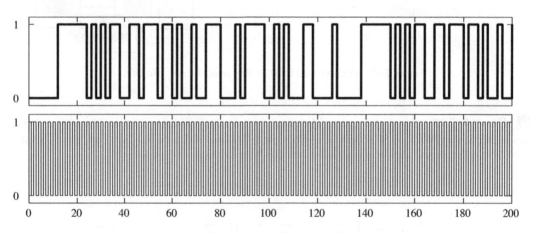

Figure 7.43   PRBS signal and clock signals.

7.3.4   Modeling of Power Electronics Circuits

The modeling of power electronics circuits with low-level blocks is rather complicated. If the blocks in SimPowerSystems are used, the modeling process is very simple and straightforward. An example of power electronic modeling will be illustrated in this section.

Example 7.11   *A thyristor based rectifier circuit can be modeled with Simulink blocks. From the Power Electronics group, it can be seen that low-level blocks such as* **Thyristor** *and higher level* **Universal Bridge** *blocks are provided. For certain applications,* **Universal Bridge** *based modeling is much easier than low-level modeling techniques.*

The **Universal Bridge** *can be used directly in modeling. The* **Synchronized 6-Pulse Generator** *block can be extracted from the* **Control Blocks** *group to construct the rectifier model shown in Fig. 7.44(a), and the AC source can be constructed as three-phase blocks in Y connection. The thyristor parameter dialog box can be opened as shown in Fig. 7.44(b).*

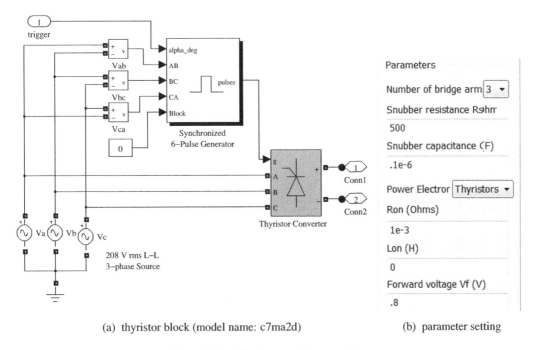

(a) thyristor block (model name: c7ma2d)   (b) parameter setting

Figure 7.44   Thyristor rectifier modeling.

To demonstrate the thyristor based rectifier system, the block diagram shown in Fig. 7.45 can be constructed. The input is a constant block, representing the expected angle of the trigger. This parameter can be set to 30° or other values. Note that to the output terminal of the system, load should also be connected, otherwise the loop may not work properly.

In the simulation model, if the trigger angle is set to 30°, the waveforms of the input and output are obtained as shown in Fig. 7.46(a). If the trigger angle is set to 50°, the waveforms are obtained as shown in Fig. 7.46(b).

7.3.5   Embedding Spice Models in Simulink

Spice is an electronic circuit simulation language and PSpice is the product of MicroSim, with a graphical user interface designed to support Spice languages. It is one of the most powerful and widely used simulation languages for electronic circuits [3]. If the circuit is only part of the whole system, while other components such as transfer functions or mechanical components are also

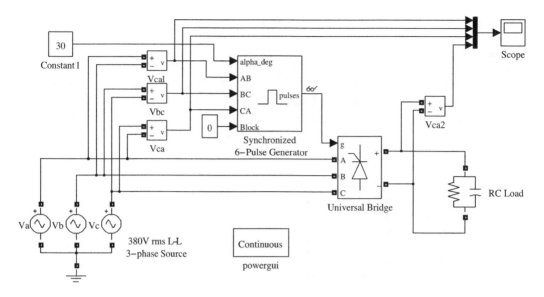

Figure 7.45   Thyristor simulation model (model name: c7ma2d2).

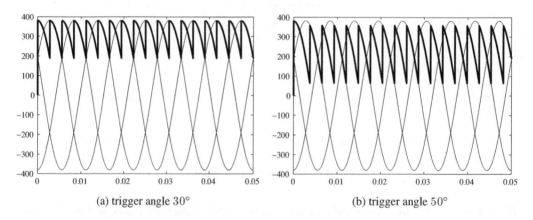

(a) trigger angle 30°   (b) trigger angle 50°

Figure 7.46   Rectifier output for different trigger angles.

involved, Spice cannot be used for modeling the whole system. Embedding a Spice model in a Simulink model is very helpful for constructing simulation models. In this section, Spice model to Simulink conversion is illustrated, so that Spice can be used in Simulink modeling.

There are several solutions to the problem. One is to use a dynamic library technique to run a Spice model from MATLAB and transfer the data back to the MATLAB workspace. Unfortunately, this method only works for off-line simulation and you cannot embed the model in Simulink. The interface can be freely downloaded from http://ave.dee.isep.ipp.pt/~jcarlos/matlab.

Another way is to use SLSP (Simulink Spice Interface). The .cir file can be embedded into a Simulink model. The interface is a commercial product, and details of the interface can be found from http://www.bausch-gall.de/prodss.htm.

SimElectronics can now translate a Spice circuit into a Simulink block. The leading command should be initiated with the keyword SUBBLK. Then the netlist2sl() function can be used to do the conversion, and a Simulink block can be automatically generated.

Example 7.12 *Consider again the circuit with a transistor studied in Example 7.6. The circuit is redrawn in Fig. 7.47(a). The numbers in the circuit are the node numbers used for Spice modeling.*

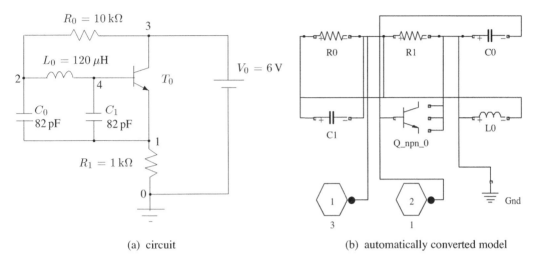

(a) circuit   (b) automatically converted model

Figure 7.47 The circuit with a transistor and the converted model.

The circuit can be described with Spice language as follows, and the program can be saved to the file c7ftri1.cir.

```
.TRAN 5ns 2us 0s 5ns UIC  * simulation parameter setting
  V0 3 0 DC 6V             * DC voltage  V0, from node 3 to 0, 6V
  R0 2 3 10Kohm            * resistor  R0, from nodes 2 to 3, with 10kΩ
  R1 1 0 1Kohm             * resistor  R1, from node 1 to 0, with 1kΩ
  C0 4 1 82pF              * capacitor  C0, from node 4 to 1, with 82pF
  C1 1 2 82pF              * capacitor  C1, from node 1 to 2, with 82pF
  Q_npn_0 3 4 1 Qn2_2N2222A * NPN transistor, connected to nodes 3,4,1, see model
  L0 2 4 120uH            * inductor  L0, from node 2 to 4, with 120μH
.PRINT TRAN V(1)           * prepare for the output data for later use
.MODEL Qn2n_2N2222A NPN(Is=11.6fA BF=200 BR=4 Rb=1.69ohm Re=423mohm
+Rc=169mohm Cjs=0F Cje=19.5pF Cjc=9.63pF Vje=750mV Vjc=750mV Tf=454ps
+Tr=102ns mje=333m mjc=333m VA=113V ISE=170fA IKF=410mA Ne=2)
.END
```

According to the syntax of Spice language, the transient analysis command .TRAN should have five parameters. The former two are respectively the step size and termination time, the third is the start time and the fourth is the maximum allowed step size, while the final paramter is the initial condition UIC.

If we want to embed the Spice model into a Simulink block diagram, the original Spice program should be rewritten as a submodel. The easiest way to do so is to insert a leading sentence " .SUBCKT *cir 3 1" to the *.cir program, where 3 and 1 are respectively the input and output nodes of the embedded model. Also, the DC voltage source statement can be removed, so that external signals can be used to drive the model. The submodel can be saved into file c7mtri2.cir.*

*The following command can be used to convert the Spice submodel into a Simulink model, shown in Fig. 7.47(b). The name of the Simulink model is assigned with the .*SUBCKT *command, and in this example,* cir*. It should be noted that, although the automatically translated Simulink model is not good in layout, the simulation model is reliable and can be used directly in simulation.*

```
>> netlist2sl('c7mtri2.cir','myLib')
```

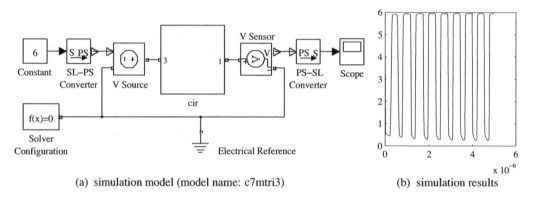

(a) simulation model (model name: c7mtri3)   (b) simulation results

Figure 7.48   Simulation of of circuit with the embedded Spice block.

With the converted Simulink model cir*, the system shown in Fig. 7.48(a) can be constructed, where the input signal is generated with a constant block in Simulink, with an amplitude of 6V. The output signal is then shown in Fig. 7.48(b).*

The method presented above can be used to translate more complicated Spice models. The model can then be embedded into Simulink block diagrams. In this way, electronic circuit modeling and simulation capabilities are significantly enhanced, and existing Spice models can also be reused in Simulink systems.

7.4   Simulation of Motors and Electric Drive Systems

Before exact modeling and simulation tools for motor drive systems were available, linear models were used to approximate motor drive systems. For instance, the motors are usually represented by a second-order transfer function, while the thyristor rectifiers are approximated by a first-order lag model. The approximate results may have significant differences from the actual systems. Thus physical modeling methodology should be adopted to accurately model motor drive systems.

7.4.1   Simulation of DC Motor Drive Systems

A double loop speed regulating system for DC motors is used as an example to present physical modeling techniques. Consider the typical DC motor drive system shown in Fig. 7.49.

The mathematical models of the filters – the current filter $F_c(s)$ and speed filter $F_s(s)$ – can be written as

$$F_c(s) = \frac{\beta}{\tau_c s + 1}, \ F_s(s) = \frac{\alpha}{\tau_s s + 1}, \tag{7.4}$$

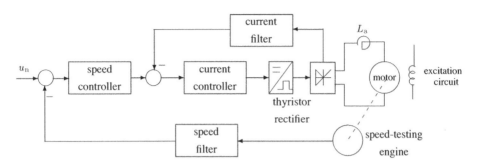

Figure 7.49   Double loop speed regulating system of DC motor drive.

where τ_c and τ_s are small filter constants, such as 0.005 s or 0.001 s. The current and speed controllers are both PI controllers and their mathematical models are respectively

$$G_c(s) = K_c \frac{T_c s + 1}{s}, \ G_s(s) = K_s \frac{T_s s + 1}{s}. \tag{7.5}$$

It can be seen from the block diagram of the control system, that the thyristor rectifier component and the DC motor component are needed. In traditional control systems, to exactly describe them is a very complicated task. Thus the double loop DC motor drive system given in Example 4.4 uses linear components to approximate them. For instance, the motor is modeled by a second-order transfer function, while the thyristor rectifier is modeled by a first-order lag system. In simulation, especially with the powerful Simulink and SimPowerSystems blockset, the thyristor, trigger and motor blocks are all provided. With these blocks, an accurate model can be constructed. In this section, a thyristor based rectifier system is demonstrated, and later, an exact model of a double loop DC motor drive system is constructed.

Example 7.13   *Assume that the parameters on the nameplate of a motor are given by [4]: the rated voltage is $U_N = 220$ V, rated current is $I_N = 136$ A, rated speed is $n_N = 1500$ r/min, $K_e = 0.228$ V/(r/min), $\lambda = 1.5$. The thyristor amplifier gain is $K_s = 62.5$, total armature resistance is $R_a = 0.863 \ \Omega$, current feedback coefficient is $\beta = 0.028$ V/A. The coefficient feedback of rotation speed is $\alpha = 0.0041$ V/(r/min), filter constants are $\tau_c = \tau_s = 0.005$ s. If the parameters of the two controllers are respectively $K_c = 1.15$, $T_c = 0.028$ s, $K_s = 20.12$, $T_s = 0.092$ s, the simulation model can be established.*

*A **Universal Bridge** block can be used to model the thyristor rectifier, and its dialog box is shown in Fig. 7.50(a). A **DC Machine** block can be used to model the motor, with the dialog box shown in Fig. 7.50(b). In the block, the armature resistance R_a is given on the nameplate. The inductance can be specified as well, and the mutual inductance can be calculated from*

$$L_f = \frac{(U_N - P R_a / U_f) R_f}{n_N U_f} = 0.637 \text{ H}.$$

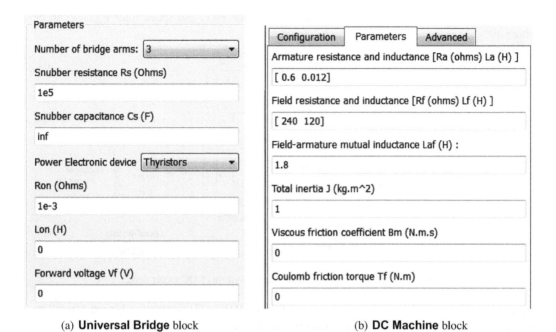

(a) **Universal Bridge** block   (b) **DC Machine** block

Figure 7.50   Parameter setting dialog boxes of the blocks.

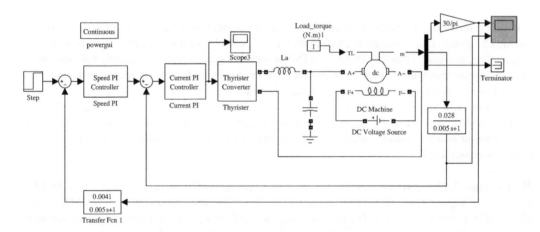

Figure 7.51   DC motor drive system (model name: c7mdcm).

The DC motor drive system can be constructed as shown in Fig. 7.51. The internal structure of the **DC Machine** *block, with the* **Look under Mask** *menu item from the shortcut menu, obtained by right clicking the* **DC Machine** *block, can be displayed as shown in Fig. 7.52. It can be seen that the electrical model and mechanical model are both constructed, and friction is also considered in the model. Thus this model is more accurate than the approximate transfer function models.*

In practical DC motor drive systems, the structures of the two PI controllers are slightly different, as shown in Figs 7.53(a) and (b). In the current controller, a bias block is used.

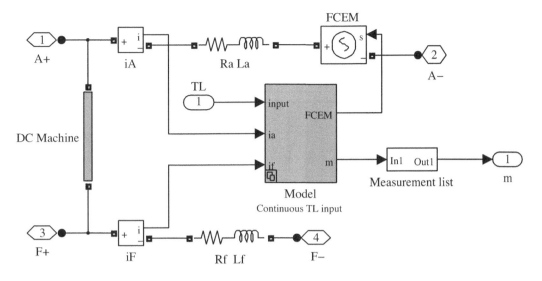

Figure 7.52   Internal structure of the **DC Machine** block.

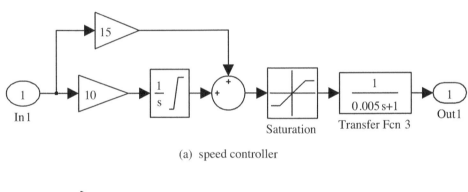

(a) speed controller

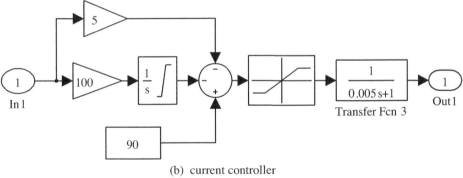

(b) current controller

Figure 7.53   Cascade PI controllers.

With Simulink models the system can be simulated and the speed and current curves can be read again, as shown in Fig. 7.54(a). If the parameters of the speed PI controller are changed by a trial-and-error method to $K_p = 15$, $K_i = 10$, the simulation results are then obtained as shown in Fig. 7.54(b). It can be seen that the control performance is improved significantly.

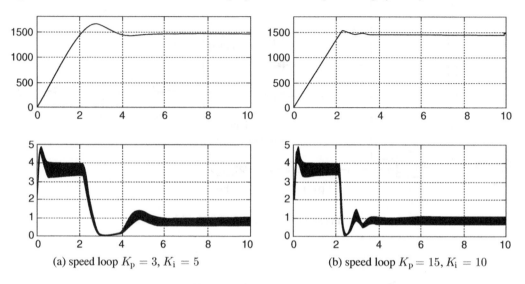

(a) speed loop $K_p = 3$, $K_i = 5$ (b) speed loop $K_p = 15$, $K_i = 10$

Figure 7.54   Control results.

*With the simulation framework, similar problems can also be studied. For instance, if we want to get the system responses when there is a load change – say the load is changed to a certain value at $t = 6\ s$ – a **Step** block can be used to describe the load change. The new simulation block diagram can be constructed as shown in Fig. 7.55.*

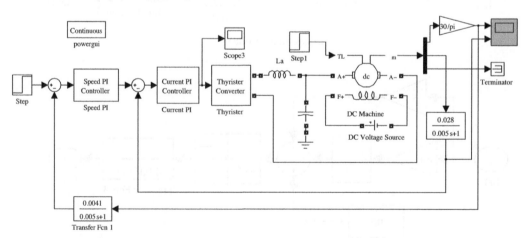

Figure 7.55   DC motor drive with load changes (model name: c7mdcm1).

Assume that the initial load is 1 N·m, and at time $t = 6\ s$, the load is changed to 15 N·m or 50 N·m. The initial value of the block can be set to 1, the step time can be set to 6, and the final value can be set to 15 or 50, respectively. The simulation results are then obtained as shown in Figs 7.56(a) and *(b). It can be seen that with the PI controllers, the capability of disturbance rejection is quite good, since the speed will return to the rated speed very quickly.*

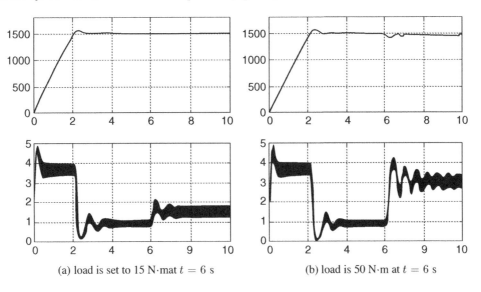

(a) load is set to 15 N·m at $t = 6$ s   (b) load is 50 N·m at $t = 6$ s

Figure 7.56 Speed regulating results with load variations.

7.4.2 Simulation of AC Motor Drive Systems

The SimPowerSystems blockset provides three asynchronous machine blocks, as shown in Fig. 7.57(a): **Asynchronous Machine pu Units**, **Asynchronous Machine SI Units** and **Single Phase Asynchronous Machine**.

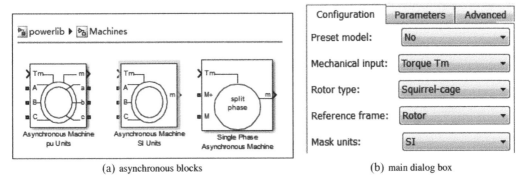

(a) asynchronous blocks   (b) main dialog box

Figure 7.57 Asynchronous blocks and parameters.

In this section, only the **Asynchronous Machine SI Units** block is used for presenting the model of AC motors. In the block, there are four input ports and four output ports. The last three input ports are used to accept three-phase stator voltages at ports **A**, **B** and **C**. The three-phase voltage, connected either in a star or triangle (Y or delta), can be connected to the ports, and the first input port is used to connect the external load signal. The first output port is a vector **m** port, consisting of

21 signals in the motor block shown in Table 7.3. The other three signals are the three-phase voltage of the rotors **a**, **b** and **c**.

Table 7.3   Twenty-one output signals in Asynchronous machine block.

Output	Output signal description
1~3	rotor current $i'_{ra}, i'_{rb}, i'_{rc}$
4~9	rotor signal in q-d-n coordinates, the currents in q- and d-axis i'_{qr}, i'_{dr}, the flux ψ'_{qr}, ψ'_{dr}, and the voltages v'_{qr}, v'_{dr}
10~12	the stator currents i_{sa}, i_{sb}, i_{sc}
13~18	the stator signals in q-d-n coordinates, $i_{qs}, i_{ds}, \psi_{qs}, \psi_{ds}, v_{qs}, v_{ds}$
19~21	rotational speed ω_m, mechanical torque T_m, rotor angle displacement θ_m

Double click the motor block. The main dialog box is shown in Fig. 7.57(b), where the **Rotor type** listbox provides the options **Wound** and **Squirrel-cage**. If the latter is selected, the output ports **a**, **b** and **c** are no longer displayed. They are short connected internally. Also, the **Reference frame** listbox provides the options **Stationary**, **Rotor** and **Synchronous**. Normally the first option is selected.

An example is given below to show the modeling and simulation of the **Asynchronous Machine SI Units** block in Y connection.

Example 7.14   *Several commonly pre-assigned model parameters are provided in the **Asynchronous Machine SI Units** block. These parameters can be used directly to simulate specific motors. The user can also set parameters with the dialog box shown in Fig. 7.57(b). In the **Parameters** pane, the motor parameters can be entered as shown in Fig. 7.58.*

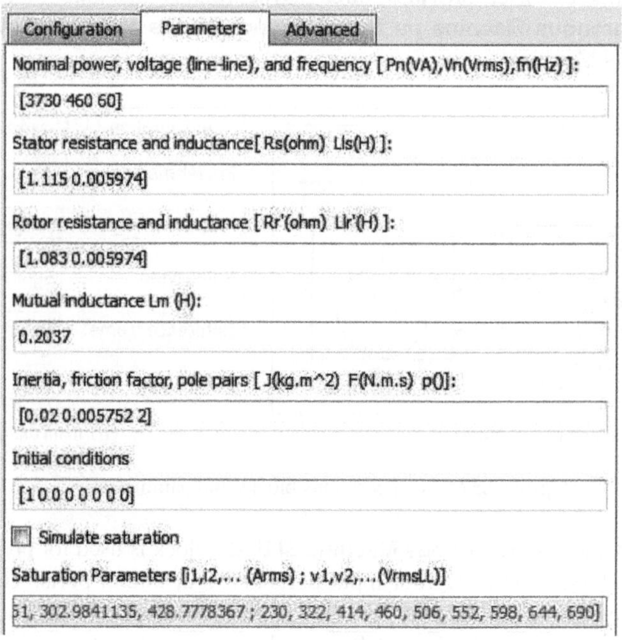

Figure 7.58   Motor parameter dialog box.

Assume that the following nameplate data of a motor can be used [4].

- *rated power $P_n = 5.5$ kW, line voltage $V_n = 380$ V, frequency $f_n = 50$ Hz.*
- *stator resistance $R'_s = 0.0217$ Ω, leakage inductance $x'_{ls} = 0.039$ Ω.*
- *rotor resistance $R_r = 0.0329$ Ω, leakage inductance $x_{lr} = 0.0996$ Ω.*
- *mutual inductance $L_m = 3.6493$ Ω.*
- *inertia $J = 11.4$ kg·m², friction coefficient $F = 0$ N·m·s, number of pole-pairs $P = 2$.*

The inductances are expressed in impedance, and $L = x/(2\pi f)$ can be used to calculate the values of the inductances.

The commonly used connection of an asynchronous motor block supports Y connection and delta connection, shown respectively in Figs 7.59(a) and (b). In this example, Y connection is used. Under such a structure, the initial phase of the three blocks are 0, 120 and 240, respectively.

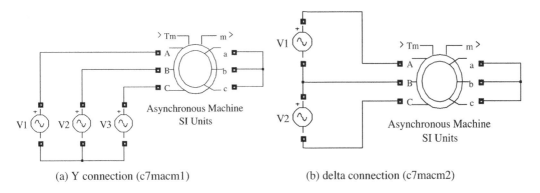

(a) Y connection (c7macm1)   (b) delta connection (c7macm2)

Figure 7.59   Motor connections.

*Thus the simulation model can be constructed as shown in Fig. 7.60. Y connections are made to the voltage sources with phase voltage of 220V, and the phases of blocks **A**, **B** and **C** are defined as 0, 120 and 240.*

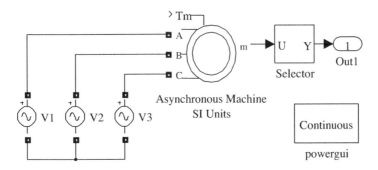

Figure 7.60   Simulink model of asynchronous motor (model name: c7macm4).

*An idle **Current Measurement** block should be placed in the model, as required by SimPowerSystems, since at least one measurement block is needed. Also a **Selector** block can be used to extract from **m** port the desired signals, as shown in Fig. 7.61.*

Parameters

Number of input dimensions: 1

Index mode: One-based ▼

	Index	Option	Index	Output Size
1	Index vector (dialog) ▼		[1 10 19 20]	Inherit from "Index"

Input port size: 21

Figure 7.61   **Selector** block setting dialog box.

*From the **Selector** block, it can be seen that the signals numbered 1, 10, 19, 20 are selected, and from Table 7.3, the selected signals are rotor current i'_{ar} of phase a, stator current i_{as} of phase **a**, rotation speed ω and output torque T. The rotation speed should be multiplied by $-30/\pi$ to get the unit of rpm. If the simulation termination time is selected as 3 s, and the algorithm is **ode15s**, with relative error tolerance of 10^{-7}, the system output can be obtained as shown in Fig. 7.62(a).*

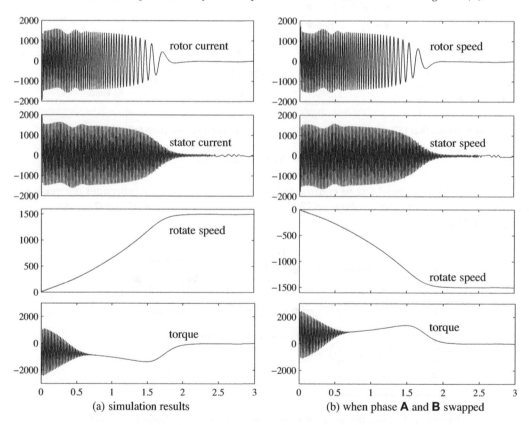

Figure 7.62   Asynchronous motor simulation results.

*If the phases **A** and **B** are swapped, the initial phases of **A** and **B** can be changed respectively to 120 and 0. The simulation results can be obtained as Fig, 7.62(b). It can be seen that when phases*

A and B are swapped, the motor rotates in the reverse direction. Other connections of sources can also be simulated.

■ **Example 7.15**   *The AC motor start-up system shown in Fig. 7.63 can be established. Assume that the load is applied at t = 0.2 s.*

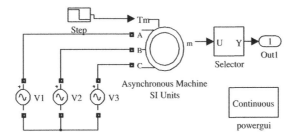

Figure 7.63   Simulink model with load (model name: c7macm5).

*The **Preset Motors** listbox in Fig. 7.64 (a) shows many commonly used motors. If the first preset asynchronous motor is used, the rated voltage is 460 V, frequency 60 Hz, rated speed 1700 rpm and the parameters of the motor dialog box as shown in Figs 7.64(b) and (c). For this system, the startup*

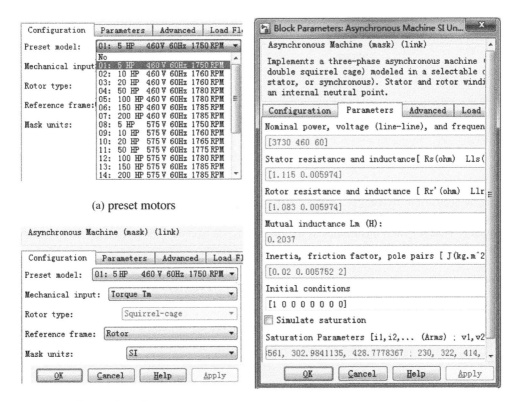

(a) preset motors

(b) other setting of preset motors

(c) motor parameters

Figure 7.64   Prototype model dialog box.

curves with idle load are shown in Fig. 7.65 (a). When a load of 100 N·m is added at t = 0.2 s, the
curves are as shown in Fig. 7.65(b).

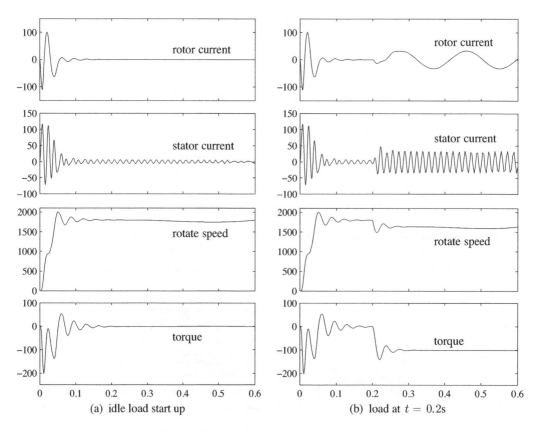

(a) idle load start up (b) load at $t = 0.2$s

Figure 7.65 Simulation of asynchronous motor.

It can be seen from the simulation results that with this simulation technique, we are able to observe the signals in the system within a very short period. This cannot be achieved with hardware experiments. Thus in power electronics as in other fields, the approach of soft experimentation has particular advantages and cannot be replaced by other methods.

7.5 Modeling and Simulation of Mechanical Systems

It has been pointed out that the Foundation Library of the Simscape blockset can be used to model mechanical systems. As well as the Foundation Library, more sophisticated SimMechanics can also be used. Simple mechanical systems are demonstrated first, and then SimMechanics based methods are explored.

7.5.1 Simulation of Simple Mechanical Systems

Simple mechanical systems can be modeled with the Foundation Library of Simscape. An illustrative example follows, to model spring–damper systems.

Example 7.16 *Consider the spring–damper system discussed in Example 7.2. For convenience, the original model is shown again here in Fig. 7.66. Assume that the elastic coefficient of the spring is 1000 N/m, and the damping ratio is 100 N·s/m, the mass is M = 1 kg. The simulation model of such a system is shown in Fig. 7.67. To connect the Simulink port and physical signal port, the converters **SL-PS** and **PS-SL** blocks should be used. For instance, the Simulink block **Pulses** can generate a square wave signal. This can be converted to a physical signal through **SL-PS** block, and the signal is applied to the **Ideal Force Source** block, and applied to the mass. The displacement and speed can be measured with the **Ideal Translational Motion Sensor** block, then they are converted with a **PS-SL** block to Simulink signals.*

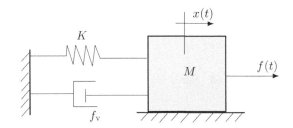

Figure 7.66   Spring–damper system.

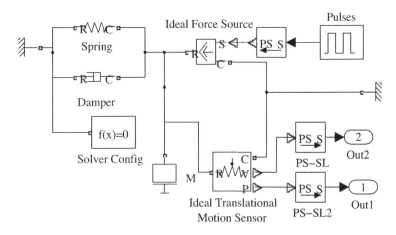

Figure 7.67   Simulink model (model name: c7mdamp1).

Assume that the external force is given by a square wave, with amplitude of 1 N, and a period of 5 s, with duty factor of 40%. The displacement and speed curves of the mass can be obtained, as shown in Fig. 7.68(a). If the elastic coefficient is changed to 100 N/m, the displacement and speed curves of the mass are obtained as shown in Fig. 7.68(b).

It should be noted that it is assumed in this example that there is no friction in the system. If friction is considered, the description is more complicated, since the direction of motion of the mass constantly changes, and static friction should also be considered. A good way to describe friction is the use of the Stateflow technique, which will be presented in Chapter 8.

Example 7.17 *With Newton's law, the simple spring–damper system can be modeled easily. The mathematical model can also be described, and simulation can then be performed. Of course, Simscape modeling is simpler and more straightforward. Now consider the multiple spring–damper*

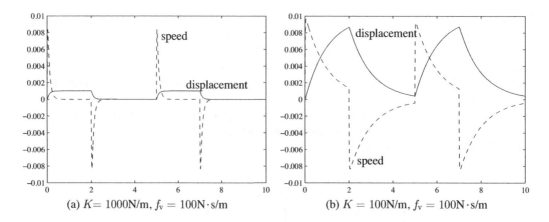

(a) $K = 1000\text{N/m}$, $f_v = 100\text{N}\cdot\text{s/m}$ (b) $K = 100\text{N/m}$, $f_v = 100\text{N}\cdot\text{s/m}$

Figure 7.68 Speed and displacement curves.

system shown in Fig. 7.69. Newton's law modeling may be rather complicated. With Simscape, the model can be established easily.

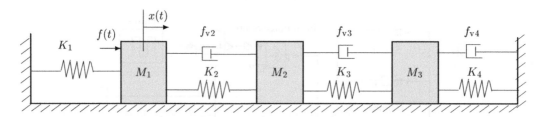

Figure 7.69 Multiple spring–damper system.

Assume that the damping ratios of the three dampers are specified respectively as $f_{v1} = f_{v2} = f_{v3} = 100$ N·s/m. The elastic coefficients are $K_1 = K_2 = 1000$ N/m, $K_3 = K_4 = 400$ N/m. The masses are $M_1 = M_2 = M_3 = 1$ kg, and the force $f(t)$ is selected as a square wave with an amplitude of 4 N. The Simulink model of the system can easily be constructed as shown in Fig. 7.70. The displacement curve of the system can be obtained as shown in Fig. 7.71.

Linearization can be performed with the following statements and the zero-pole-gain model can be obtained as

$$G(s) = \frac{100(s + 195.9)(s + 10)(s + 4.083)}{(s + 318.2)(s + 145.2)(s + 8.486)(s + 4.082)(s^2 + 24.01s + 699.6)}.$$

```
>> G=linearize('c7mdamp2'); G=minreal(G)
```

7.5.2   Introduction to the SimMechanics Blockset

SimMechanics Blockset was released in 2001. With the powerful MATLAB/Simulink and three-dimensional animation facilities, mechanics systems can be modeled and simulated with building blocks. This is an innovative achievement of MATLAB in the area of physical modeling. SimMechanics can be used to model various bodies and joints based on the concept of Newtonian mechanics, and analysis and design are implemented for the established mechanisms.

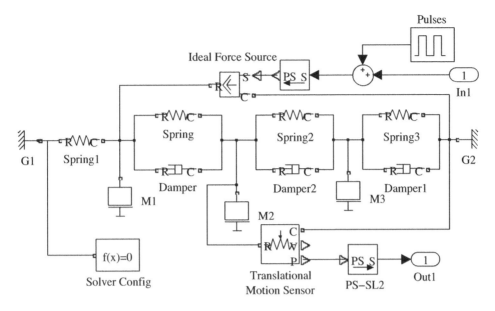

Figure 7.70 Simulink model of the multiple spring–damper system (model name: c7mdamp2).

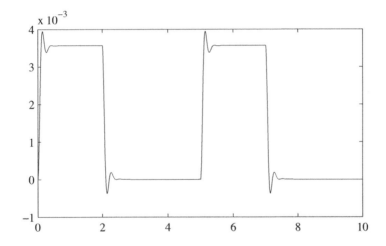

Figure 7.71 Displacement curve of the multiple spring–damper system.

SimMechanics blockset can be opened from the Simulink block library or with the MATLAB command `mechlib`. The blockset interface is shown in Fig. 7.72, with the following groups:

- **Bodies** group. Double click the icon and, as shown in Fig. 7.73, the group provides four blocks. The **Ground** block has one port, and the **Body** has two ports, labeled **B** (for base) and **F** (for follower). From a system point of view, **B** can be understood as the input while **F** is the output.

  Also, the other two blocks **Machine Environment** and **Shared Environment** are provided. They can be used to specify the mechanical and solver parameters of the system.

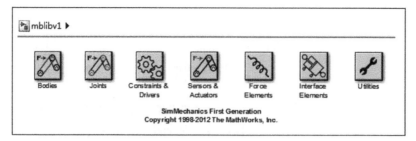

(a) standard SimMechanics — first generation version

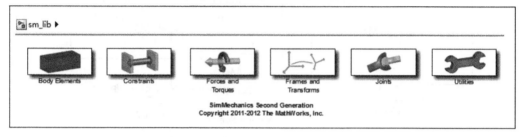

(b) SimMechanics — second generation version

Figure 7.72   SimMechanics blocksets.

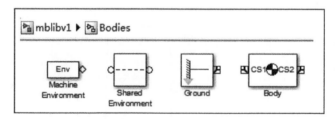

Figure 7.73   Body group.

- **Interface Elements** group is shown in Fig. 7.74. Two blocks **Prismatic-Translational Interface** and **Revolute-Rotational Interface** are provided and they are used for interfacing SimMechanics and Simscape Foundation Library blocks.

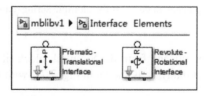

Figure 7.74   Interface Elements group.

- **Force Elements** group is shown in Fig. 7.75. The blocks **Body Spring & Damper** and **Joint Spring & Damper** are provided. The former is a spring–damper mechanism connected to a planar body, and the latter is a spring–damper connected to a joint.

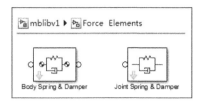

Figure 7.75   Force Elements group.

- **Constraints & Drivers** group with blocks is shown in Fig. 7.76. In this group, statics constraints elements such as **Gear Constraint, Parallel Constraint** and **Point-Curve Constraint** are provided. Drivers such as **Distance Driver, Angle Drive, Linear Driver** and **Velocity Driver** blocks are also provided.

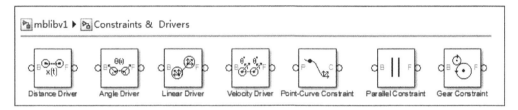

Figure 7.76   Constraints & Drivers group.

- **Sensors & Actuators** group is shown in Fig. 7.77. The blocks in the group are used to exchange information with Simulink signals. For instance, **Body Sensor** block can be added to the extra port of the **Body** blocks such that the information of position, velocity, acceleration in translational and rotational motion are detected, and the output signal can be connected to Simulink blocks such as **Scope**. The actuators can be used to receive Simulink signals. For instance, if an external force is applied to a body object, the force can be connected to a **Body Actuator**, and the output can be connected to the additional port of the **Body** block. This method seems to be complicated, but connections between Simulink and SimMechanics blocks must be converted using these intermediate components.

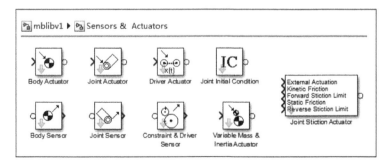

Figure 7.77   Sensors & Actuators group.

- **Utilities** group is shown in Fig. 7.78, with the blocks **Continuous Angle, Mechanical Branching Bar** and **Convert from Rotation Matrix to Simulink 3D Animation format**.

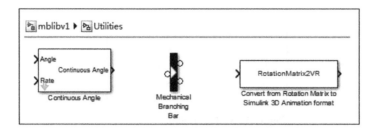

Figure 7.78   **Utilities** group.

- **Joints** group is shown in Fig. 7.79. In this group, various joints are provided, such as **Prismatic**, **Revolute**, **Spherical**, **Planar**, **Universal**, **Cylindrical**, **Weld**, **Screw** and **Six-DoF**.

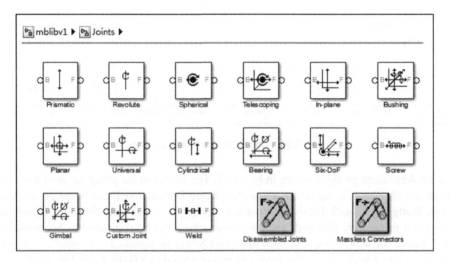

Figure 7.79   **Joints** group.

7.5.3   *Examples of Mechanical System Simulation*

In robot control systems, the mechanical part can be modeled with SimMechanics, and the control part can be modeled with MATLAB/Simulink and other blocksets, such that the whole system can be modeled and simulated using the same framework. This makes the whole system modeling much easier and more effective.

A four-bar mechanism system is used as an example to demonstrate the modeling and simulation process. SimMechanics is used in the modeling process. With the introductory modeling techniques, the readers should be able to model other mechanical systems.

▣ **Example 7.18**   *Consider the four-bar mechanism shown in Fig. 7.80 [5], with the sizes of the bars specified. It can be seen from the sketch that there are two fixed standers, three bars linked to each other or to the standers with four revolute pairs with one degree of freedom. In the mechanism, the dotted line AD is usually regarded as a bar, such that the mechanism is usually referred to as a four-bar mechanism.*

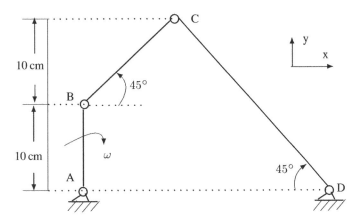

Figure 7.80   Illustration of the four-bar mechanism.

Assume that $\omega = 0$. *For a four-bar mechanism, the following procedure can be used in the modeling and simulation:*

1) Block diagram construction

With the blocks provided in SimMechanics, the prototype Simulink model can be constructed as shown in Fig. 7.81. In the block diagram, the two standers can be modeled by the **Ground** *block, and the bars are modeled with the* **Body** *blocks. The* **Revolute** *block from the* **Joints** *group can be copied into the model window to represent the revolute pairs; four of these blocks are needed. The* **Body** *block from the* **Body** *group can also be copied into the model window, three are needed. These blocks can then be connected in the order shown in Fig. 7.80. In the prototype model, the relevant blocks are copied and connected directly. The parameters and the interface with MATLAB/Simulink are not yet considered.*

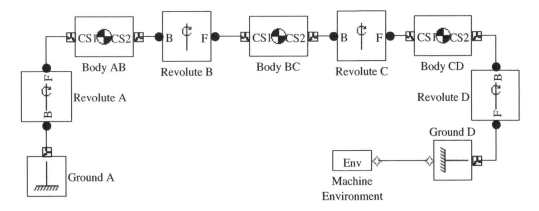

Figure 7.81   Prototype model of a four-bar mechanism (filename: c7mmech4.md).

The world coordinate setting is quite similar to the one used for virtual reality shown in Fig. 5.74(a). The orientation of the axes is also the same as that shown in Fig. 5.74(b), following the right hand principle. In the four-bar mechanism, the coordinates of the x and y axes are assumed to be as shown in Fig. 7.72, while the z axis is assumed to be coming out from the page.

The coordinate system is defined in this way, and the four revolute pairs should then be rotating in a positive direction around the z axis.

2) Model parameters setting

*a) **Ground parameter setting**. Assume that point A is the origin of the world coordinate. Thus the position of the left stander is (0,0,0) cm, while the right one is at (30,0,0) cm.*

*Double click that **Ground D** block. The dialog box shown in Fig. 7.82 will be displayed, and the **Location [x,y,z]** edit box can be assigned to vector variable* D, *that is, the vector* [30, 0, 0], *and the unit is assumed to be **cm**. Also, the **Machine Environment** block can be used, and the checkbox **Show Machine Environment port** should be selected.*

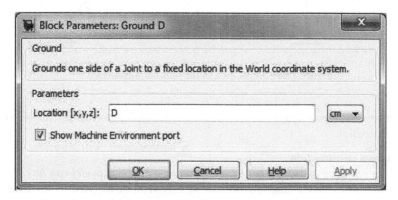

Figure 7.82   ***Ground*** parameter setting dialog box.

*b) **Environmental variable parameter setting**. Double click the **Machine Environment** block. The dialog box shown in Fig. 7.83 will be displayed, allowing the user to specify simulation control parameters of the mechanical system, for instance, the **Gravity vector** for the three axes. It is easily seen that the acceleration due to gravity vector is* [0, −9.81, 0], *meaning that, for x and z axes, the acceleration of gravity is zero, and for the y axis it is* −9.81 *m/s².*

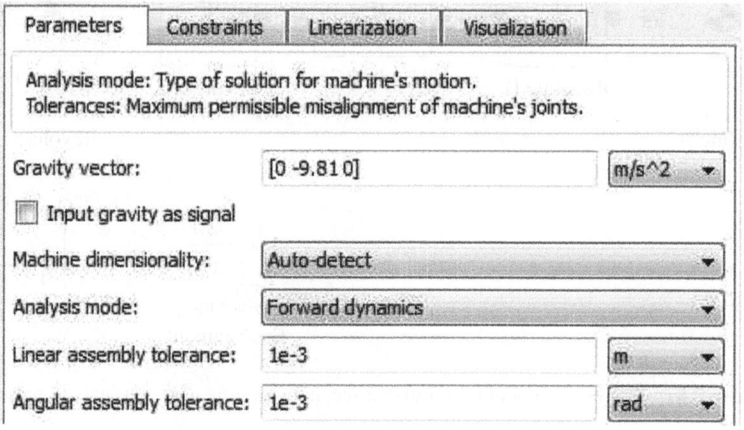

Figure 7.83   Environmental variable parameter dialog box.

*c) **Parameter setting of the revolute pairs**. Now consider the setting of the revolute pairs. Double click the **Revolute A** block and the dialog box shown in Fig. 7.84 will open. In the edit box labeled*

***Number of sensor/actuator ports**, the default value of 0 is specified. Under the **Parameters** pane, the parameters of the revolute pairs should be specified including the coordinate and the orientation vector. For simplicity, the coordinate of the four revolute pairs can all be set to the world coordinate. Since all four revolute pairs can rotate in the positive direction of the z axis, the **Axis of Action** edit box of each of them should be set to [0, 0, 1].*

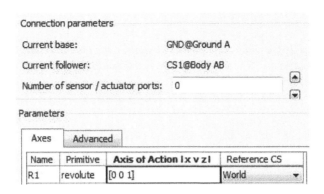

Figure 7.84   Parameter dialog box of revolute pairs.

*d) **Bar position specifications**. With simple computation from the sketch, it can be found that the bar lengths are respectively $l_{AB} = 10$ cm, $l_{BC} = 10\sqrt{2} = 14.14$ cm and $l_{CD} = 20\sqrt{2} = 28.28$ cm. The spatial positions of points B and C in the world coordinate are respectively $(0, 10, 0)$ cm and $(10, 20, 0)$ cm. Double click the dialog box of a bar. A dialog box shown in Fig. 7.85 will be opened. The following parameters are essential to bar objects.*

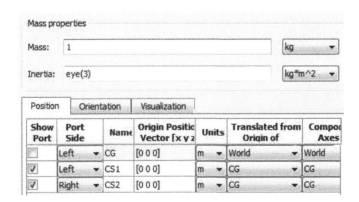

Figure 7.85   Parameter dialog box of **Body** block.

- ***Lengths of the bars**: Although in the dialog box, the lengths of the bars are not explicitly expected, they are actually needed in calculating the masses, gravitational forces and positions.*

- ***Masses of the bars**: Assume that the bars are all evenly distributed iron cylinders, with a diameter of 1 cm. The mass per unit length is $7.8\pi r^2 = 6.13$ g/cm. Since the lengths of the bars are known, the masses of the bars can be computed easily.*

- **Inertia matrix**: *The edit box labeled **Inertia tensor** accepts the inertia matrix of the body. For cylindrical iron bars, the inertia matrix is a diagonal matrix defined as*

$$T = \begin{bmatrix} mL^2/2 & & \\ & mr^2/12 & \\ & & mL^2/12 \end{bmatrix}$$

where m is the mass, r is the radius and L is the length of the bar. Note that the value should be converted to the unit of $kg \cdot m^2$. The calculation of inertia matrix for bodies of other shapes may be tedious. Sometimes numerical integration has to be evaluated. The following commands can be used to enter the necessary data into the MATLAB workspace

```
>> r=0.5; gg=7.81*pi*r^2; L1=10; L2=10*sqrt(2); L3=20*sqrt(2);
   B=[0,10,0]; C=[10,20,0]; D=[30,0,0]; M1=L1*gg*1e-9;
   M2=L2*gg*1e-9; M3=L3*gg*1e-9; T1=diag([r^2/2,L1^2/12,L1^2/12])*M1;
   T2=diag([r^2/2,L2^2/12,L2^2/12])*M2;
   T3=diag([r^2/2,L3^2/12,L3^2/12])*M3;
```

- **Center of gravity**: *The center of gravity of the bars, labeled **CG**, should be assigned in the dialog box. Since the bars are assumed to be evenly distributed, the center of gravity is in fact the middle point of the bar, and they can be easily calculated. The specification of the **CG** properties can still be described in terms of the world coordinate system, and for this example the centers of gravity of the three bars can be calculated as* B/2, (C-B)/2 *and* (D-C)/2.

- **Coordinate system**: *The coordinate system is labeled **CS**. In SimMechanics, when describing the positions, the relevant coordinate should also be given. In the c7mmech4 model, the world coordinate is mainly used. However, in practical mechanical systems, the world coordinate is not always the best choice. Different coordinates should be selected for different bodies.*

  *In the four-bar mechanism, each bar can have its own coordinate, with the base port as the origin. Now let us consider the **Body AB** block. Double click it, and the dialog box shown in Fig. 7.86 will be opened.*

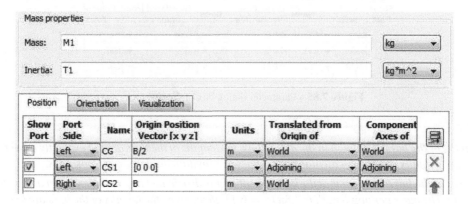

Figure 7.86   Dialog box of **Body AB** block.

*Since the base port **B** of each bar is connected directly to other blocks, in the base port setting of **CS1**, the **Translated from Original of** list should be assigned to **Adjoining**, and the **Origin Position Vector** edit box should be assigned to* [0, 0, 0]. *In the center of gravity and follower port **F** items, the world coordinate should be selected. and the center of gravity **CG** should be assigned to* `B/2`, *while for the follower port **CS2**, the vector* `B` *should be specified. Meanwhile, the inertia matrix **Inertia** edit box should be filled in with the variable name **T1**.*

*For the **Body BC** block, again the base port should be assigned as **Adjoining** and* [0, 0, 0]. *The follower port should be assigned as the world coordinate, with position* `C`, *and the center of gravity should also use the world coordinate, with position of* `(C-B)/2`, *as shown in Fig. 7.87(a). Since in the **Body CD** block, the base and follower ports are all connected to pre-specified blocks, they can both be assigned as **Adjoining** with position* `[0,0,0]`. *The center of gravity should still be assigned to the world coordinate, with the position* `(D-C)/2`, *as shown in Fig. 7.87(b).*

Position | Orientation | Visualization

Show Port	Port Side	Name	Origin Position Vector [x y z]	Units	Translated from Origin of	Components Axes of
☐	Left	CG	(C-B)/2	m	World	World
☑	Left	CS1	[0 0 0]	m	Adjoining	Adjoining
☑	Right	CS2	C	m	World	World

(a) **Body BC** dialog box

Position | Orientation | Visualization

Show Port	Port Side	Name	Origin Position Vector [x y z]	Units	Translated from Origin of	Components Axes of
☐	Left	CG	(D-C)/2	m	World	World
☑	Left	CS1	[0 0 0]	m	Adjoining	Adjoining
☑	Right	CS2	[0 0 0]	m	Adjoining	Adjoining

(b) **Body CD** dialog box

Figure 7.87 Position parameters of other bars.

3) Simulation parameters setting

*In the simulation process of SimMechanics, apart from the common control parameters specified in Simulink, some other simulation parameters should be assigned. When SimMechanics is installed, a **Simscape → SimMechanics** menu will be added to the **Simulation → Configuration Parameters** menu, as shown in Fig. 7.88. If animation display is required, the **Show animation during simulation** checkbox should be selected.*

4) Simulation and analysis of the system

*Once all the settings are completed, the menu **Simulation → Start** can be used to start the simulation process, and the simulation results can be obtained. For instance, for this example, the simulation results can be displayed as shown in Fig. 7.89. It can be seen that the seemingly complicated mechanical system can easily be simulated with SimMechanics and the MATLAB environment. With the use of three-dimensional animation facilities, the simulation results can be visually animated.*

7.5.4 Interfacing Simulink with Other CAD Tools

The mechanical system modeling method based on SimMechanics and Simscape blocksets has been presented. It should be pointed out that this method is not the best one in the field of mechanical

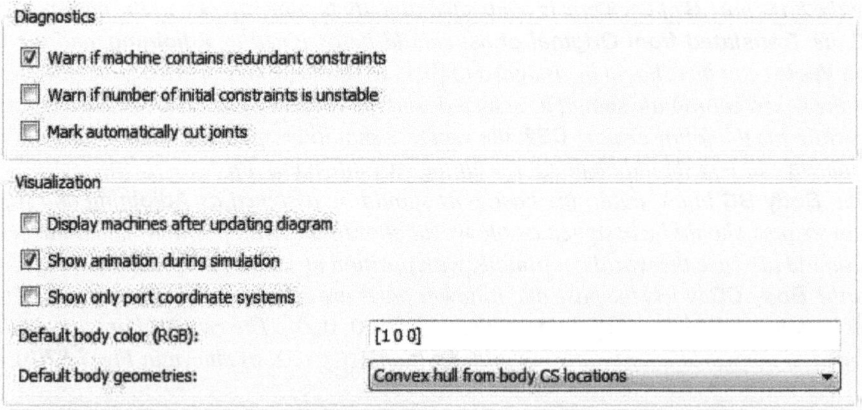

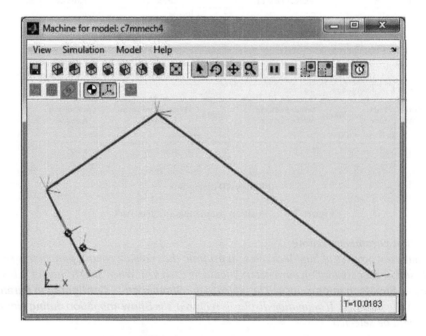

Figure 7.88   Dialog box for mechanical simulation parameters.

Figure 7.89   Animation of simulation results.

engineering. Broadly speaking, when the whole system consists of several subsystems such as electrical, control and communication subsystems, simulation under the Simulink framework is currently the only choice. This means that the simulation models with other professional software should be able to be embedded into Simulink. There are currently two options for these problems: one is to convert the models established in other professional software into Simulink subsystems, and the other is to establish a joint simulation mechanism with Simulink and the other professional software.

SimMechanics provides interfaces with several commonly used mechanical CAD systems, such as SolidWorks or Pro/ENGINEER (ProE). Readers familiar with these tools can model mechanical systems in that software, and then translate the models into SimMechanics, using the relevant commands provided in Simulink. Thus they can be converted into SimMechanics blocks easily and automatically.

■ **Example 7.19**   *A file robot.xml is provided in SimMechanics [6]. The file was created in SolidWorks: a robot arm model was designed and the data was extracted. The prototype model in SolidWorks is shown in Fig. 7.90.*

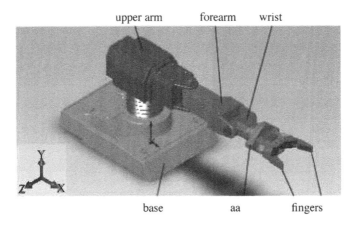

Figure 7.90   Prototype of robot arm design by SolidWorks.

With the MATLAB function `mech_import('robot.xml')`, *the prototype described in the file robot.xml can be loaded into MATLAB, and a SimMechanics model can be generated automatically, as shown in Fig. 7.91, the layout of the model having been modified slightly. Sensors and actuators can be added to the model to implement measurement and control of the robot arm.*

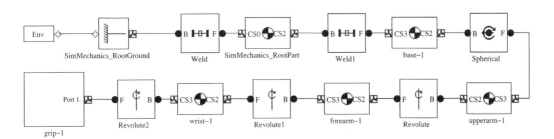

Figure 7.91   Automatically generated SimMechanics model (model name: c7_robot_t).

ADAMS [7] stands for Automatic Dynamic Analysis of Mechanical Systems, a software developed by MDI. The software can be used in modeling and virtual prototyping of mechanical systems. Since well-established three-dimensional animation facilities are provided, it can also be used in mechanical system simulation. Also, there is a good interface between ADAMS and Simulink, and the joint simulation can be performed under Simulink such that the mechanical models developed in ADAMS can be invoked [8].

Exercises

7.1 With model masking facilities in Simulink, create individual R, L, C blocks from the **R-L-C Branch** block in the SimPowerSystems blockset.

7.2 Consider the circuit shown in Fig. 7.92. Assume that $R_1 = R_2 = R_3 = R_4 = 10\,\Omega, C_1 = C_2 = C_3 = 10\,\mu\text{F}$. The input voltage is a sinusoidal function $v(t) = \sin(\omega t)\text{V}$, with $\omega = 10$ rad/s. Simulate the circuit to find the analytical solution of the output voltage $v_c(t)$. Draw the Bode plot of the circuit: (a) linearize the model, then use the bode() function to draw its Bode plot; (b) select a set of ω_i, and measure the magnitude and phase of the output voltage signal, so that the Bode plot can be drawn. Check whether the two results are consistent.

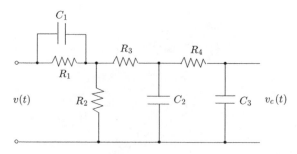

Figure 7.92   Problem (7.2).

7.3 Establish simulation models for the transistor amplifier circuits shown in Figs 7.93(a) and (b), and draw time responses under various control signals.

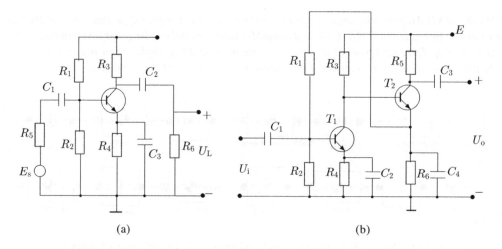

(a)   (b)

Figure 7.93   Problem (7.3).

7.4 Establish simulation models for the circuits with operational amplifiers shown in Figs 7.94(a), (b) and (c). Draw the system responses under different input signals, and extract the linearized models. By assigning the gains of the operational amplifiers to 10^4, 10^6 and infinity, compare the results.

7.5 Establish the Simulink models for digital logic circuits and then validate following equations
(a) $Z_1 = A + B\overline{C} + D$, (b) $Z_2 = AB(C + D) + D + \overline{D}(A + B)(\overline{B} + \overline{C})$.

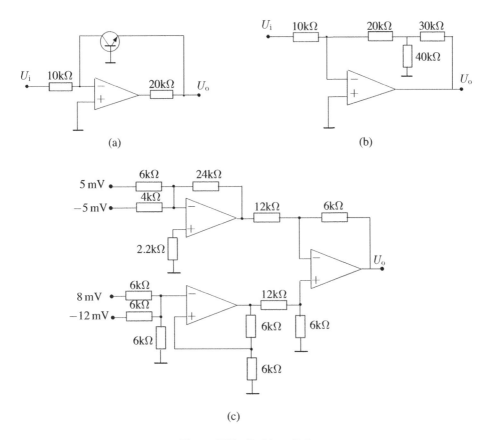

Figure 7.94   Problem (7.4).

7.6   Assume that the circuit of a trigger logic system is shown Fig. 7.95. Draw the time sequence diagram of the output signal through simulation methods.

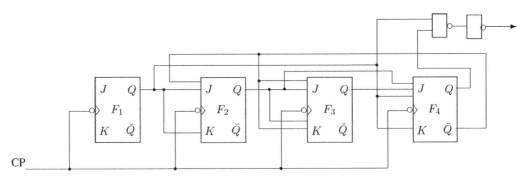

Figure 7.95   Problem (7.6).

7.7   Consider again the problem in Example 7.16. If the friction cannot be neglected, and it is known that the friction coefficient is $f_v = 0.25$, simulate the system and draw the results.

7.8 Create a three-dimensional animation world file with V-realm builder program, and simulate the c7mmech4 model with animation display.

References

[1] R C Dorf, R H Bishop. Modern control systems, 11th Edition. Upper Saddle River: Pearson, Prentice-Hall, 2008

[2] R X Peng. Fundamentals of Digital Electronics. Wuhan: Wuhan University of Technology Press, 2001. In Chinese

[3] W H Gao, H Wang. Analysis and design of analog circuits – PSpice applications. Beijing: Tsinghua University Press, 1999. In Chinese

[4] Y K He. Computer simulation of AC motors. Beijing: Science Press, 1990. In Chinese

[5] D Wang. Workbook of theoretical mechanics. Beijing: People's Education Press, 1963. In Chinese

[6] The MathWorks Inc. SimMechanics user's guide, 2010

[7] Z G Li. Introduction to ADAMS with examples. Beijing: National Defence Industry Press, 2006. In Chinese

[8] M Järviluoma, J Kortelainen. ADAMS/Simulink simulation of active damping of a heavy roller. Technical Report BTUO57-031129, VTT Technical Research Centre of Finland, 2003

8

Modeling and Simulation of Non-Engineering Systems

In the previous chapter we saw that there are various engineering blocksets available in MATLAB/Simulink, which can be used directly in modeling and simulation of engineering systems but there are also many non-engineering systems blocksets. Then there are also a lot of third-party programs and blocksets developed by scholars worldwide. Users can also develop their own toolboxes and blocksets.

In Section 8.1, modeling and simulation of pharmacokinetics systems are presented. Compartment modeling is briefly introduced, and physiology based pharmacokinetics modeling methods and nonlinear generalized predictive control of anesthetic processes are shown. Section 8.2 deals with MATLAB/Simulink based image and video processing. Image Processing Toolbox and Computer Vision System Toolbox blockset are also presented, and real-time video processing systems are explored. In Section 8.3, the finite state machine concept is explained and modeling and simulation problems with Stateflow are presented. Stateflow can often be used in complicated supervision modeling and simulation. This tool generalizes the capabilities of logical systems modeling, and can be used to describe systems with loops of conditional processes. In Section 8.4, we look at modeling and simulation of discrete event systems and a queuing system is used as an example to demonstrate the use of the SimEvents blockset.

8.1 Modeling and Simulation of Pharmacokinetics Systems

In this section the basic concepts of pharmacokinetics (PK) and pharmacodynamics (PD) are presented. Modeling and control problems are studied. In modeling, compartment models are considered first, and then physiology based modeling are covered. Simulink models of anesthetic processes are constructed and transfer function models are obtained. Finally, nonlinear generalized predictive control of anesthetic processes are explored.

8.1.1 Introduction to Pharmacokinetics

Pharmacokinetics is a discipline based on the theories of pharmacy, mathematics and kinetics, and it was established in 1980s. This discipline deals with the absorption, distribution, metabolism and excretion of drugs when they are introduced into human bodies, for example through intravenous (IV) administration. In particular, the relationships between drug concentration in plasma or in

System Simulation Techniques with MATLAB® and Simulink®, First Edition. Dingyü Xue and YangQuan Chen.
© 2014 John Wiley & Sons, Ltd. Published 2014 by John Wiley & Sons, Ltd.

organs are studied as a function of time. Pharmacodynamics is a discipline that studies mainly the physiological effects of drugs on human bodies. The mechanism of drug action, and the relationship of drug effect with concentration are also studied. The relationship between pharmacokinetics and pharmacodynamics is illustrated in Fig. 8.1, where $C_p(t)$ is the drug concentration, while $C_e(t)$ is the drug effect.

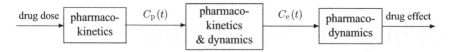

Figure 8.1 Illustration of pharmacokinetics and pharmacodynamics.

If feedback control philosophy is introduced, a controller can be designed so that the administration of drugs can be made in a closed loop, in order to achieve the best possible effect. For instance, the speed and dose of anesthetics can be controlled by a micro-injection pump so as to achieve the best performance.

8.1.2 Compartment Modeling of Pharmacokinetics Systems

It has been shown in pharmacokinetics experiments that the drug concentration in arterial blood can be written as the sum of several weighted exponentials [1]

$$c = \sum_{i=1}^{n} c_i e^{-\lambda_i t}, \tag{8.1}$$

where c_i and λ_i are undetermined coefficients, and n is the number of exponential terms. From the identifiability consideration, n cannot be too large. Normally, selecting n as 2 or 3 would be adequate [2]. A compartment model is an easy model to use for studying the relationship of the drug concentration in plasma over time. Human bodies are regarded as one or more interconnected compartments. In the model, the compartment is the basic modeling element, and there are interconnections between the compartments as shown in Fig. 8.2, where the variable k_{ij} is the flow from compartment j to compartment i, and the subscript 0 means the outside world. The mathematical model of the whole system can be written as

$$\frac{dx_i}{dt} = \sum_{j=1, j \neq i}^{n} k_{ij} x_j - \sum_{j=1, j \neq i}^{m} k_{ji} x_i - k_{0i} x_i + u_i(t), i = 1, 2, \cdots, n. \tag{8.2}$$

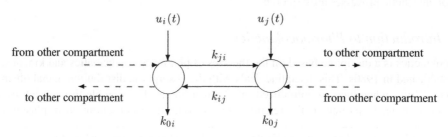

Figure 8.2 Interconnection of two compartments.

The compartment model is a special form of linear ordinary equation, and its matrix form can be written as

$$
\begin{bmatrix} \dot{x}_1 \\ \dot{x}_2 \\ \vdots \\ \dot{x}_n \end{bmatrix} = \begin{bmatrix} a_{11} & a_{12} & \cdots & a_{1n} \\ a_{21} & a_{22} & \cdots & a_{2n} \\ \vdots & \vdots & \ddots & \vdots \\ a_{n1} & a_{n2} & \cdots & a_{nn} \end{bmatrix} \begin{bmatrix} x_1 \\ x_2 \\ \vdots \\ x_n \end{bmatrix} + \boldsymbol{B}u \tag{8.3}
$$

where

$$
a_{ij} = k_{ij},\, j \neq i, \quad \text{and} \quad a_{jj} = \sum_{i=1,j\neq j}^{n} k_{ij} - k_{0j}, \tag{8.4}
$$

and the subscript 0 indicates the outside world

$$
a_{jj} \leq 0, a_{i,j} \geq 0 (i \neq j), \quad k_{0j} = 0, |a_{jj}| \geq \sum_{i=1,i\neq j}^{n} k_{ij}. \tag{8.5}
$$

Specially, if two or three compartments are involved, we can use the simplified form

$$
\begin{bmatrix} \dot{x}_1(t) \\ \dot{x}_2(t) \end{bmatrix} = \begin{bmatrix} -(c_1 + c_2) & c_3 \\ c_2 & -c_3 \end{bmatrix} \begin{bmatrix} x_1(t) \\ x_2(t) \end{bmatrix} + \begin{bmatrix} 1 \\ 0 \end{bmatrix} u(t), \quad y(t) = c_1 x_1(t), \tag{8.6}
$$

$$
\begin{bmatrix} \dot{x}_1(t) \\ \dot{x}_2(t) \\ \dot{x}_3(t) \end{bmatrix} = \begin{bmatrix} -(c_1 + c_2) & c_3 & 0 \\ c_2 & -(c_3 + c_4) & c_5 \\ 0 & c_4 & -c_5 \end{bmatrix} \begin{bmatrix} x_1(t) \\ x_2(t) \\ x_3(t) \end{bmatrix} + \begin{bmatrix} 1 \\ 0 \\ 0 \end{bmatrix} u(t), \quad y(t) = c_1 x_1(t). \tag{8.7}
$$

If these coefficients are given, the state space block in Simulink can be used directly to construct the model. A compartment based Wada blockset has been developed [3]. In the blockset, one-, two- and three-compartment blocks are provided in Fig. 8.3. The **Branch** and **Juncture** blocks are also provided where the **Branch** block can be used to control the distribution of drugs to different organs, controlled by the branch factor in the second input port. These blocks can be used directly in creating compartment models.

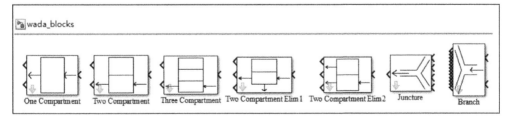

Figure 8.3   Wada blockset (blockset name: wada_blocks).

Right click the **Two-Compartment** block, and the shortcut menu will be displayed. Selecting the **Look under Mask** menu item from the shortcut menu, the internal model structure shown in Fig. 8.4(a) will be displayed. The parameter dialog box shown in Fig. 8.4(b) will then be displayed.

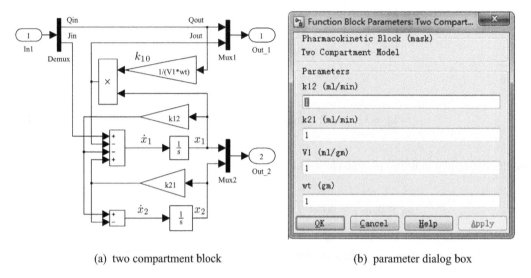

(a) two compartment block   (b) parameter dialog box

Figure 8.4   Two-compartment model and parameters setting.

From the model in Fig. 8.4(a), the mathematical model of the two-compartment model can be written as

$$\begin{bmatrix} \dot{x}_1 \\ \dot{x}_2 \end{bmatrix} = \begin{bmatrix} -(k_{12}+k_{10}) & k_{21} \\ k_{12} & -k_{21} \end{bmatrix} \begin{bmatrix} x_1 \\ x_2 \end{bmatrix} + \begin{bmatrix} 1 \\ 0 \end{bmatrix} u, \; \boldsymbol{y}_2 = \boldsymbol{x}, \tag{8.8}$$

where $k_{10} = 1/(Vw)$, w and V are respectively the mass and volume of the compartment.

The internal structures of the two-compartment models are shown respectively in Figs 8.5(a), (b); they all have additional output ports. The mathematical models are respectively

$$\begin{bmatrix} \dot{x}_1 \\ \dot{x}_2 \end{bmatrix} = \begin{bmatrix} -(k_{12}+k_{10}) & k_{21} \\ k_{12} & -k_{21} \end{bmatrix} \begin{bmatrix} x_1 \\ x_2 \end{bmatrix} + \begin{bmatrix} 1 \\ 0 \end{bmatrix} u, \; \boldsymbol{y}_2 = \boldsymbol{x}, \; y_3 = k_{20}x_2, \tag{8.9}$$

$$\begin{bmatrix} \dot{x}_1 \\ \dot{x}_2 \end{bmatrix} = \begin{bmatrix} -(k_{12}+k_{10}) & k_{21} \\ k_{12} & -k_{21} \end{bmatrix} + \begin{bmatrix} x_1 \\ x_2 \end{bmatrix} + \begin{bmatrix} 1 & k_{10} \\ 0 & 0 \end{bmatrix} \begin{bmatrix} u_1 \\ u_2 \end{bmatrix}, \; \boldsymbol{y}_2 = \boldsymbol{x}, \; y_3 = k_{10}x_1, \tag{8.10}$$

and it can be seen that the latter has two output ports, and the excretion and metabolism are defined differently.

If the compartment model is used alone, since the physiological structure of body is not being considered, and the state variables do not have good physical meaning, the control effect may not be satisfactory. Researchers are more interested in physiology based modeling techniques. The

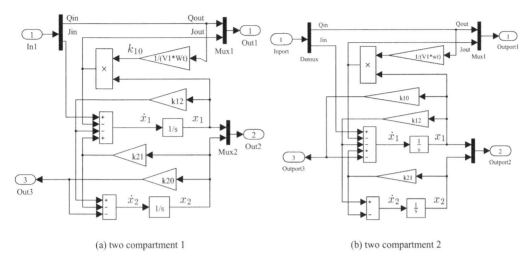

(a) two compartment 1   (b) two compartment 2

Figure 8.5   Two-compartment models with excretion and metabolism.

physiology based model can also be established with Wada's compartment blockset. Details of this will be given later.

8.1.3   *Physiology based Pharmacokinetic Modeling with Simulink*

8.1.3.1   **Wada Model with Transport Delays**

In ordinary compartment models, transport delays are not considered. If transport delays are considered, the model shown in Fig. 8.6 can be adopted. It can be seen that the model consists of two parts: the one in the forward path is called the cardiopulmonary subsystem, and the one in the feedback path is referred to as the systemic subsystem. The state variables for the system can be selected as

$$\boldsymbol{x}_{\mathrm{p}} = [x_{\mathrm{LH}}, x_{\mathrm{LB}}, x_{\mathrm{LT}}, x_{\mathrm{RH}}]^{T}, \boldsymbol{x}_{\mathrm{s}} = [x_{\mathrm{VRG}}, x_{\mathrm{M}}, x_{\mathrm{F}}, x_{\mathrm{R}}]^{T}, \tag{8.11}$$

the state space model with delay can be written as [4]

$$\begin{cases} \dot{\boldsymbol{x}}_{\mathrm{p}} = \boldsymbol{A}_{\mathrm{p}}\boldsymbol{x}_{\mathrm{p}} + \boldsymbol{B}_{\mathrm{p}}[Q\boldsymbol{c}_{\mathrm{v}} + u(t - \tau_{\mathrm{i}})] \\ c_{\mathrm{a}}(t) = \boldsymbol{C}_{\mathrm{p}}\boldsymbol{x}_{\mathrm{p}}(t - \tau_{\mathrm{p}}) \end{cases} \quad \begin{cases} \dot{\boldsymbol{x}}_{\mathrm{s}} = \boldsymbol{A}_{\mathrm{s}}\boldsymbol{x}_{\mathrm{s}} + \boldsymbol{B}_{\mathrm{s}}Qc_{\mathrm{a}} \\ c_{\mathrm{v}}(t) = \boldsymbol{C}_{\mathrm{s}}\boldsymbol{x}_{\mathrm{s}}(t - \tau_{\mathrm{r}}), \end{cases} \tag{8.12}$$

where the state variables $\boldsymbol{x}_{\mathrm{p}}$ and $\boldsymbol{x}_{\mathrm{s}}$ are respectively the drug mass cardiopulmonary subsystem, while the one in the feedback path is referred to as the systemic subsystem. The state variables for the system can be selected as the drug masses of the cardiopulmonary and systemic subsystems,

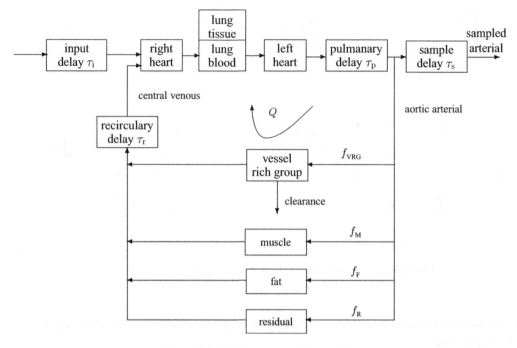

Figure 8.6   Wada transport delay model.

respectively. The volumes of the organs are expressed by V. The matrices in the state space model are then

$$
A_{\mathrm{p}} = \begin{bmatrix} -\dfrac{Q}{V_{\mathrm{RH}}} & 0 & 0 & 0 \\[2ex] \dfrac{Q}{V_{\mathrm{RH}}} & -\left(k_{\mathrm{LB,LT}}+\dfrac{Q}{V_{\mathrm{lung}}}\right) & k_{\mathrm{LB,LT}} & 0 \\[2ex] 0 & k_{\mathrm{LB,LT}} & -k_{\mathrm{LT,LB}} & 0 \\[2ex] 0 & \dfrac{Q}{V_{\mathrm{lung}}} & 0 & -\dfrac{Q}{V_{\mathrm{LH}}} \end{bmatrix}, \quad B_{\mathrm{p}} = \begin{bmatrix} 1 \\ 0 \\ 0 \\ 0 \end{bmatrix}, \quad C_{\mathrm{p}}^{T} = \begin{bmatrix} 0 \\ 0 \\ 0 \\ \dfrac{1}{V_{\mathrm{LH}}} \end{bmatrix}, \quad (8.13)
$$

and

$$
A_{\mathrm{s}} = \begin{bmatrix} -\dfrac{Qf_{\mathrm{VRG}}+Cl}{V_{\mathrm{VRG}}} & 0 & 0 & 0 \\[2ex] 0 & -\dfrac{Qf_{\mathrm{M}}}{V_{\mathrm{M}}} & 0 & 0 \\[2ex] 0 & 0 & -\dfrac{Qf_{\mathrm{F}}}{V_{\mathrm{F}}} & 0 \\[2ex] 0 & 0 & 0 & -\dfrac{Qf_{\mathrm{R}}}{V_{\mathrm{R}}} \end{bmatrix}, \quad B_{\mathrm{s}} = \begin{bmatrix} f_{\mathrm{VGR}} \\ f_{\mathrm{M}} \\ f_{\mathrm{F}} \\ f_{\mathrm{R}} \end{bmatrix}, \quad C_{\mathrm{s}}^{T} = \begin{bmatrix} \dfrac{f_{\mathrm{VGR}}}{V_{\mathrm{VGR}}} \\[2ex] \dfrac{f_{\mathrm{M}}}{V_{\mathrm{M}}} \\[2ex] \dfrac{f_{\mathrm{F}}}{V_{\mathrm{F}}} \\[2ex] \dfrac{f_{\mathrm{R}}}{V_{\mathrm{R}}} \end{bmatrix}. \quad (8.14)
$$

Wada presented a set of actual parameters [4]

```
>> T_i=5/60; T_s=5/60; T_p=5/60; T_r=30/60; Q=5.5; V_RH=0.26;
   V_lung=0.45; V_LH=0.26; k_LBLT=8.0; k_LTLB=2.3;   Cl=0.64;
   rho=0.63; V_VRG=5.5; f_VRG=0.7; V_M=13.2; f_M=0.05; V_F=36.7;
   f_F=0.04; V_R=14.9; f_R=0.21;
   A_P=[-Q/V_RH, 0, 0, 0; Q/V_RH, -(k_LBLT+Q/V_lung), k_LTLB, 0;
        0, k_LBLT, -k_LTLB, 0; 0, Q/V_lung, 0, -Q/V_LH];
   B_P=[1; 0; 0; 0]; C_P=[0,0,0,1/V_LH];
   A_S=diag([-(Q*f_VRG+Cl)/V_VRG, -Q*f_M/V_M, ...
             -Q*f_F/V_F, -Q*f_R/V_R]);
   B_S=[f_VRG; f_M; f_F; f_R];
   C_S=[f_VRG/V_VRG, f_M/V_M, f_F/V_F, f_R/V_R];
   T=[0,1,1.01,1.2,1.21,5]'; U=[50,50,29,29,10,9]';
```

From the above mathematical description, the Simulink model can be established as shown in Fig. 8.7. Users can assign the above MATLAB commands to the **PreLoadFcn** property of the model such that each time the model is opened, the data are loaded into the MATLAB workspace automatically. The system can then be simulated, and the input–output curves can be obtained as shown in Figs 8.8(a) and (b), with the following statements

```
>> stairs(T,U); figure; [t,x,y]=sim('wada_dly',[0,10]); plot(t,y)
```

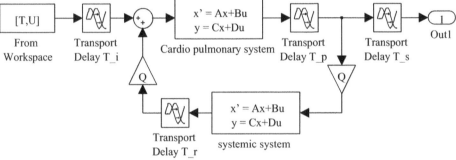

Figure 8.7   Wada model with transport delay (model file: wada_dly).

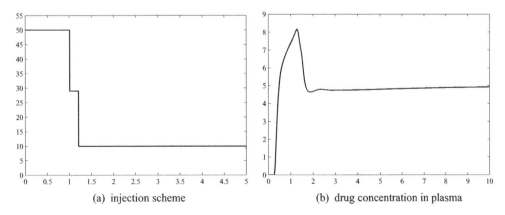

(a) injection scheme   (b) drug concentration in plasma

Figure 8.8   Drug injection scheme and concentration in plasma.

8.1.3.2   Pharmacokinetics Toolbox and Simulink Models

A pharmacokinetics toolbox and Wada blockset is a practical tool in the modeling of anesthetic processes [3], and it can be initiated with the command `wada_model`. The blockset is shown in Fig. 8.3. Also, a physiology based pharmacokinetics Simulink model is offered, as shown in Fig. 8.9. In the model, each group of organs is modeled with a compartment, and the compartment blocks can be used directly to create the model. The toolbox provided can be used to set the parameters of the compartments. For instance, rat modeling under fentanyl can be directly assigned so a fentanyl.m file, while the data in human bodies can be assigned to a file fent_hum.m.

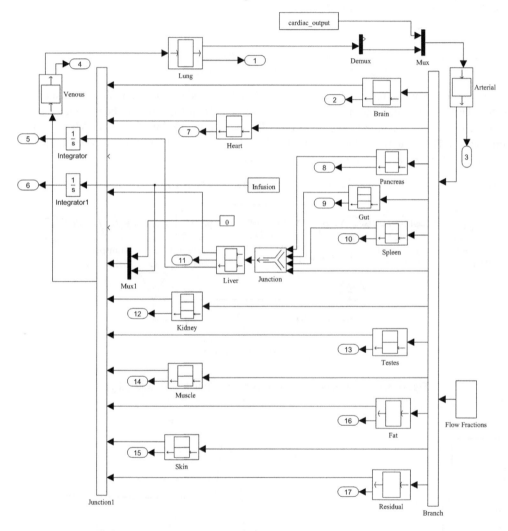

Figure 8.9   Physiological pharmacokinetics model (model name: wada_model).

8.1.3.3   Mapleson Model and Programs

In this section, the effects of anesthetics upon the human body are studied, and control problems are considered. Another physiology based pharmacokinetics model was proposed by Mapleson [1], and there are differences from the Wada model. In this model, the effect of organs upon the drug can be shown in Fig. 8.10, where K_x is the branch factor for each branch, and excretion and metabolism are conducted in kidney and liver, respectively. Higgins extended the Mapleson model and specified data in the Mapleson model, and wrote a program to assign data in the model when fentanyl is administrated [5].

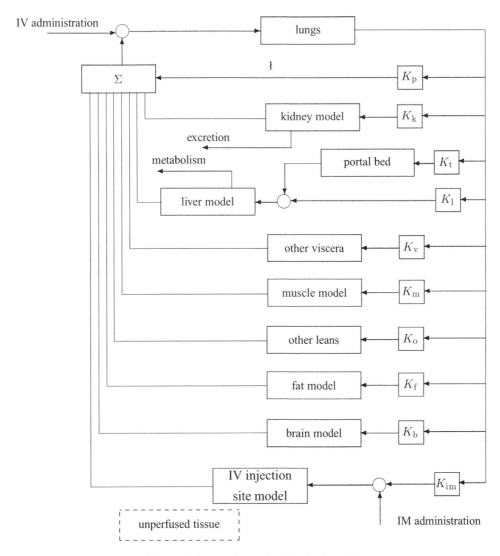

Figure 8.10 Mapleson physiological model.

Unfortunately, the Higgins model was written in C, so a MEX interface to the C code was written such that the program can be executed under the MATLAB environment [6, 7]. The approximate linear transfer functions for each organ were obtained and finally the Simulink system based on the Mapleson model was constructed as shown in Fig. 8.11. In the block diagram the coefficient 336.6 was obtained to convert the unit μmol in the original C code into the unit ng. It can be seen from the

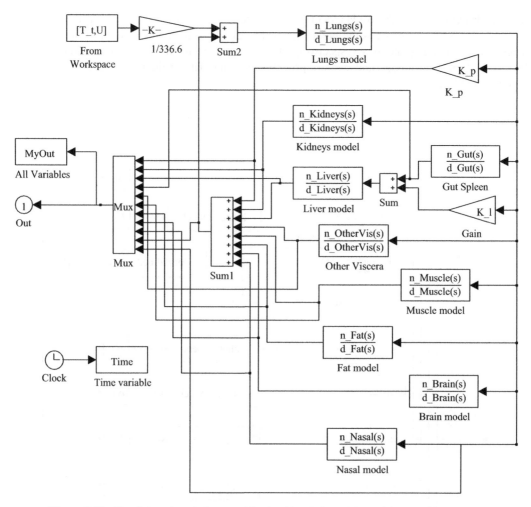

Figure 8.11 Physiology based pharmacokinetics Simulink model (model name: hig_simu0).

block diagram that, since transfer function blocks were used, the structure and parameters are much simpler than the compartment blocks in Wada's model. It has been shown [6, 7] that the parameters in the subtransfer function models in each organ for different weights and cardiac output can be interpolated. Thus the system model parameters for different weights and cardiac outputs can be established in the new model. Here only syntaxes are shown for the modeling and analysis; the code can be downloaded from the book's website.

```
input parameters:  bd_wt, t_f, DT, card_out, FenDose, DoseDur
get_higgins_new,  % call higginsm.mexw32 function in simulation
ident_tf             % identify subtransfer functions from simulation results
```

where the variable bd_wt is the weight (kg), card_out is the cardiac output (L/min), t_f is the final simulation time (s), DT is the step size (s), and FenDose and DoseDur are vectors describing the injection dose (μg) and time durations (s), respectively.

📖 **Example 8.1** *Assuming the weight of a person is 60 kg, the cardiac output 6.8 L/min, simulation duration 3600 s, step size 1 s, total dose 100 μg and injection period is 60 s, the following commands can then be used to assign the data in the MATLAB workspace. The modified Higgins program in MEX C form can be called so that the drug concentration in the organs can be calculated and the transfer functions can be identified.*

```
>> bd_wt=60; card_out=6.8; FenDose=100; DoseDur=60; t_f=3600; DT=1;
   get_higgins_new; ident_tf, G_Brain, G_Lungs
```

The subtransfer functions can be identified. For instance, the brain model and lungs model are respectively

$$G_{\text{brain}}(s) = \frac{6.14 \times 10^{-5}}{s + 0.2988}, \quad G_{\text{lungs}}(s) = \frac{0.3995s + 153.8}{s^2 + 9.966s + 31.16}.$$

Apart from the subtransfer functions, other drug concentrations can also be calculated. For instance, `ArtPool(:,1)` *stores the drug concentration in artery in μmol/L. The drug concentration in other organs can also be returned to the MATLAB workspace. The variable names returned can be listed with the MATLAB command* `who`.

With the subtransfer functions, the overall transfer function can then be evaluated directly with the relevant MATLAB commands. For instance, the overall transfer function from the IV administration input to the drug concentration in the artery pool can be obtained with the following MATLAB statements

```
>> G_Fdbk=K_p+G_Kidneys+(G_Gut+K_l)*G_Liver+G_OtherVis+G_Muscle+...
         G_Fat+G_Nasal+G_Brain;    % overall transfer function in feedback path
   G_Sys=zpk(feedback(G_Lungs,G_Fdbk,1))
```

$$G(s) = \frac{\begin{array}{c} 0.39945(s+385.1)(s+1.425)(s+1.26)(s+0.8064)(s+0.8043)(s+0.2988) \\ (s+0.1894)(s+0.1757)(s+0.1625)(s+0.1069)(s+0.008874)(s+0.001536) \end{array}}{\begin{array}{c} (s+1.261)(s+1.015)(s+0.8103)(s+0.8063)(s+0.2984)(s+0.1927)(s+0.1626) \\ (s+0.1495)(s+0.06935)(s+0.004999)(s+0.001113)(s^2+10.42\,s+30.86) \end{array}}.$$

With optimal model reduction techniques [8], a fifth-order model can be found

$$G_1(s) = \frac{12.973(s + 0.3171)(s + 0.1201)(s + 0.008881)(s + 0.001536)}{(s + 1.956)(s + 0.2878)(s + 0.06987)(s + 0.004935)(s + 0.001098)},$$

```
>> G1=opt_app(G_Sys,4,5,0);
   for i=1:3, G1=opt_app(G_Sys,4,5,0,G1); end
```

where `opt_app()` *function [9] can be downloaded for the book's website. If the initial values of the reduced-order model are not selected properly, loops can be used to further search for the optimum reduced-order models.*

With the following statements, the artery concentration curve, together with linearized model and reduced fifth-order model can be obtained as shown in Fig. 8.12. It can be seen that the identified model is very close to the Higgins model in terms of simulation results. Also, it can be seen that the reduced fifth-order model can approximate the simulation model very well.

```
>> u=100/60*(t<=1); y=ArtPool(:,1)*336.6; % artery concentration in ng/ml
   y1=lsim(G_Sys,u,t); y2=lsim(G1,u,t); plot(t,y,t,y1,'--',t,y2,':')
```

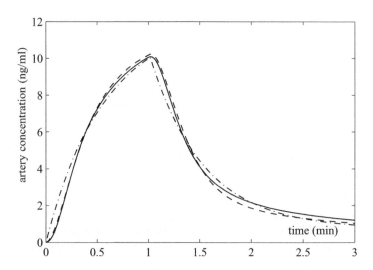

Figure 8.12   Artery concentration with different methods.

It can be seen from the Wada model and the Higgins model that the former method can only be used for fixed parameters. If the weight and cardiac output are changed, the Wada model cannot be used, while Higgins code can be used to solve the problems easily. The conversion from Higgins code to Simulink model, and the deficiency in the Higgins model are overcome. The Simulink interpretation can also be obtained with the new code.

8.1.4   Pharmacodynamic Modeling

It has been stated that pharmacokinetics can be used to set up the relationship between the drug dose and the drug concentration. In pharmacodynamics, drug effect is studied. The drug concentration in the sample brain $C_b(t)$ of the Higgins model is used to bridge the two disciplines. Suppose that the concentration of the sample brain can be written as $C_b(t)$, then the drug effect can be defined by either of the following two quantities:

$$C_{e1}(t) = \frac{C_b^\gamma(t)}{EC_{50}^\gamma + C_b^\gamma(t)}, \quad C_{e2}(t) = E_0 - \frac{C_b^\gamma(t)}{EC_{50}^\gamma + C_b^\gamma(t)} E_{max} \tag{8.15}$$

where EC_{50} is the concentration at 50% drug effect. For a different drug, the values are all different. For fentanyl, in different papers the values of EC_{50} are 6.9, 7.8 and 8.1 ng/ml. The equation $C_{e1}(t)$ is also known as the Hill equation, and γ is often known as the steepness of the Hill equation. The concentration $C_{e1}(t)$ can be directly measured with electroencephalogram (EEG), where $C_b(t)$ is the drug concentration in the sample brain. $C_{e2}(t)$ is mainly the frequency response concept, where the drug effect is also known as the spectral edge. The constant R_0 is defined as $E_0 = 20$ Hz, and the frequency at the maximum drug effect is often given by $E_{max} = 15$ Hz.

Based on the two pharmacodynamics criteria, the Simulink model can be constructed as shown in Fig. 8.13(a). The parameter input dialog box for the masked block is shown in Fig. 8.13(b), and the masked block is shown in Fig. 8.13(c). The block can be modeled directly with pharmacokinetics and pharmacodynamic modeling.

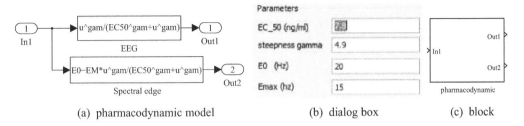

(a) pharmacodynamic model   (b) dialog box   (c) block

Figure 8.13 Masked pharmacodynamic model (model name: c8mpdblk).

8.1.5 *Nonlinear Generalized Predictive Control of Anesthesia*

A MEX C control system program was developed, where the Higgins pharmacokinetics, pharmacodynamics and generalized predictive controller routines are embedded. The executable file nln_gpcx.mexw32 can be generated and it can be executed in MATLAB directly. Meanwhile, a graphical user interface has been written to implement the modeling, simulation and control facilities. The filename is higmodel.m. This program can be executed in MATLAB directly. The command `higmodel` can be used to start the program, and the interface shown in Fig. 8.14(a) will be displayed. The menu **Model** → **Parameters** can be used to open the dialog box shown in Fig. 8.14(b), and the parameters can be entered in the dialog box. The button **Confirm** can be clicked to accept the parameters. System modeling and analysis can then be completed with the interface.

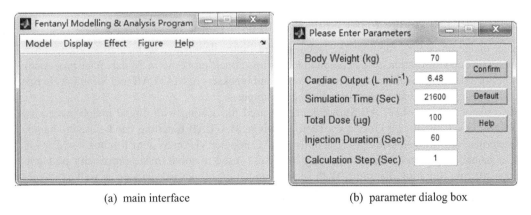

(a) main interface   (b) parameter dialog box

Figure 8.14 Graphical user interface of anesthetic process.

The parameters can be entered, then the menu **Model** → **Run algebraic model** can be used to calculate the drug concentrations in different organs, with the Higgins model. Simulation can then be performed. The menu **Model** → **Identification** can be used to identify the subtransfer functions.

Identification results can be displayed with the **Display** menu. Users can selectively display the drug concentration, subsystems and overall models. Also, the reduced-order models can be displayed.

Simulink model parameters can either be identified from the simulation results, or by interpolation methods . Here, the **Model** → **Establish Simulink model** command can be used to set the parameters in the model. The identification method is suggested such that accurate results can be obtained.

The **Effect** → **EEG** menu can be used to draw drug effect curves with pharmacodynamic curves, as shown in Fig. 8.15(a). The **Effect** → **NLGPC control** menu can be used to simulate nonlinear generalized predictive control to achieve drug effect, by computer-controlled injection with the IV administration pump. The control results can be obtained as shown in Fig. 8.15(b). It can be seen that the control results are satisfactory.

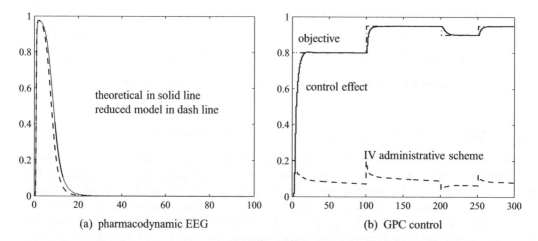

Figure 8.15   Pharmacodynamic and control results.

8.2   Video and Image Processing Systems

Digital image processing includes the techniques of image denoising, segmentation, recognition, compression and restoration, using computer technology. Basic problems in digital image processing can be solved directly with the relevant toolboxes and blocksets in MATLAB and Simulink. In this section, the topics are discussed mainly through examples.

Two major tools in MATLAB/Simulink can be used for dealing with digital image processing problems. One is the Image Processing Toolbox, where MATLAB functions can be used to handle image processing problems directly. The other is the Computer Vision System Toolbox – which was earlier named Video and Image Processing Blockset – used to solve image processing problems by constructing Simulink diagrams. With such a simulation framework, videos as well as images can be handled directly in Simulink, and real-time processing system of images and videos can be established.

With the command `viplib`, the main interface of the Computer Vision System Toolbox can be opened as shown in Fig. 8.16, where the groups such as **Sources**, **Analysis & Enhancement**, **Sinks**, **Text & Graphics**, **Conversions**, **Transforms**, **Morphological Operations** and **Filtering** groups are provided. Also, plenty of illustrative examples are provided in the **Demos** group, and they can be reused easily for solving related problems.

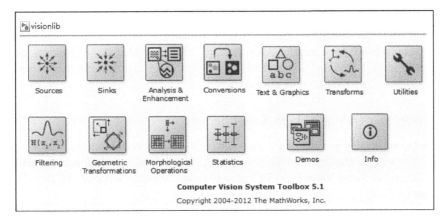

Figure 8.16 Computer Vision System Toolbox.

8.2.1 Importing Pictures and Videos into MATLAB

Various blocks are provided in the blockset for importing images and videos into the Simulink environment. These blocks are provided in the **Sources** group, and are displayed as shown in Fig. 8.17, when the group icon is double clicked. The commonly used blocks in the group are summarized below

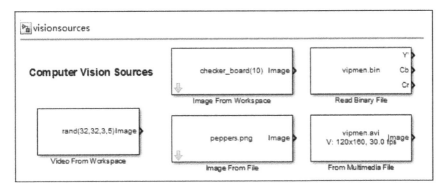

Figure 8.17 **Sources** group for image file input.

- **Image file inputting**: Images in files and in MATLAB workspace can be loaded into Simulink environments, with the blocks **Image From File** and **Image From Workspace**. The currently supported file types are bmp, jpg, jpeg, png, tif and tiff, covers most commonly used image types. These image blocks can be used directly in Simulink modeling as an input source. Static image files can also be read with the imread() function provided in the Image Processing Toolbox.

- **Multimedia inputting**: The multimedia files can be loaded into the Simulink environment with the **Video From File** and **Video From Workspace** blocks. The currently supported file suffixes are avi, mp4, wmv, wav, wma and mp3. The loaded images and sound can be processed directly in Simulink. With the MATLAB functions mmreader(), multimedia files can also be loaded into MATLAB directly. It should be noted that to load the whole of a multimedia file, especially large-scale multimedia files, a large amount of memory is needed, if those in the blockset need to process the video frame by frame. The block diagram approach is much more economical in large-scale video processing tasks.

- **Real-time acquisition from cameras**: Apart from the block provided in the blocksets, an Image Acquisition Toolbox is provided. In the blockset, only one block is provided to load the real-time video from cameras into the Simulink environment.

8.2.2   *Display and Output of Videos and Images*

The image and video display group **Sinks** is shown in Fig. 8.18, and the images and videos can either be saved into files with **To Multimedia File** or the MATLAB workspace with **Video To Workspace**, or displayed directly in **Video Viewer** and **To Video Display**. Also, the video can be stored with the **To Binary File** block.

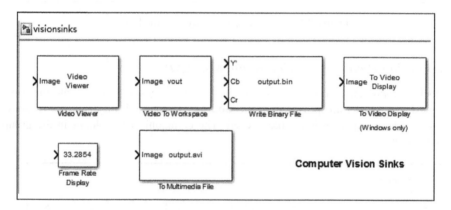

Figure 8.18   Sinks group.

Example 8.2   *Now consider a simple example, where we want to play a video file stored in the computer. For instance, the high-definition movie file wildlife.wmv saved in the windows folder, the Simulink model shown in Fig. 8.19(a) can be constructed. In this model, the **From Multimedia File** block from the **Sources** group, and the **View Viewer** block from the **Sinks** group can be used in constructing the Simulink model. Double click the input block, and the dialog box shown in Fig. 8.20 is displayed. The filename should be specified in the dialog box to establish the link between the model and the file. The **Image signal** listbox can either be assigned to the default **One multidimensional***

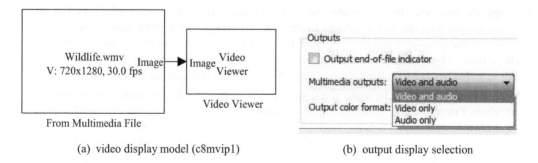

(a) video display model (c8mvip1)   (b) output display selection

Figure 8.19   Simple video display system.

*signal, or to the **Separate color signal** option. For colored video files, the latter option will provide three output ports, returning the red, green and blue (RGB) components of the color videos.*

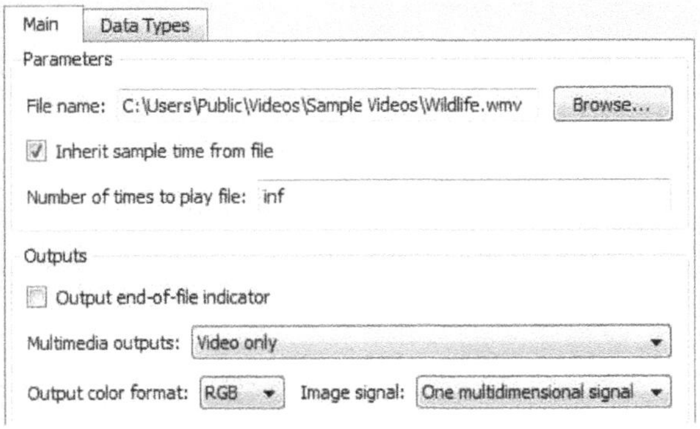

Figure 8.20   Filename specification dialog box.

*Unfortunately, the above model is not a complete one, since the sound in the multimedia file cannot be played simultaneously. The **Video Viewer** block can only display video, and no sound is played. From the multimedia source block, select the **Multimedia output** listbox – it can be seen that besides the default **Video only** option, other options are available: for instance, the **Video and audio** option can be selected as shown in Fig. 8.20(b). If this option is selected, two output ports in the block will be displayed. One is still connected to the **Video Viewer** block, and the other is idle. The **To Audio Device** block in the **Sinks** group in the DSP System Toolbox blockset can be used to connect to the port. The new multimedia file player model can be constructed as shown in Fig. 8.21.*

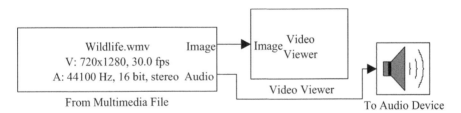

Figure 8.21   Multimedia playing model (model name: c8mvip1a).

The MATLAB function `mmreader()` *can also be used to load the movie file into the MATLAB workspace. However, since the high-definition file is very large, a lot of memory space is required. Here a smaller video file vipmen.avi can be used instead to demonstrate the problem. A structured array can be established in the MATLAB workspace. The **NumberOfFrames** property of the variable is the total number of frames in the video file. The movie file can be displayed with the* `movie()` *function with the following statements.*

```
>> W=mmreader('vipmen.avi'); nF=W.NumberOfFrames;
   W1(1:nF)=struct('cdata',...
                   zeros(W.Height,W.Width,3,'uint8'),'colormap',[]);
   for i=1:nF, W1(i).cdata=read(W,i); end
   h=figure; set(h,'Position',[100 100 W.Width W.Height])
   movie(h,W1,1,W.FrameRate);
```

The MATLAB function `imtool()` in the Image Processing Toolbox can be used to display and process the images. Other facilities can be used to process the picture. For instance, the resizing of the picture can be implemented in the function.

▉ **Example 8.3**   *The* `imread()` *function can be used to load the tiantan.jpg file into the MATLAB workspace. The function* `imtool()` *can be displayed, as shown in Fig. 8.22(a).*

```
>> W=imread('tiantan.jpg'); imtool(W)
```

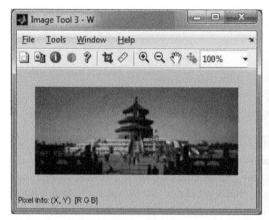
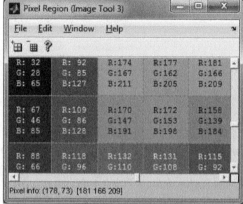

(a) image display   (b) pixel display

Figure 8.22   Image viewer and pixel display.

It can be seen that there is a toolbar in the display window, from which the processing of the image file is very easy. For instance, the 🔲 *button can be used, and the window shown in Fig. 8.22(b) can be displayed. The color values of the pixels in the selected area can be displayed. The selected area can be moved with mouse drag action so that the pixel colors of the area of interest can be displayed. This technique is useful, since first-hand knowledge can be obtained and it might be helpful for edge detection work, which will be addressed later.*

8.2.3   Fundamental Blocks for Video and Image Processing

A large number of blocks are provided in the Computer Vision System Toolbox blockset, and the important commonly used blocks are summarized below:

- **Image analysis and enhance group** has the blocks shown in Fig. 8.23, including **Histogram Equalization, Template Matching, Edge Detection, Median Filter** and **Corner Detection**. With these blocks, processing of images and videos is very easy and straightforward. Also, parameters can be assigned to the blocks so that different algorithms can be adopted in the analysis and enhancement of images. In the MATLAB Image Processing Toolbox, functions such as `edge()`, `histeq()`, `medfilt2()` and `imadjust()` can also be used to fulfill the above-mentioned tasks.

- **Image conversion group** is shown in Fig. 8.24, where the blocks **Chroma Resampling, Color Space Conversion, Gamma Correction** are provided. For instance, the **Color Space Conversion** block can be used to convert the images with different color spaces from each other. The available

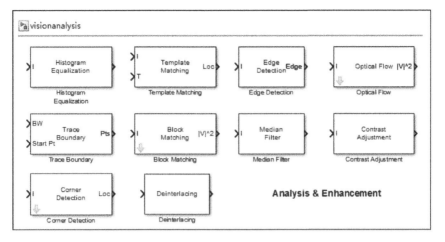

Figure 8.23   Analysis & Enhancement group.

color space models include RGB, HSV, grayscale and YCrCb. These models have areas and characteristics for their own particular applications. The **Autothreshold** block can be used to convert a grayscale image into black and white , and the threshold can be selected automatically according to the images.

For image variables, the functions `rgb2gray()` and `rgb2hsv()` can be used to convert color images, while the `im2bw()` function can be used to convert an image into black and white. However, the threshold should be given as a constant. The function `imadjust()` can be used to correct images, including the use of the γ correction method.

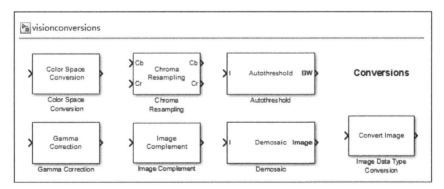

Figure 8.24   Conversion group.

- **Image transformation group**: the **Transforms** group is displayed as shown in Fig. 8.25. Various transformations can be made to images, including **2D FFT, 2D-IFFT, 2D-DCT, 2D-IDCT** and Hough transforms. Filters are used to filter out various of kinds of noise, and the Hough transform can be used to identify lines from black-and-white images.

In MATLAB and its Image Processing Toolbox, two-dimensional Fourier transforms and Fourier cosine transforms can be obtained with `fft2()`, `ifft2()`, `dct2()` and `idct2()` functions, while the Hough transform can be carried out with the `houghlines()` and `hough()` functions.

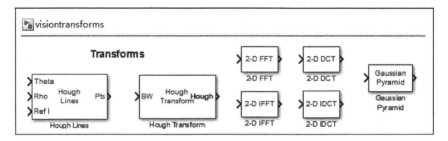

Figure 8.25 Image transform group.

- **Morphological group**: the **Morphological Operations** are displayed in Fig. 8.26, including **Dilation** and **Erosion**. They can be used for certain image processing problems. Suitable geometric structured dilation may review certain characteristics in a blurred image, while the erosion method can extract the skeleton from black-and-white images. Also for black-and-white images, the functions `imdilate()` and `imerode()` can be used to perform dilation and erosion tasks as well.

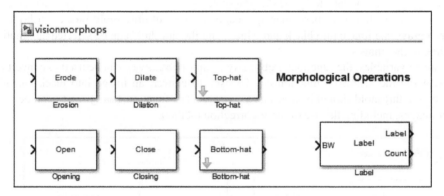

Figure 8.26 **Morphological** group.

- **Image filters group**: the contents of the group are shown in Fig. 8.27, where **2-D FIR Filter**, **2-D Convolution**, **Median Filter** and **Kalman Filter** blocks are provided. These blocks can be used to process directly images and videos.

  In the Image Processing Toolbox, the functions `convmtx2()`, `ftrans2()`, `fwind1()` and `fwind2()` can be used, while median filters can be directly handled with function `medfilt2()`.

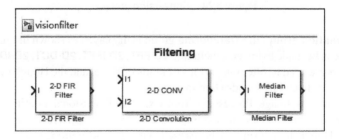

Figure 8.27 **Filtering** group.

- **Adding graphics and text to the images**. The **Text & Graphics** group is shown in Fig. 8.28, where **Draw Shapes**, **Insert Text** and **Draw Markers** blocks are provided, which allows the superimposition of objects on top of the existing images. The **Compositing** block can be used in image post-processing.

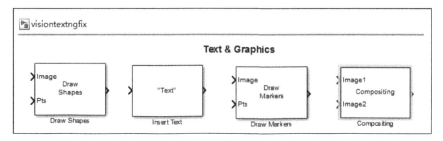

Figure 8.28   **Text & Graphics** group.

- **Demos and applications group**: A large number of demonstration models are provided in the Computer Vision System Toolbox blockset, including autonomous driving, image stitching, corner detection, object recognition and tracking and video surveillance recording. The frameworks of the demo models can be used directly to solve other application problems with slight modifications.

8.2.4   Processing of Video and Images through Examples

With the blocks provided in the Computer Vision System Toolbox blockset, Simulink models can be constructed easily. Various working examples are given, and they can be used in practice as well. In this section, several simple examples are given to demonstrate the modeling method for video processing tasks.

Example 8.4   *Many image processing algorithms can only be used in processing grayscale images, rather than colored images. If color images are to be processed, the original image can be converted to grayscale first. Consider the video file vipmen.avi provided with MATLAB. Since this is already in the MATLAB search path, there is no need to set the path name to the file. It can be used directly in block setting.*

*A blank model window can be opened first. The **From Multimedia File** block can be copied into the window. Open the dialog box and specify the source video filename vipmen.avi in the relevant edit box. Copy the **Color Space Conversion** block into the model window and select from the listbox the item **RGB to Intensity**; the output of the block can then be set to grayscale signals. An **Edge Detection** block can be connected to the block, such that the binary edge of the video file can be obtained, which can be displayed with the **Video Viewer** block. The Simulink model can be constructed as shown in Fig. 8.29.*

The last frame image in the original video file can be obtained as shown in Fig. 8.30(a). With the Sobel operator, the edges in the image can be detected as shown in Fig. 8.30(b). If the Canny operator is used instead, the new edge detection results can be obtained as shown in Fig. 8.30(c). It can be seen that the Canny operation keeps too much of detailed information, and if the Roberts operator is accepted, the image in Fig. 8.30(d) can be obtained.

*To display the original grayscale images together with the edges detected as binary images, the **composition** block in the **Text & Graphics** group can be used to accept the two images, and the new Simulink model can be established as shown in Fig. 8.31.*

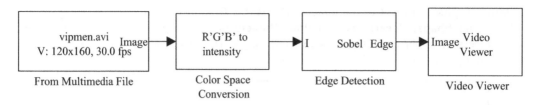

Figure 8.29   Video processing system (model name: c8mvip2).

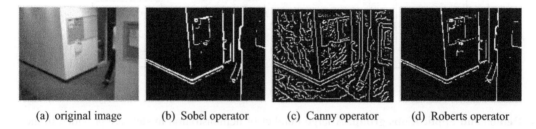

(a) original image   (b) Sobel operator   (c) Canny operator   (d) Roberts operator

Figure 8.30   Edge detection result in the last frame of the video.

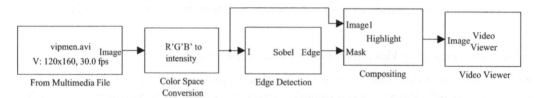

Figure 8.31   New video processing system (model name: c8mvip4).

*In order to display directly the composited videos, double click the **Composition** block. A dialog box will then be displayed as shown in Fig. 8.32(a). The **Operation** selection can be set to **Highlight selected pixels**, so that the system can be simulated easily. The results are shown in Fig. 8.32(b). It can be seen that the detected edges are highlighted on the original data.*

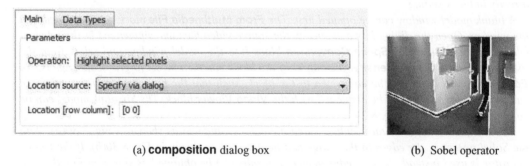

(a) **composition** dialog box   (b) Sobel operator

Figure 8.32   Video composition system.

To superimpose the detected edges on the original color image, the Simulink model shown in Fig. 8.33 can be employed. This model can be used directly for the compositions of color images and videos.

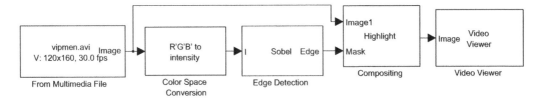

Figure 8.33   Color video edge detection with composition (model name: c8mvip6).

Static images can also be processed with the related functions in the Image Processing Toolbox. For instance, the function `imread()` *can be used to load data files into the MATLAB workspace, and the* `rgb2gray()` *file can be used to convert the original color images to grayscale images. The function* `edge()` *can then be used to detect the edges of the grayscale images and function* `imtool()` *can be used to display the images obtained. For video files, the* `mmreader()` *function can be used to load the images frame by frame, and that the video file can be processed with loop structures. Again, due to the memory issues, only small-scale videos can be processed in this way.*

```
>> W=mmreader('vipmen.avi'); W1=read(W,W.NumberOfFrames); % last frame
   W2=rgb2gray(W1); W3=edge(W2,'Prewitt'); imtool(W3)
```

The borders in each frame of the video can be extracted with MATLAB commands using a loop structure. However, the best way to process video files is using the Simulink model.

Example 8.5   *In photography, overexposure and underexposure phenomena often happen. In this case, the histograms of the images may be clustered closely in a condensed interval, and the composition of the image cannot be seen clearly. This phenomenon happens because the histograms may not be well distributed. With the use of histogram equalization methods, the original image can be corrected. Even when the histogram is well distributed, the method can still be used for local processing.*

Consider the original image shown in Fig. 8.34(a) which is the surface of Mars, reproduced from [10]. The lower-right corner of the image is too dark to see any useful information. Image processing needs to be used on this image.

The histogram of the image can be obtained with the `imhist()` *function in the Image Processing Toolbox, as shown in Fig. 8.34(b). It can be seen that the grayscale values of the pixels are clustered in a condensed way in the interval of 0~30. Details of the image cannot be distinguished. Histogram equalization needs to be carried out with the following statements*

```
>> W=imread('c8fvid1.tif'); imtool(W); figure; imhist(W)
```

With the use of the Computer Vision System Toolbox, the Simulink model is shown in Fig. 8.35(a). In this model, the image file is specified in a source block, and it is connected with a **Histogram Equalization** *block, and the result is displayed in a* **Video Viewer** *block. The processed image is obtained as shown in Fig. 8.36(a), and the histogram of the new image is shown in Fig. 8.36(b), using the* `histeq()` *function. It can be seen that details can be better displayed in the new image.*

```
>> W1=histeq(W); imshow(W1); figure; imhist(W1)
```

It can also be seen that although the equalized image is better than the original one, the result is not perfect. The adaptive histogram equalization function `adapthisteq()` *in the Image Processing*

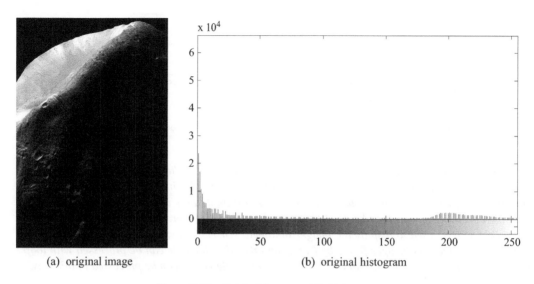

(a) original image (b) original histogram

Figure 8.34 Original image and its histogram.

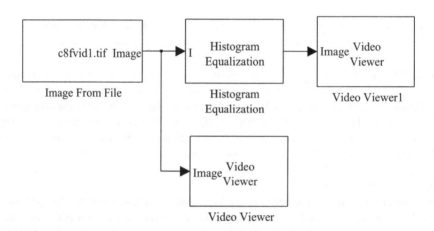

Figure 8.35 Simulink model for histogram equalization system (model name: c8mvip3).

Toolbox can be used to find better images. With the following statements, the improved image can be obtained as shown in Fig. 8.37(a), and the new histogram is shown in Fig. 8.37(b). It can be seen that the improved image is much better and the new histogram is well distributed.

```
>> W2=adapthisteq(W,'Range','original',...
          'Distribution','uniform','clipLimit',.2)
   imtool(W2); figure; imhist(W2)
```

The Simulink blocks can be used directly in the video processing tasks. The adaptive histogram equalization algorithm has not been implemented yet in blocks. Users are advised to write a piece of code for themselves to implement such a block, if needed.

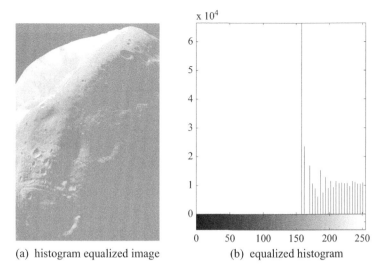

(a) histogram equalized image   (b) equalized histogram

Figure 8.36   Histogram equalization of the image.

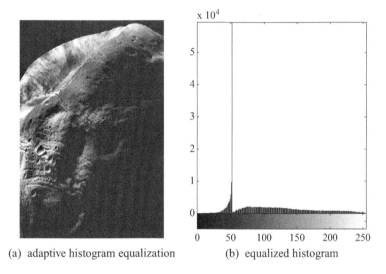

(a) adaptive histogram equalization   (b) equalized histogram

Figure 8.37   Image processing with adaptive histogram equalization.

Example 8.6   *Consider the original image shown in Fig. 8.38(a) [10]. The original image is not quite clear. The dilation method can be used to process the original image. The Simulink image dilation model can be constructed as shown in Fig. 8.38(b), where the dilation element [0, 1, 0; 1, 1, 1; 0, 1, 0] can be specified in the* **Structuring element** *edit box in the dialog box of the block. The dilated result is shown in Fig. 8.38(c). It can be seen that the processed image is better and clearer.*

  The dilation and erosion of images can also be performed by the functions imerode() *and* imdilate() *in the Image Processing Toolbox. If the dilation matrix is selected as A = [0, 1, 0; 1, 1, 1; 0, 1, 0], the following MATLAB statements can be used to process the original*

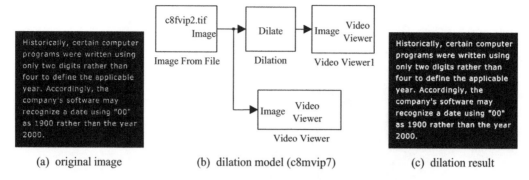

(a) original image   (b) dilation model (c8mvip7)   (c) dilation result

Figure 8.38   Image dilation system and results.

image, and the inverted image is obtained as shown in Fig. 8.39. The inverted image shows the dilation results more clearly.

```
>> W=imread('c8fvip2.tif'); imtool(~W);          % show inverted image
   A=[0 1 0; 1 1 1; 0 1 0]; W1=imdilate(W,A);    % dilation
   figure; imtool(~W1)                            % show inverted image
```

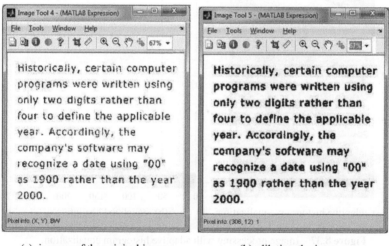

(a) inverse of the original image   (b) dilation the inverse

Figure 8.39   Dilation processing of the original inverse.

As explained earlier, these functions can only be used to handle single image files. For video files, loops can be considered for separately processing each frame. However, the process is much more complicated than the Simulink modeling process.

■ **Example 8.7**   *The morphological function* bwmorph() *can be used in processing black-and-white images. This function can be used in skeleton extraction, edge detection and border* *extraction problems for black-and-white images. Consider the original black-and-white image shown in Fig. 8.40(a). The function* `bwmorph()` *can be used to extract the boundary and skeleton of the original image, as shown in Figs 8.40(b) and (c). The skeleton extracted is not very good due to the disturbance of the borders in the image. The function* `bwmorph()` *can be called with the* `'thin'` *option repeatedly in loops for 50 times, and a better skeleton can be extracted as shown in Fig. 8.40(d). There are other options in the* `bwmorph()` *function, and more on the options can be found with the command* `doc bwmorph`.

```
>> W=imread('c8fvip3.bmp'); imtool(W)      % display original image
   W1=bwmorph(W,'remove');  imtool(~W1)    % boundary pixels extraction
   W2=bwmorph(~W,'skel',inf); imtool(~W2)  % skeleton extraction
   W3= ~W; for i=1:50, W3=bwmorph(W3,'thin'); end,
   imtool(~W3)  % final image by repeated thinning process
```

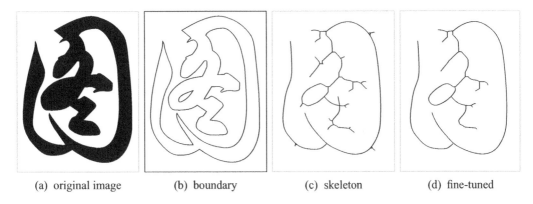

(a) original image   (b) boundary   (c) skeleton   (d) fine-tuned

Figure 8.40   Morphological based image processing.

In the Image Processing Toolbox and the Computer Vision System Toolbox, a lot of demonstration programs and models are provided. These models can be a reference for users to construct their own models. More information on digital image processing can be found in [10, 11].

8.2.5   Real-time Processing of Videos and Images

The Image Acquisition Toolbox provided in MATLAB allows the user to extract images and videos in real time from cameras connected to the computer. The only block provided in the toolbox is labeled **From Video Device**, and it can be used to act as an input block.

Example 8.8   *Consider again the video processing system constructed in Fig. 8.33. If the video file input is replaced by a real-time input, for instance the camera input represented by the* **From Video Device** *block, a new Simulink model can be established, as shown in Fig. 8.41.*

On double clicking the **From Video Device** *block, a dialog box is displayed as shown in Fig. 8.42(a). The camera connected to the current computer – the* **winvideo1 (Built-in iSight)** *camera – has been recognized automatically. The user can also select the camera resolution with the* **Video format** *listbox. Further settings of the camera can be made if the* **Edit properties** *button is clicked. The dialog box shown in Fig. 8.42(b) can be displayed. The effect can also be tested with the* **Preview** *button.*

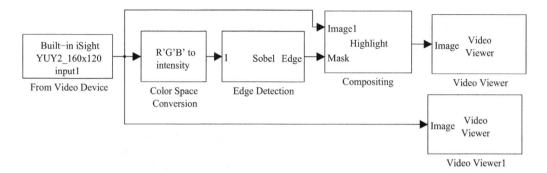

Figure 8.41   Real-time edge detection and composition system (model name: c8mvip8).

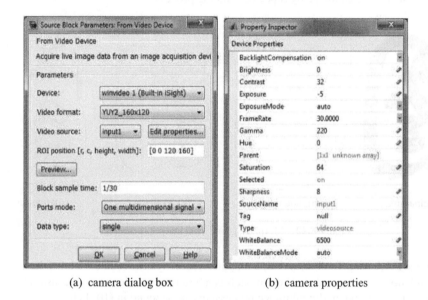

(a) camera dialog box   (b) camera properties

Figure 8.42   Dialog box and properties of cameras.

Start the simulation block diagram and real-time video can be acquired from camera and edge detection and composition can be applied immediately.

8.3   Finite State Machine Simulation and Stateflow Applications

The theoretical foundation of Stateflow simulation is the theory of the finite state machine (FSM). In the finite state machine, the number of the states is finite. When a certain event happens, the state of the system transits to another state. Finite state machines are also called event driven systems. In Stateflow, the state transition conditions can be designed such that the state can transit from one to another.

Stateflow is the graphical implementation tool for finite state machines. It can be used in solving complicated supervisory logic problems. The state transition can be defined with the graphical tool. The Stateflow chart can be embedded into Simulink models.

8.3.1 Introduction of Finite State Machines

Finite state machines can be designed in the graphical user interface provided in Stateflow. The finite states can be constructed, and graphical methods can be used to draw the state transition and transition conditions, such that the overall finite state machine can be drawn. Thus in Stateflow, state and state transition are the two most important components. An illustration of a finite state machine is sketched in Fig. 8.43. The word "finite" used here indicates that the number of states is finite. In the sketch, there are four states, and the transitions between them are conditional. The transitions of states sometimes are conditional. In finite state machines, the conditions and events for state transitions should be declared.

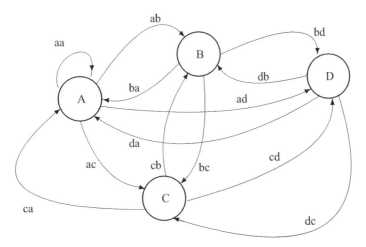

Figure 8.43   Sketch of a finite state machine.

Stateflow charts should normally be embedded in Simulink models. Stateflow charts are event driven, and the events may come from the same Stateflow chart or from Simulink models.

8.3.2 Fundamentals of Stateflow

Under the MATLAB command window, the `stateflow` command can be issued to launch the Stateflow environment. The Stateflow interface is shown in Fig. 8.44. The left window is a Stateflow library and the right one is a Simulink window with a blank Stateflow **Charts** block. Stateflow charts can be embedded in Simulink models.

Double click the Stateflow **Chart** block and a Stateflow editing interface is opened, as shown in Fig. 8.45. The Stateflow chart can be edited in this interface. Powerful editing facilities are provided in the Stateflow editor, and complicated logical expressions can be edited in the interface as well. It should be noted that the output ports of a Stateflow chart can be the states, and the number of them is finite.

In the Stateflow editing interface, the shortcut menu can be shown by right clicking the mouse, as shown in Fig. 8.46(a). If the **Properties** menu item is selected, the dialog box shown in Fig. 8.46(b) can be displayed. The properties of the whole chart can be defined.

On the left hand side of the Stateflow editing interface, there is an object palette, and many important Stateflow chart components can be copied from the palette. The Stateflow interface can be used to edit Stateflow charts in a graphical way. The procedures of Stateflow chart modeling are:

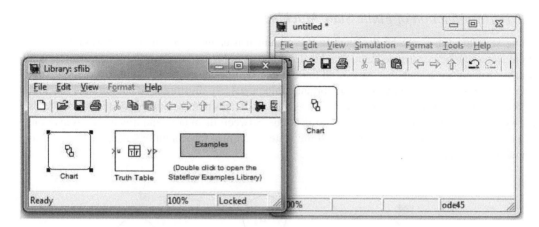

Figure 8.44   Stateflow startup interface.

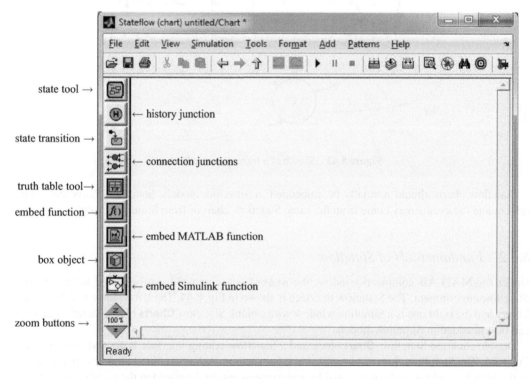

Figure 8.45   Stateflow graphical edit interface and tools.

• **Use state tool**. The states are the operational status of the systems. In Stateflow, there are two statuses, active and inactive. The state tool can be used to add states in the chart. Click the 🖼 button and a state can be drawn in the Stateflow editing interface. There is a question mark on the state icon. The user can add the name and status to the state. For instance, the name can be assigned as **on**. All the needed states can be drawn, as shown in Fig. 8.47.

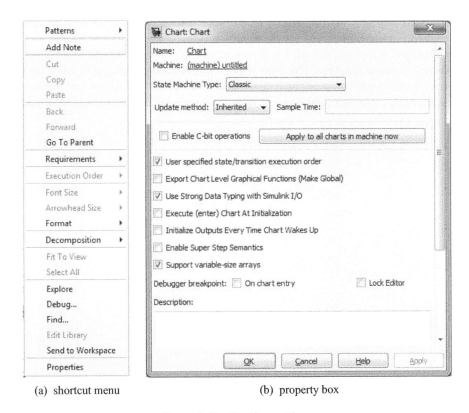

(a) shortcut menu   (b) property box

Figure 8.46   Stateflow settings.

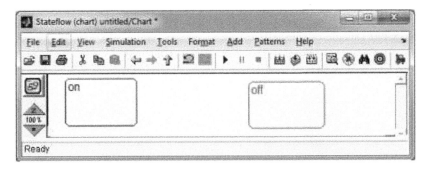

Figure 8.47   New states in the Stateflow window.

As well as the names of the states, other information can be added to the states. For instance, a statement led by `entry:` can be provided, indicating the statements to execute when entering the state. The keywords `exit:` and `during:` can be used to indicate respectively the statements to run when exiting the states, or remaining in the states. Note that, each state can have only one assigning statement.

The button 🔳 can also be used to add new states, and the states can be regarded as temporary states with no names.

- **State transition setting**. Click the border of one state, and drag mouse to another state icon, and then release mouse button, and a state transition curve with an arrow is displayed. On the state transition curve, a question mark **?** is displayed, allowing the user to specify state transition conditions and callback functions. Click the transition curve, and a shortcut menu is displayed. Choose from it the **Properties** menu item, and a dialog box is opened as shown in Fig. 8.48. The state transition conditions and callback functions can be written in the edit box labeled by **Label**.

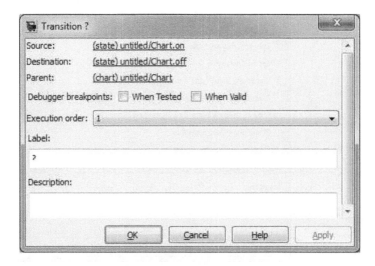

Figure 8.48   Stateflow state transition setting.

- **Event and data setting**. Stateflow provides an **Add** menu, as shown in Fig. 8.49(a). Proper event and data definitions can be selected. For instance, an event can be entered from Simulink. The Stateflow chart also allows the user to exchange data with the Simulink environment, and the menu in Fig. 8.49(b) should be selected.

(a) event menu   (b) data menu

Figure 8.49   **Add** menus.

- **Input and output setting**. Selecting the **Add** → **Data** → **Input from Simulink** menu item, a dialog box shown in Fig. 8.50(a) will open. Input and output ports can be added to the Stateflow charts. The names and data types of the ports can be assigned. If the menu item **Add** → **Data** → **Constant** is selected, the dialog box shown in Fig. 8.50(b) will be displayed.
- Stateflow also allows MATLAB code, Stateflow subcharts and Simulink models to be embedded in the logical chart. Varieties of flow control with state transitions can also be defined.

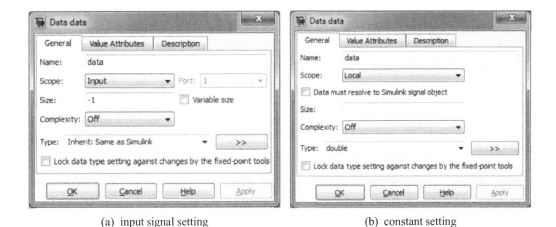

(a) input signal setting   (b) constant setting

Figure 8.50   Dialog boxes of input signals and constants.

The blocks and parameters in the Stateflow chart can be set in a dialog box, as shown in Fig. 8.51, with the **Tools** → **Explorer** menu item.

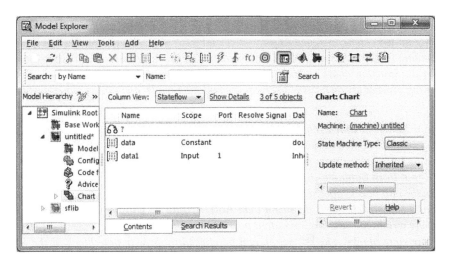

Figure 8.51   Stateflow analysis interface.

8.3.3   Commonly Used Commands in Stateflow

It was shown earlier that the `stateflow` command can be used to open the main interface of Stateflow. The following statements can also be used for the Stateflow facilities and tools.

- `sfnew` command can be used to create a new Simulink model with a blank Stateflow chart. The command `sfnew fnm` can create a Simulink model with the name `fnm`, where a blank Stateflow chart is included.

- `sfexit` command can be used to close all the Stateflow chart models, and exit from the Stateflow environment.

- `sfsave` command can be used to save the Stateflow chart in a file.
- `sfprint` command can be used to print the Stateflow model.

8.3.4 Application Examples with Stateflow

Several examples are given to show the modeling and application procedures. The switching differential equation is demonstrated with Stateflow. The switching conditions can be modeled as state transition conditions. Friction force problems are also demonstrated using Stateflow modeling techniques. Finally, the Stateflow modeling and simulation example of a table tennis refereeing system is demonstrated.

▉ Example 8.9 *Consider the switching differential equation shown in Example 5.22, with $\dot{x} = A_i x$ and*

$$A_1 = \begin{bmatrix} 0.1 & -1 \\ 2 & 0.1 \end{bmatrix}, \quad A_2 = \begin{bmatrix} 0.1 & -2 \\ 1 & 0.1 \end{bmatrix}$$

*and the initial state variables are $x_1(0) = x_2(0) = 5$. The switching condition is that if $x_1 x_2 < 0$, the system is switched to A_1, but if $x_1 x_2 \geq 0$, the system is switched to A_2. Such a switching system can also be modeled using finite state machine methods in Stateflow. Assume that there are two **states** in the finite state machine.[1] We can assign one output port sys for the Stateflow chart, and the two corresponding **states** can be assigned as sys = 1 and sys = 2. The two **states** can be switched when the conditions $x_1 x_2 < 0$ and $x_1 x_2 \geq 0$ are satisfied, respectively. Meanwhile, two input ports, x_1 and x_2 for the Stateflow chart are used to describe the switching conditions.*

The following procedure can be used to create a simulation model containing Stateflow finite state machines:

*1) **Create a Stateflow block**. Open a blank model window, and open the Stateflow group. A Stateflow block can be copied to the Simulink model window.*

*2) **Define the input and output ports in the Stateflow block**. As mentioned, when the menu **Add** → **Data** → **Input from Simulink** shown in Fig. 8.49(b) is selected, the dialog shown in Fig. 8.50(a) is opened. In the edit boxes labeled **Name**, the variable names x1 and x2 can be entered, such that two input ports can be defined. If the menu item **Add** → **Data** → **Output to Simulink** is selected, and you specify sys in the edit box, an output port can be defined.*

*3) **Draw the state blocks**. The ▣ button can be clicked to draw a state block in the Stateflow chart. Right click the state icon, and in the shortcut menu, select the **Properties** menu item, and the dialog box shown in Fig. 8.52 will open. The properties of the states can be entered. Double clicking the state block also allows you to edit the icon. In the **Label** edit box, the following information can be assigned to the two state icons respectively*

```
con1/                                con2/
entry: sys=1;                        entry: sys=2;
```

where con1 and con2 are the names of the states. The entry keywords that follow the statements are executed when entering the state, and the values of sys are set to 1 and 2, respectively.

---

[1] The confusing keyword "state" is used again, since the "state" here means the status in Stateflow, that is, **state 1** and **state 2**, while in state space equations, the word state means the "state variables" x_1, x_2. We shall use **state** in this example for status.

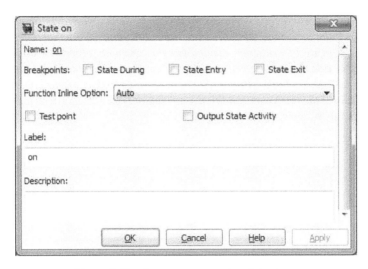

Figure 8.52   State property setting dialog box.

4) Defining state transition conditions. Move the mouse to the border of a state box, and the cursor is changed to +. Drag the mouse to another state and release mouse button, and a state transition curve with an arrow will be drawn automatically. There will be a question mark shown on top of the transition curve. For instance, to express the condition $x_1 x_2 < 0$, such that the system is transited from state 1 to state 2, double click the question mark, and the transition condition can be set to [x1*x2 < 0].

5) Default initial state. Clicking the ⬚ *button, the default state can be entered. Please note that this procedure cannot be omitted.*

The final Stateflow chart can be constructed as shown in Fig. 8.53(a). The Simulink model with switch block and the Stateflow chart can be established as shown in Fig. 8.53(b), where the threshold of the **Switch** *block can be set to 1.5. The model can then be used in modeling switching differential equations.*

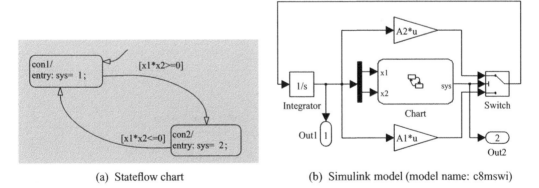

(a) Stateflow chart   (b) Simulink model (model name: c8mswi)

Figure 8.53   Simulation model of switching differential equations.

When simulating such a system, the state variables can be obtained as shown in Fig. 8.54. The results obtained are the same as those obtained in Example 5.22. The expression $x_1 x_2$ can also be *shown in the figure as well, where the switching process can be demonstrated. In simulation, the state blocks are blinking all the time, indicating the state transition status. Note that different simulation algorithms may result in slightly different results. For this example, algorithm **ode23s** is the best one to use; if **ode45** or **ode15s** are selected, inaccurate results may be obtained. Compared with the results in Example 5.22, the default setting of Stateflow is not satisfactory for continuous systems, since the zero-crossing of the transition condition $x_1 x_2$ cannot be accurately detected. Zero-crossing detection for continuous systems should be set.*

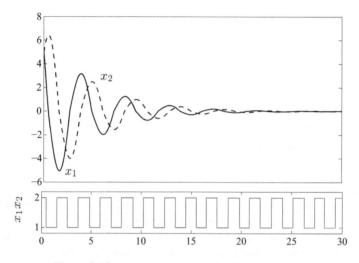

Figure 8.54   Simulation results of switching system.

*Right click on the Stateflow chart, and select **Properties** in the shortcut menu that appears and the dialog box shown in Fig. 8.55 will be displayed. The **Update method** listbox has several options. The **Continuous** option can be selected. Then check the **Enable zero-crossing detection** checkbox. Exact zero-crossing can be detected, and correct simulation results can be obtained.*

Name: Chart
Machine: (machine) c8mswi

State Machine Type: Classic   ▼

Update method: Continuous ▼   Sample Time:

☑ Enable zero-crossing detection

☐ Enable C-bit operations   Apply to all charts in machine now

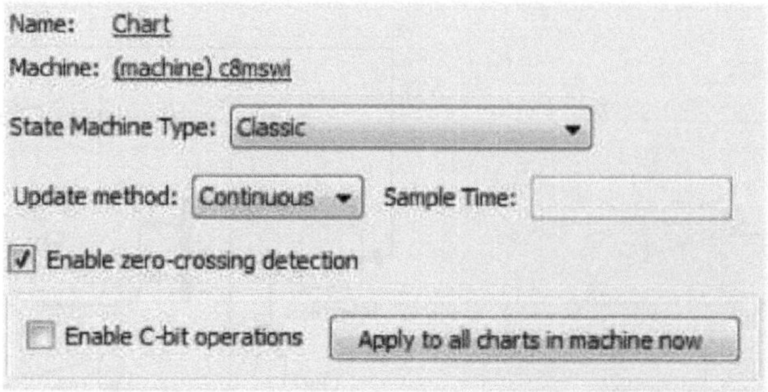

Figure 8.55   Stateflow switch condition.

A conventional Simulink model can be constructed as shown in Fig. 8.56. The model can be simulated with a conventional algorithm, such that exact simulation results can be obtained.

■ **Example 8.10**   *Consider the mechanical model in Fig. 8.57 [12]. A similar problem has been studied in Example 7.2, but friction was not considered in that case. If friction force is considered,*

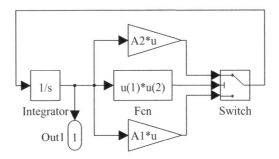

Figure 8.56 Another model (model name: c8mswi1).

both static and kinetic friction need to be involved, and the size and direction of the forces are not fixed, so many cases need to be considered.

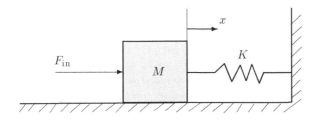

Figure 8.57 Mechanical model.

Here Stateflow has to be used to describe the friction force, in several cases. In Fig. 8.57, the mechanical model is shown. The mass M is moving under the external force F_{in}. Thus, the mass M subjects to the following forces: the external force F_{in}, friction force F_f and the pulling force of F_{str} of the spring, so that resultant force is then

$$F = F_{in} - F_f - F_{str}. \tag{8.16}$$

From Newton's second law, the following equations can be written

$$M\ddot{x} = F = F_{in} - F_f - F_{str}, \tag{8.17}$$

where x is the displacement, thus $\ddot{x}$ is the acceleration. The spring tension is $F_{str} = Kx$, where K is the elastic coefficient, and the friction force F_f can be calculated from

$$F_f = \begin{cases} \text{sign}(\dot{x})\mu F_n, & |F_{sum}| > \mu F_n \\ F_{sum}, & \dot{x} = 0 \text{ and } |F_{sum}| \leq \mu F_n \end{cases} \tag{8.18}$$

where μ is the friction coefficient, F_n is the normal force, $\dot{x}$ is the speed of the mass. When $\dot{x} = 0$, the mass is at rest. F_{sum} is the force at rest, satisfying $F_{sum} = F_{in} - F_{sliding}$. When the speed is nonzero, an external force can make it stay at zero. When the speed of the mass is zero, then F_{sum} is the force *keeping the acceleration at zero. The friction force can further be classified as static and kinetic friction forces, which depend upon whether the mass is at motion or not, that is,*

$$\mu F_n = \begin{cases} \mu_{\text{static}} F_n = F_{\text{static}}, & \dot{x} = 0 \\ \mu_{\text{sliding}} F_n = F_{\text{sliding}}, & \dot{x} \neq 0 \end{cases} \tag{8.19}$$

where μ_{static} and μ_{sliding} are respectively the static friction coefficient and the kinetic friction coefficient. Considering the above two conditions, the following formula can be used to compute the friction force

$$F_f = \begin{cases} \text{sign}(\dot{x}) F_{\text{sliding}}, & \dot{x} \neq 0 \\ F_{\text{sum}}, & \dot{x} = 0 \text{ and } |F_{\text{sum}}| < F_{\text{static}} \\ \text{sign}(F_{\text{sum}}) F_{\text{static}}, & \dot{x} = 0 \text{ and } |F_{\text{sum}}| \geq F_{\text{static}}. \end{cases} \tag{8.20}$$

It can be seen that the mass M has two states, motion and rest. A state stuck *can be assigned to describe the status. It can be seen from (8.18) that when $|F_{\text{sum}}| > F_{\text{static}}$, the mass is in the motion state, and the variable* stuck *is set to 0. If the above conditions are not satisfied, and the mass is at rest, the flag variable* stuck *is set to 1. In this way, the state of the friction can be defined. Based on this logical description, the Stateflow chart can be constructed as shown in Fig. 8.58. In the Stateflow chart, the resultant force* Fsum, *the zero speed detecting signal* novelocity *and the static friction force* Fstatic *are the signals from Simulink, and the state* stuck *is the signal transmitted to Simulink.*

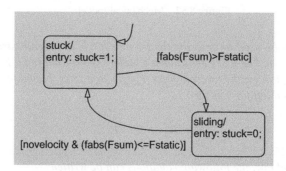

Figure 8.58   Stateflow logical model.

Also, from (8.20), the subsystem representing the friction force F_f can be constructed as shown in Fig. 8.59. Here, two switch blocks can be used to describe the three conditions.

*From these two models, the overall Simulink model can be established, as shown in Fig. 8.60. In this model the **Hit Crossing** block is used to detect the zero-crossing point of the speed signal. The mathematical model in (8.17) is described by double integrator blocks. Note that the model constructed is different from that given by [12], and the structure given here is easier to understand.*

*In the Simulink model, the speed signal is from the start port of the first integrator, not from its output port, so as to avoid an algebraic loop. This can be done by ticking the **Show state port** checkbox of its parameter dialog box. Also, the* stuck *signal can be used to reset the integrator.*

*In the simulation model, a triangular waveform is used as the external force input signal. The **Repeating Sequence** block in the **Sources** group is used to generate this signal, with its parameter*

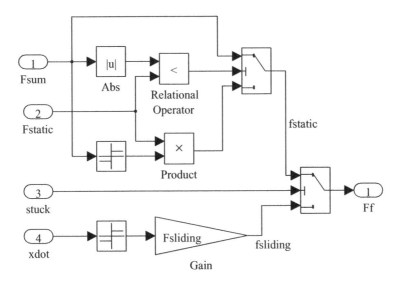

Figure 8.59 Friction force subsystem.

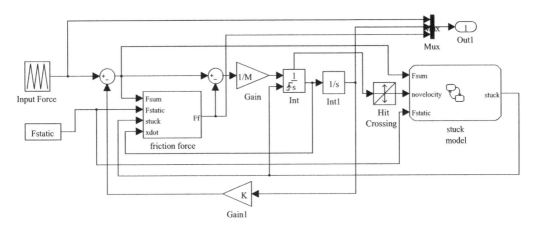

Figure 8.60 Simulink model of the friction system (model name: c8fstr1).

dialog box as shown in Fig. 8.61. The parameters in the dialog box mean that the period of the periodic signal is 10 s, and that the peak value of 5 is reached at 5 s.

Since this system is a stiff equation, the algorithm **ode15s** *should be selected, and the* **Relative tolerance** *option assigned to* 10^{-6}. *In this way, the original system can be reliably simulated.*

When the system model is established, the following statements can be used to input the relative parameters into the MATLAB workspace

```
>> M=0.01; K=1; Fsliding=1; Fstatic=1;
```

The input, output and friction force can be obtained as shown in Fig. 8.62(a), and the input signal is a triangular waveform.

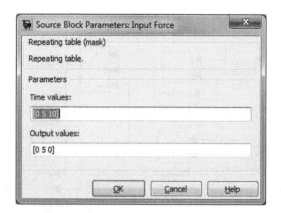

Figure 8.61   Dialog box of input signal parameter setting.

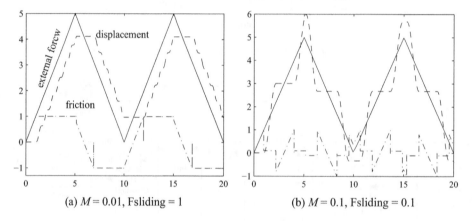

(a) $M = 0.01$, Fsliding = 1   (b) $M = 0.1$, Fsliding = 0.1

Figure 8.62   Input, output curves with different parameters.

Also, the parameters of the system can be altered. For instance, the following MATLAB statements can be used to enter the new parameters, and then, on restarting the simulation process, the curve obtained is shown in Fig. 8.62(b).

```
>> M=0.1; Fsliding=0.1;
```

Example 8.11   *The previous two examples are continuous systems, and they can be modeled and simulated without Stateflow, so the modeling process can be simple and straightforward. But let us now consider a real event-driven example – a vision based referee system for table tennis matches. From the vision subsystem, four signals are generated. These signals can be used to drive the Stateflow chart model of the referee system [13].*

The four input signals driving the Stateflow chart are respectively, P_l, P_r, P_n and sA. The two output signals of the Stateflow chart are score_A *and* score_B. *The input signal P_l represents the number of hits on the left hand side table in this round, while P_r is the number on the right. P_n indicates whether it is a net ball, only applicable to serving. The signal sA is the key representing the server side, with 1 for Side A serving, 0 for side B serving. The output signal* score_A *is for the score of Side A, while* score_B *is for Side B's score. In order to describe the state transition conditions* *easily, two new signals Q_s and Q_r can be determined, representing respectively the number of hits on the server table and the receiver table.*

Six states, S_i, $i = 1, 2, \cdots, 6$, can be defined in the referee system, as shown in Table 8.1. Based on these six states, a state transition sketch can be constructed as shown in Fig. 8.63.

Table 8.1 state specifications.

State	State description	State names
S_1	serving	***serve***
S_2	strike back by the receiver	***receiver_return***
S_3	strike back by the server	***server_return***
S_4	score by the server	***server_score***
S_5	score by the receiver	***receiver_score***
S_6	end of game	***end_game***

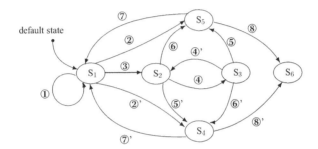

Figure 8.63 State transition chart.

The state transition regulations are listed in Table 8.2. From the state transition conditions, it can be seen that some of the processes can easily be implemented, while others may not be easy to construct. For state transition condition ⑦, a Substateflow chart or embedded M-functions can be used.

The following procedures are used to establish the Stateflow charts:

*1) **Stateflow chart drawing**. From the above analysis, the logical flow expression defined in Table 8.1 can be used to draw the main Stateflow chart, as shown in Fig. 8.64. In the flow chart, the callback functions can be specified for conditional state transitions with the syntax* `[st_conditions]callbacks`*, where the string inside the square brackets* [] *is used to express the conditions for state transitions while the string in curly braces* {} *can be used to express the callback functions, or embedded subcharts. If MATLAB functions are used, the name of the function should be prefixed by* `ml`*.*

*From the main Stateflow chart, it can be seen that a state block can be connected with several state transition lines. For instance, the state **server_score** has two outgoing state transition lines, labeled automatically **1** and **2**. In the first transition line, a transition condition is specified, while there is no transition condition in the second line. This means that when* `[(score_A >= 11 | score_B >= 11) & ml.abs(score_B-score_A) >= 2]` *condition is true, that is, one of the sides reached score of 11, and the score difference is greater than 2, the state is transferred to the*

Table 8.2   State transition condition descriptions.

Order	State	Descriptions	Transition conditions
①	$S_1 \Rightarrow S_1$	serve again	$P_n = 1 \;\&\; Q_s = 1 \;\&\; Q_r \geq 1$
②	$S_1 \Rightarrow S_5$	serve fault	$Q_s \neq 1 \mid Q_r = 0$
②′	$S_1 \Rightarrow S_4$	serve score	$P_n \neq 0 \;\&\; Q_s = 1 \;\&\; Q_r \geq 2$
③	$S_1 \Rightarrow S_2$	serve successful	$P_n \neq 0 \;\&\; Q_s = 1 \;\&\; Q_r = 1$
③	$S_2 \Leftrightarrow S_3$	reseiver strike back	$Q_s = 1 \;\&\; Q_r = 0$
④′	$S_3 \Leftrightarrow S_2$	server strike back	$Q_s = 0 \;\&\; Q_r = 1$
⑤	$S_2 \Rightarrow S_5$	receiver score	$Q_r = 0 \;\&\; Q_s \geq 2$
⑤′	$S_2 \Rightarrow S_4$	receiver fault	$Q_r \geq 1 \mid Q_s = 0$
⑥	$S_3 \Rightarrow S_5$	server loss point	$Q_s \geq 1 \mid Q_r = 0$
⑥′	$S_3 \Rightarrow S_4$	server score	$Q_s = 0 \;\&\; Q_r \geq 2$
⑦	$S_4 \Rightarrow S_1$	next serve	condition ⑧ no satisfied
⑦′	$S_5 \Rightarrow S_1$	next serve	⑧′not satisfied
⑧	$S_4 \Rightarrow S_6$	receiver win	$M(A, B) \geq 11 \;\&\; |A - B| \geq 2$
⑧′	$S_5 \Rightarrow S_6$	server win	$M(A, B) \geq 11 \;\&\; |A - B| \geq 2$

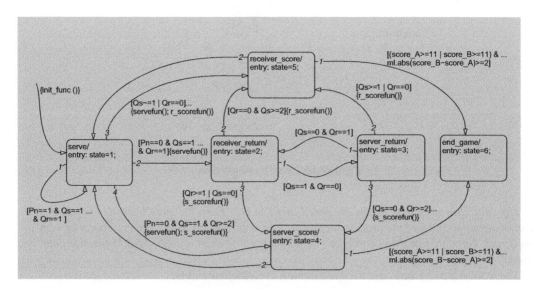

Figure 8.64   Main Stateflow chart.

end_game *state to end the simulation process. However, if the condition does not hold, the transition line **2** will be activated, and the state is transferred to the **server** state, to start the serve process.*

*2) **Embed function drawing and programming**. The default state transition should be assigned first. On the transition line, the function* `init_func()` *can be called first, and the function can be described by an embedded chart. Clicking the button 𝒇₀ from the object palette of the Stateflow editing interface, the embedded chart can be drawn in the same editing interface. The chart shown in Fig. 8.65 can be used, where the small circle indicates the temporary state, created by clicking the ⛭ button; its action is similar to the ⊞ button, and the difference is that the name of the state cannot be specified.*

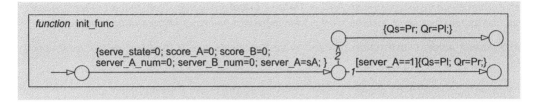

Figure 8.65   Initialization function `init_func()` flow chart.

As well as the initialization embedded chart, other charts can also be drawn in a similar way, as shown in Fig. 8.66. For instance, in Fig. 8.66(c), the server chart `servefcn()` can be used to show the map from variable P to Q, and the modeling is much simpler in this way.

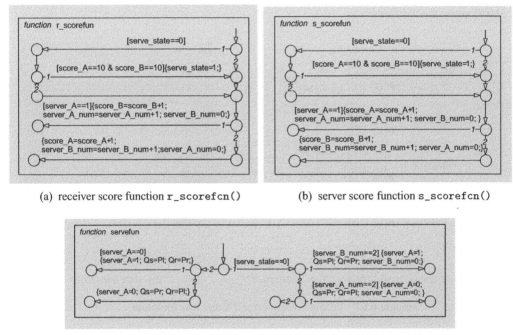

(a) receiver score function `r_scorefcn()`   (b) server score function `s_scorefcn()`

(c) serve function `servefcn()`

Figure 8.66   Other embedded functions.

3) Setting of local variables and interface variables. There are several interfaces between the Stateflow chart and the Simulink model. In this system, four interfaces are designed from Simulink to Stateflow, and they are respectively Pl, Pr, Pn *and* sA. *The first three correspond to the signals* P_l, P_r *and* P_n, *while* sA *is the server of the game. Three interfaces can also be designed from the Stateflow chart to the Simulink model, where* score_A *and* score_B *represent the scores of the sides A and B. The third output,* state, *is the value of the state. These interfaces can be assigned from the menus* **Add** → **Data** → **Input from Simulink** *and* **Output to Simulink**. *Other variables such as* `serve_state` *can be assigned with* **Add** → **Data** → **Local** *menu item. Note that all other variables should be assigned as local variables, otherwise there will be errors in simulation processes.*

4) **Simulink model construction**. *The following statements can be used to generate a set of test data*

```
>> t=sort(unique(0.01*round(rand(1000,1)*2000))); % event time vector
   u1=round(2*rand(size(t))); u2=round(2*rand(size(t)));
   u3=round(0.6*rand(size(t))); U1.time=t; U1.signals.values=u1;
   U2.time=t; U2.signals.values=u2; U3.time=t; U3.signals.values=u3;
```

Based on this data, a Simulink model for a table tennis referee testing system can be established as shown in Fig. 8.67(a). The data in MATLAB workspace can be used to drive the Stateflow chart, and the simulation results can be obtained as shown in Fig. 8.67(b). Random numbers can be used to drive the system. **Manual Switch** *can be used to select the server side. The simulation algorithm* **Discrete (no continuous states)** *can be selected, and the* **File** → **Model Properties** → **Callbacks** → **StartFcn** *menu can be used.*

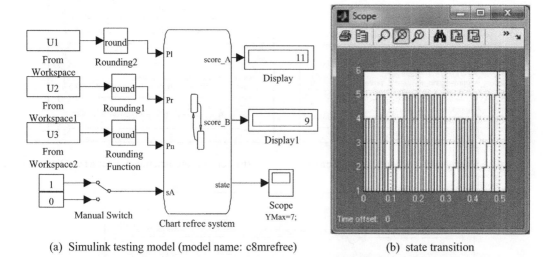

(a) Simulink testing model (model name: c8mrefree)   (b) state transition

Figure 8.67   Testing simulation system and state transition.

Example 8.12   *Embedded Stateflow subcharts were used in the previous example to describe the functions for initialization, server score, receiver score and serving. These functions can also be implemented with embedded M-functions. By clicking the ▥ button in the object palette of the Stateflow editing interface, the embedded M-function can be defined. Double click the* **eM** *sign and a MATLAB function editor will open automatically. The user can write in the editor the callback M-function. The newly constructed Stateflow chart is provided in the model file c8mrefree1.mdl, as shown in Fig. 8.67(a), and the four embedded M-functions are listed below.*

```
function init_func
serve_state=0; score_A=0; score_B=0; server_A_num=0;
server_B_num=0; server_A=sA;
if server_A==1,Qs=Pl; Qr=Pr; else, Qs=Pr; Qr=Pl; end
```

```
function s_scorefun
if serve_state==0 & score_A==10 & score_B==10, serve_state=1; end
if server_A==1,
   score_A=score_A+1; server_A_num=server_A_num+1; server_B_num=0;
else
   score_B=score_B+1; server_B_num=server_B_num+1; server_A_num=0;
end
function r_scorefun
if serve_state==0 & score_A==10 & score_B==10, serve_state=1; end
if server_A==1,
   score_B=score_B+1; server_A_num=server_A_num+1; server_B_num=0;
else
   score_A=score_A+1; server_B_num=server_B_num+1;server_A_num=0;
end
function servefun
if serve_state==0
   if server_B_num==2, server_A=1; Qs=Pl; Qr=Pr; server_B_num=0;
   elseif server_A_num==2, server_A=0;Qs=Pr;Qr=Pl; server_A_num=0; end
else, server_A= server_A;
   if server_A==1, Qs=Pl; Qr=Pr; else, Qs=Pr; Qr=Pl; end
end
```

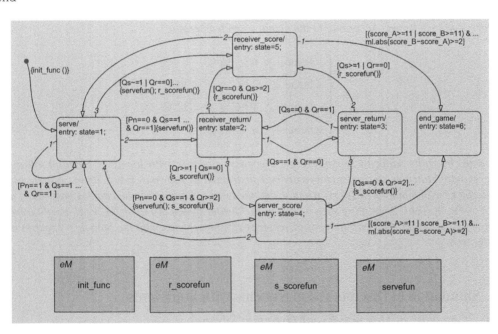

Figure 8.68 Modified main Stateflow chart.

*Note that the comments on the embedded M-functions are that there is no *.m corresponding to them, and the functions are embedded in the chart already. Also the variables used have already been defined in the **Add** → **Data** menu, so there is no need to use input and output arguments in the function call. The results under the new system are the same as the original ones.*

8.3.5 Describing Flows with Stateflow

It is very easy to implement flow control structures in MATLAB statements. However, it might be rather complicated to construct models for them with Simulink blocks, although similar blocks have been provided in subsystems. With the use of Stateflow charts, the loops and conditional branches can easily be constructed.

The conditional branch structure

> `if` conditions 1, statements 1, `else`, statements 2, `end`

can also be expressed in Stateflow charts as shown in Fig. 8.69(a). This chart was drawn by mouse clicks and may not look very standardized. The 🗷 button can be clicked to draw temporary state icons, and the state transition relationship can also be as represented in Fig. 8.69(b).

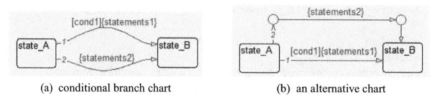

(a) conditional branch chart   (b) an alternative chart

Figure 8.69 Stateflow charts for conditional branch structures.

🖥 **Example 8.13** *Assume that a room temperature regulation system implements several temperature control schemes. According to the actual room temperature*

$$
y = \begin{cases}
\text{scheme 1,} & u \geq 30°\text{C} \\
\text{scheme 2,} & 22°\text{C} \leq u < 30°\text{C} \\
\text{scheme 3,} & 10°\text{C} \leq u < 22°\text{C} \\
\text{scheme 4,} & u < 10°\text{C}.
\end{cases}
$$

*It might be rather difficult to express the piecewise function with low-level Simulink blocks. However, with Stateflow, the chart can easily be constructed as shown in Fig. 8.70(a). Assume that the interfaces with Simulink are ports **u** and **y**, and the testing Simulink model can be constructed as shown in Fig. 8.70(b). Simulating such a system, the results shown in Fig. 8.70(c) can be obtained. It can be seen that the temperature control schemes can be selected with the Stateflow chart in an intuitive way.*

8.4 Simulation of Discrete Event Systems with SimEvents

8.4.1 Concepts of Discrete Event Dynamic Systems

In everyday life, we often encounter problems that are not easily modeled by the continuous variable systems presented earlier. For example, a person enters an elevator and presses the button for the fifteenth floor. This is an independent event, since it is independent of the elevator system, and the selections of other persons in the same elevator. The solutions to this kind of systems are well beyond the concepts of the continuous variable systems discussed so far. This kind of system is driven by events, thus they are referred to as discrete event dynamic systems (DEDS) or simply discrete event

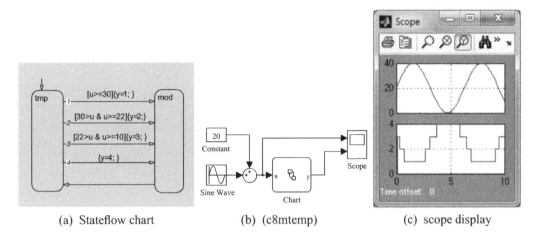

(a) Stateflow chart   (b) (c8mtemp)   (c) scope display

Figure 8.70   Illustration of Stateflow implementation of conditional branch.

systems (DES) [14]. In the real world, many systems are discrete systems driven by events, such as service systems, scheduling systems, computer network systems or traffic management systems.

There are various methods and tools for analyzing discrete event systems. The systems can be analyzed with the finite state machine methods with Stateflow, or by theoretical approaches with minimax algebra and Petri nets. SimEvents is a simulation based framework for analyzing discrete event systems. In this section, some commonly used concepts about discrete event systems are presented. A server–queue problem is used as an example of simulating discrete event systems using the SimEvents blockset.

There are two essential components in discrete event systems: the entity and the event. An entity describes an object in the system, while an event describes the driving conditions when the state of the system changes. Entities can be classified as temporary entities or permanent entities. In queuing serving systems, the service desk can be regarded as the permanent entity residing in the system, while the customers in the queue can be regarded as temporary entities. Temporary entities enter the system at particular time instances, and reside in the system for a period of time and, after reacting with other entities, they may leave the system. Permanent entities reside in the system all the time. The arrival and departure of temporary entities and the interactions of the entities change the internal states of the system [15]. Each entity has its own states and properties.

Events are the conditions that cause the change in the states in the system. For queuing systems, the often encountered events are the arrival of entities, start of services, end of services and the departure of the entities.

Discrete event systems can be modeled directly with Stateflow – as discussed in the previous section – since the states and state transition conditions can be modeled easily with Stateflow. The states of the whole simulation system can be driven by the events. Discrete event dynamic systems can also be constructed with the blocks provided in the SimEvents blockset. In this section, modeling and simulation problems by SimEvents are discussed.

8.4.2   Introduction to SimEvents

In the Simulink model library, double click the SimEvents icon, or type `simevents` command in the MATLAB command window, then the SimEvents blockset will be opened, as shown in Fig. 8.71.

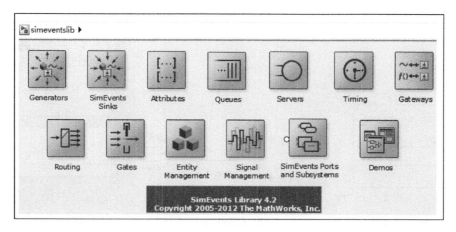

Figure 8.71   SimEvents blockset.

It can be seen that there are several groups in the blockset, such as **Generators, SimEvents Sinks, Entity Management, Attributes, Queues, Signal Management, Servers, SimEvents Ports & Subsystems, Timing** and **Routing**. The blocks in these groups can be used to model and simulate discrete event systems.

Double clicking the **Generators** group, a subgroup shown in Fig. 8.72(a) will be opened. The three subgroups contained in **Generators** are **Entity Generators**, **Event Generators** and **Signal Generators**, where the **Entity Generators** subgroup is shown in Fig. 8.72(b), including the two blocks **Time-Based Entity Generator** and **Event-Based Entity Generator**.

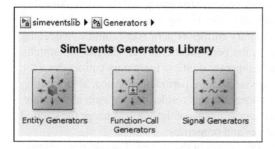

(a) SimEvents generator group

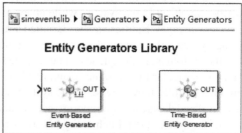

(b) **Entity Generators Library**

Figure 8.72   SimEvents generators library.

The **Event Generator** and **Signal Generator** groups can be opened as shown in Figs 8.73 (a) and (b), where the blocks **Signal-Based Function-Call Event Generator, Entity-Based Function-Call Event Generator, Event-Based Random Number** and **Event-Based Sequence** are provided. These blocks can be used to generate the events and signals needed to drive the simulation model.

The SimEvents **Sinks** group provides the blocks shown in Fig. 8.74, including **Entity Sink, Signal Scope, X-Y Signal Scope, Attribute Scope, X-Y Attribute Scope, Discrete Event Signal to Workspace**. These blocks can be used to show simulation results.

The SimEvents **Queues** group is shown in Fig. 8.75(a). The blocks in this group include **FIFO Queue** (standing for first-in, first-out), **LIFO Queue** (last-in, first-out) and **Priority Queue**.

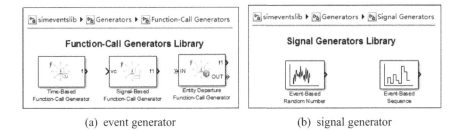

(a) event generator   (b) signal generator

Figure 8.73   Event and signal generators.

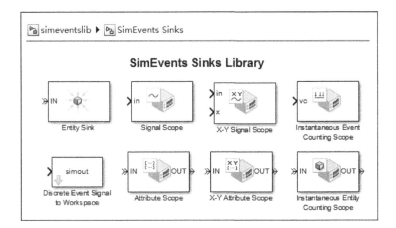

Figure 8.74   SimEvents Sinks group.

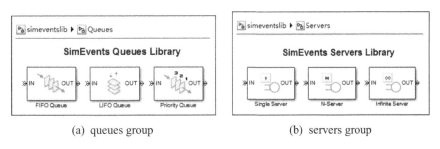

(a) queues group   (b) servers group

Figure 8.75   Queues and servers groups.

The SimEvents **Servers** group is shown in Fig. 8.75(b), including **Single Server, N-Server** and **Infinite Server** blocks.

It can be seen from the SimEvents blockset that there are usually two types of signal ports in SimEvents blocks. One is the ordinary Simulink port, and the other one, labeled with ⭐, is the SimEvents signal used to describe entities. This type of signal can be detected with entity detection blocks. SimEvents also provides **Routing** as shown in Fig. 8.76. The blocks include **Replicate, Output Switch, Input Switch** and **Path Combiner**.

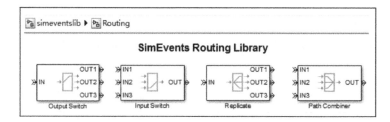

Figure 8.76   SimEvents routing library.

8.4.3   Modeling and Simulation of Queuing Systems

In this section, the ideas of the previous example can now be used for modeling and simulation of queuing models. The simplest queuing model is first demonstrated, and ways of extending this and doing further studies of the queuing systems are also demonstrated.

Example 8.14   *In a barbershop, assume that the arrival intervals of customers satisfy a uniform distribution in the interval* (0, 20) *minutes. Each barber completes his haircut job in 15 minutes. SimEvents can be used to study how many barbers are needed in the barbershop.*

For SimEvents modeling, we can assume that the arrival time of the entities – that is, customers in this example – can be modeled with the **Time-Based Entity Generator** *block. An entity is generated and arrives and enters a first-in first-out queue to wait to be serviced. If the barber is idle, the next entity in the queue can be serviced. Based on this idea, the Simulink model can be constructed as shown in Fig. 8.77.*

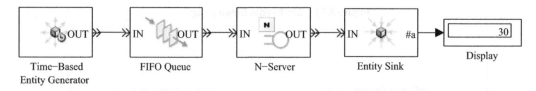

Figure 8.77   SimEvents model (model name: c8mqueue1).

Double clicking the **Time-Based Entity Generator** *block, a dialog box is displayed. The listbox labeled* **Generate entities with** *can be opened, as shown in Fig. 8.78(a). The random numbers can be generated with the options* **Intergeneration time from dialog** *and* **Intergeneration time from port t**; *the latter option can be used to automatically add a* **t** *port to accept signals from Simulink. This signal can be used to control the time spacing.*

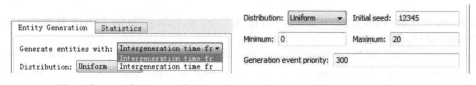

(a)   random number generation   (b)   uniform distributed random numbers

Figure 8.78   Entity generator parameters dialog boxes.

*The dialog box can also be used to define random numbers. The listbox **Distribution** provides options **Constant**, **Uniform** and **Exponential**, which can be used to specify time spacing. If the option **Uniform** is selected, as shown in Fig. 8.78(b), the interval of the random numbers can be generated.*

*In the simulation model, the **FIFO Queue** block can be used to represent the first-in first-out queue. The **N-Server** server describes N servers, with the dialog box shown in Fig. 8.79(a). The number of servers can be directly specified in the dialog box. The service time can be set by external signals. Here we set the service time to constant 15.*

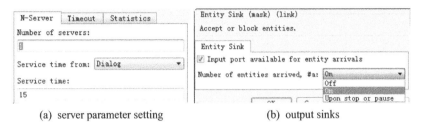

(a) server parameter setting (b) output sinks

Figure 8.79   Server and output parameter setting dialog box.

*Double click the **Entity Sink** block, and the dialog box in Fig. 8.79(b) is shown. The **Number of entities arrived, #a** listbox can be set to **on**. An extra output port can be added automatically to the block, and the number of entities arrived can be output from this port.*

It can be seen from the simulation results that if there is one barber in the shop, 30 customers can be served within 8 hours. If two or three barbers are working in the barbershop, 48 customers can be served within 8 hours. It is concluded that two barbers are sufficient for the barbershop.

By using simulation methods, different schemes can be assessed. However, some important parameters such as average waiting time are not obtained with the current model. Modifications should be made to the model for further analysis.

Example 8.15   *Apart from the default input and output ports of the SimEvents blocks, many more options can be used to provide more input and output ports. Double click the **FIFO Queue** block, and select the **Statistics** pane, as shown in Fig. 8.80(a), with each checkbox corresponding to an output port. The relevant ports include **Number of entities departed, #d**, **Number of entities in the queue, #n**, **Average wait, w**, **Average queue length, len** and **Number of entities time-out**,*

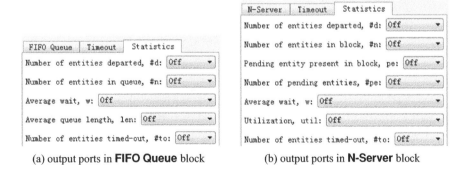

(a) output ports in **FIFO Queue** block   (b) output ports in **N-Server** block

Figure 8.80   Dialog boxes in the queue and server blocks.

#to. *Note that the term "average" means the average value up to the current time. The actual average value can be retrieved as the final value of this port when the simulation is completed, and this quantity has statistical meanings.*

*Double click the **N-Server** block to open its dialog box, and select the **Statistics** pane, as shown in Fig. 8.80(b), where each option still corresponds to an output port, and the output ports include **Number of entities departed, #d, Number of entities in block, #n, Pending entity present in block, pe, Average wait, w, Utilization, util** and **Number of entities time-out, #to**.*

With the selected output ports, the reconstructed simulation model is shown in Fig. 8.81. Some other simulation results can be obtained as shown in Fig. 8.82.

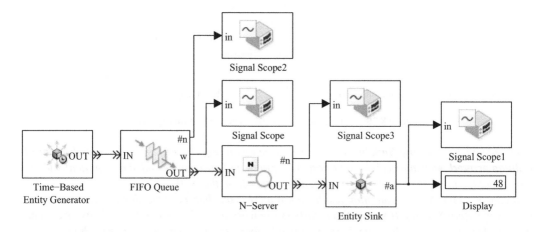

Figure 8.81   New simulation model (model name: c8mqueue2).

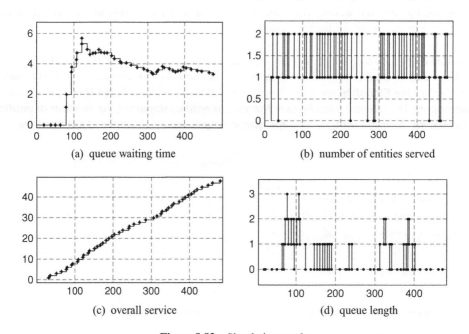

(a) queue waiting time

(b) number of entities served

(c) overall service

(d) queue length

Figure 8.82   Simulation results.

It can be seen from Fig. 8.82(a) that the average queuing time is 3.35 minutes. If three barbers are employed, the average waiting time is reduced to 0.3 minutes. However, the total number of serviced customers is still 48.

Example 8.16   *The serving time for each customer was assumed to be a fixed 15 minutes. If the service time is a random number uniformly distributed in the interval (12,18) minutes, the Simulink model in Fig. 8.83 can be constructed. In this model, the **N-Server** has one more input port **t**. This port is added by using the dialog box in Fig. 8.84(a): in the listbox **Service time from**, the **Signal port t** option is selected. A signal generator block **Event-Based Random Number Generator** can then be used, and the contents of the block are shown in Fig. 8.84(b). If the listbox **Uniform** is set to **Distribution**, and the interval of the random number is set to **(12,18)** for the service time.*

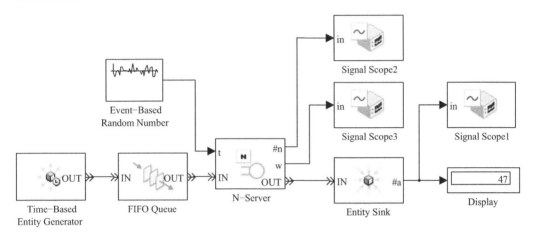

Figure 8.83   New simulation model (model name: c8mqueue3).

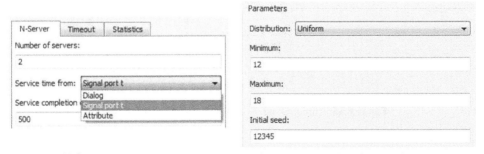

(a) **N-Server** block parameters   (b) parameters of random number generator

Figure 8.84   Parameter setting dialog box.

In the new simulation system, the number of service customers is 47, as shown in Fig. 8.85(a), since the service time is generated randomly. The number of served entities are also obtained as shown in Fig. 8.85(b). It can be discovered from simulation results that if the number of barbers is increased, the total serviced customer number is still 47.

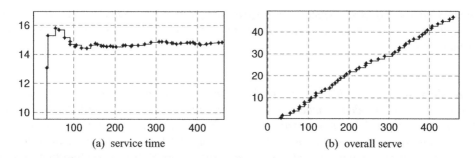

(a) service time   (b) overall serve

Figure 8.85   Simulation results.

The simulation presented earlier analyzed multiple server problems with one block. With the use of SimEvents, multiple servers with multiple queue problems can also be analyzed. A simple example will now be demonstrated.

Example 8.17   *Assume that the customers enter the barbershop uniformly and join two queues, waiting for the two barbers for service. The simulation model can be established as shown in Fig. 8.86. An **Output Switch** block can be used to administrate the two queues. Double clicking the block, the dialog box shown in Fig. 8.87 will be displayed, which enables the user to specify the **Number of entity output ports**, and then in the **Switching criterion** listbox, select the option **Equiprobable**. It can be seen that the simulation results are the same as those obtained earlier.*

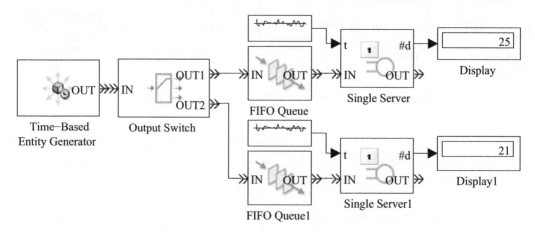

Figure 8.86   Simulink model (model name: c8mqueue4).

In addition, the priority of the queues and entities can also be set. More detailed information can be found in [16] and [14, 15].

Exercises

8.1   Write down the mathematical models of the single-compartment model, ordinary two-compartment and three-compartment model in the Wada's blockset.

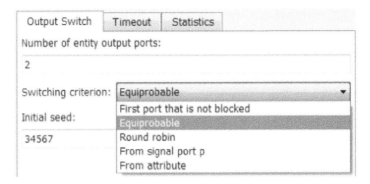

Figure 8.87   Dialog box of the switch block.

8.2 For the anesthetic control and simulation interface, the generalized predictive controller was written in C, and was not rewritten in MATLAB. In fact, the code can be rewritten as an S-function, so that the Simulink model for simulation can be established. Rewrite the S-function representation of the generalized predictive controller, and draw the simulation block diagram.

8.3 It can be seen from Example 8.5 that when dealing with certain images, an adaptive histogram equalization approach may behave much better than the ordinary histogram equalization approach. Unfortunately, there is no such MATLAB block provided. The MATLAB function `adapthisteq()` can only be used to process static images. Try to extend this function to a Simulink block so that videos can also be processed with this block.

8.4 Construct the loop structure and switch structure by the use of Stateflow chart drawing techniques.

8.5 A seven-segment LCD decoder was studied earlier. The kernel part of the circuit is the truth table block. Try to establish a truth table using Simulink and perform the simulation again and see whether the same results are obtained.

References

[1] D J Waters, W W Mapleson. Exponentials and the anaesthetist. Anaesthesia, 1964, 19:274–293

[2] N R Davis, W W Mapleson. A physiological model for the distribution of injected agents, with special reference to pethidine. British Journal of Anaesthsia, 1993, 70:248–258

[3] D R Wada, D R Stanski, W F Ebling. A PC-based graphical simulator for physiological pharmacokinetic models. Computer Methods and Programs in Biomedicine, 1995, 46:245–255

[4] D R Wada, D S Ward. Open-loop control of multiple drug effects in anaesthesia. IEEE Transactions on Biomedical Engineering, 1995, 42(7):666–677

[5] M J Higgins. Clinical and theoretical studies with opioid analgestic fentanyl. Master's thesis, University of Glasgow, 1990

[6] D Xue. Experimentation of linear dynamic fitting analysis for the Higgins model with fentanyl drug administrations. Technical Report, Department of Automatic Control and Systems Engineering, Sheffield University, Sheffield, UK, 1999

[7] M Mahfouf, D A Linkens, D Xue. A new generic approach to model reduction for complex physiologically-based drug models. Control Engineering Practice, 2002, 10(1):67–82

[8] D Xue, D P Atherton. A suboptimal reduction algorithm for linear systems with a time delay. International Journal of Control, 1994, 60(2):181–196

[9] D Xue. Computer aided design of control systems – MATLAB languages and applications (2nd edition). Beijing: Tsinghua University Press, 2006. In Chinese

[10] R C Gongzalez, R E Woods. Digital image processing with MATLAB. Englewood Cliffs: Prentice-Hall, 2nd Edition, 2002
[11] R C Gongzalez, R E Woods. Digital image processing. Englewood Cliffs: Prentice-Hall, 2nd Edition, 2002
[12] The MathWorks Inc. Simulink/Stateflow technical examples – Using Simulink and Stateflow in automotive applications, 1998.
[13] X H Ma. Research on computer assisted table tennis referee system based on digital image processing. Shenyang: MSc Thesis, Northeastern University, 2010. In Chinese
[14] D Z Zheng, Q C Zhao. Discrete event dynamic systems. Beijing: Tsinghua University Press, 2001. In Chinese
[15] Q T Gu. Modeling and simulation of discrete event systems. Beijing: Tsinghua University Press, 1999. In Chinese
[16] The MathWorks Inc. SimEvent user's manual, 2009

9

Hardware-in-the-loop Simulation and Real-time Control

In the previous chapters, Simulink was used to model complicated systems. Single variable and multivariable systems, linear and nonlinear systems, continuous, discrete and hybrid systems, time invariant and time varying systems, engineering and non-engineering systems can all be simulated by Simulink. With the use of S-function and Stateflow techniques, complicated systems, as well as discrete event systems can also be modeled and simulated.

So far, numerical simulation approaches in Simulink have been studied. However, the interaction with the real world has yet to be considered. For many practical processes, accurate mathematical models cannot be obtained, so exact Simulink models cannot be constructed. Sometimes due to the complexity of actual systems, the Simulink models established may not be accurate. The actual system should somehow be embedded in the simulation loop to get more accurate simulation results. This kind of simulation is usually referred to as a hardware-in-the-loop simulation. Since this kind of simulation is often performed in real-time, it is sometimes referred to as real-time simulation.

Real-Time Workshop provided by MathWorks can translate the Simulink models into C code, and the standalone executable files can also be generated using this tool, so that real-time control can be performed. Also, third-party software and hardware also provide interfaces to Simulink. Good examples of these products are dSPACE, with its Control Desk and Quanser plus WinCon (which can be used to implement hardware-in-the-loop simulation and real-time control experiments). MATLAB and Simulink support many products from well-known hardware manufacturers such as Motorola, Texas Instruments etc, and can directly generate executable code for them from Simulink models. A low-cost NIAT tool can also be used to implement hardware-in-the-loop simulation and experiments. Moreover, a much lower cost platform using Arduino is introduced. In this chapter, these tools will be demonstrated with applications in real-time control.

9.1 Simulink and Real-Time Workshop

9.1.1 Introduction to Hardware-in-the-loop Techniques

In hardware-in-the-loop simulation systems, part of the simulation loop is composed of computer software, while the rest is the actual hardware systems.

In practical control systems, the hardware in the loop can either be the controllers or the plant. In aerospace and military systems, the hardware in the system is usually the controllers. However, in

---

System Simulation Techniques with MATLAB® and Simulink®, First Edition. Dingyü Xue and YangQuan Chen.
© 2014 John Wiley & Sons, Ltd. Published 2014 by John Wiley & Sons, Ltd.

process control systems, the hardware is usually the plant, and the controllers can be the Simulink models.

One advantage of hardware-in-the-loop simulation is that the validation of control results is very straightforward. In industrial control, hardware-in-the-loop simulation techniques can significantly reduce the time required to design controllers and can increase the reliability of the systems.

Fast prototype design is a new approach in controller design and implementation. The controllers can be constructed with Simulink and Stateflow, and the executable code for the controllers can be generated easily, and the parameters of the controllers can be tuned on-line, according to the actual control behavior. The designed controller can be regarded as a prototype, and once the control results are satisfactory, the code can be downloaded to the real controllers and the control actions can be carried out without the use of MATLAB or Simulink.

MathWorks provides the following tools to support Simulink controllers:

- **Real-Time Workshop, RTW**: Optimized C and Ada code can be generated automatically from the Simulink model. The executable file is much faster than the Simulink model.

- **Real-Time Workshop Embedded Coder**: This can be used to develop embedded C programs.

- **Real-Time Windows Target** and **xPC Windows Target**: The controller described by Simulink can be used to access the input and output ports of I/O boards, such as A/D and D/A converters, so that real-time control systems can be established.

9.1.2   Standalone Code Generation

Sometimes the simulation process of Simulink model can be very slow, so we may need to speed up the process. On the other hand, we may sometimes need the simulation process to be executed independently, without support from MATLAB environment. Thus, standalone executables need to be considered.

In real-time simulation, fixed step size simulation algorithms have to be used. The details of setting up Simulink algorithms can be found in Fig. 4.34(b) in Chapter 4. Examples are given below to demonstrate the use of the real-time tools.

Example 9.1   *Consider the Simulink model in Example 8.60, with the model file c8fstr1.mdl. Selecting a fixed step size of 0.001 s, and the algorithm **ode5(Dormand-Prince)**, the following statements can be used for assigning the values of the parameters*

```
M=0.01; K=1; Fsliding=1; Fstatic=1;
```

*These commands can be written into the **PreLoadFcn** property of the model, and saved in a new file c9fstr1.mdl. Running the model*

```
>> tic, [t,x,y]=sim('c9fstr1'); toc
```

*it can be seen that the execution time required is 11.32 s on the testing machine. The menu **Tools** → **Real-Time Workshop** → **Build Model** can be used, and an executable file c9fstr1.exe generated. This file is a standalone file, which can be executed without MATLAB. The following commands can be used, and the execution time can be measured as 0.37 s. This is much faster than the model running under MATLAB.*

```
>> tic, !c9fstr1
   toc % note that this command cannot be used in the previous line
```

The executable file saves the results to the data file c9fstr1.mat. With the `load` *command, the data obtained can be loaded into the MATLAB workspace. The two variables* `rt_tout` *and* `rt_yout` *are returned, which store respectively the time and output signal of the system. For instance, the simulation curves can be drawn with the following statements, as shown in Fig. 9.1.*

```
>> load c9fstr1; plot(rt_tout,rt_yout)
```

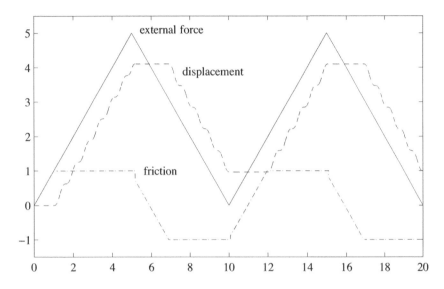

Figure 9.1   Simulation results obtained with the executable file.

It should be noted that the simulation results are obtained with the fixed step algorithm, but the results are the same as the variable step algorithm. With the **Simulation** → **Configuration Parameters** *menu in the Simulink model window, the* **Real-Time Workshop** *pane of the dialog box is shown in Fig. 9.2. If only the C code is needed, then the* **Generate code only** *checkbox should be selected. The* **Tools** → **Real-time workshop** → **Build Model** *menu item can be selected to generate the source code. The files *.c, *.mk and *.h can be generated in the* c9fstr1_grt_rtw *folder. The executable file can then be constructed. Compiling parameters can be selected from the dialog box. If the executable file is to be generated, deselect the* **Generate code only** *checkbox.*

If standalone executable is not needed, and the only requirement is to speed up the simulation process, there is no need to use fixed step algorithms. The menu **Simulation** → **Accelerator** *can be used to construct dynamic link library files. The model can then be saved to a new file c9fstr2.mdl. Executing the file, the time required is 0.35 s. So it can be seen that the efficiency of the simulation model has again been significantly improved.*

```
>> tic, [t,x,y]=sim('c9fstr2'); toc
```

The Simulink models with S-functions written in C language can be converted into standalone executables, while the one containing the MATLAB version of S-functions cannot be manipulated in this way.

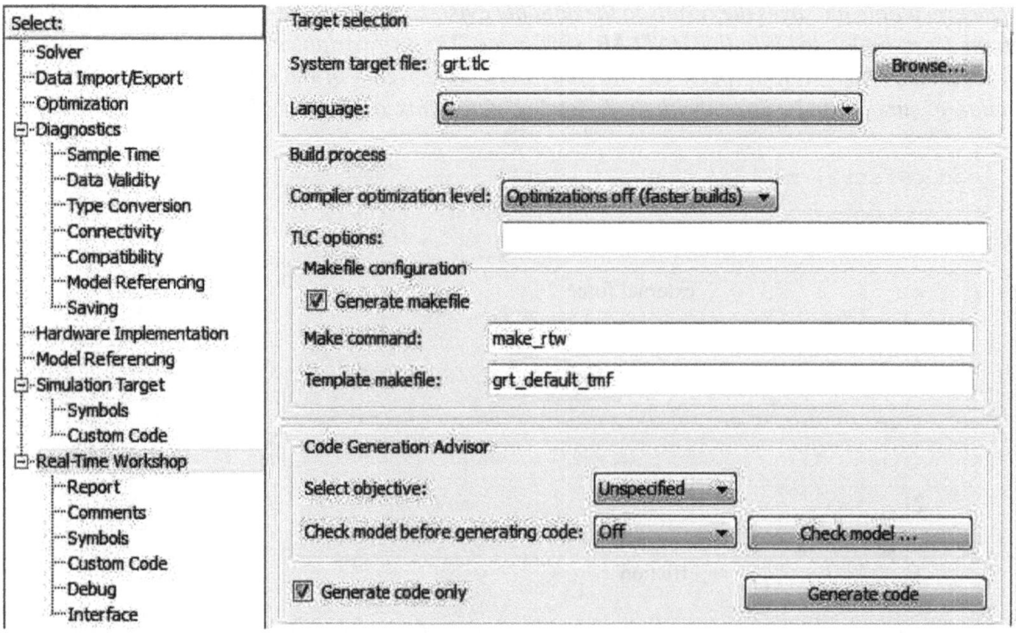

Figure 9.2   Simulink parameter setting dialog box.

9.1.3   Real-time Simulation and Target Computer Simulation

Hardware-in-the-loop simulation and fast prototype design can be performed directly with Real-Time Windows. This allows the use of a computer as a host and a target in simulation [1]. When the Simulink model has been constructed, external simulation mode can be used to generate executable files.

In ordinary off-line control system design, the mathematical models of the plant model should be extracted first. Then, based on the mathematical model of the plant, a controller can be designed. However, when the controller is used in real-time control, the behavior of the system may not be as good; indeed, the control results may be very poor. This is because, in numerical simulation, many of the factors of the plant models may have to be neglected, such as the accuracy of the mathematical models, external disturbances, model parameter variations, measurement noise and so on. In actual systems, all these may affect the behavior of the practical systems.

Thus, hardware-in-the-loop simulation techniques are very important. Since the controller designed can be used to control the actual plant, the control behavior can be assessed directly. xPC is a low-cost hardware-in-the-loop simulation software package, and the major input and output boards are supported.

Here we shall concentrate on presenting the real-time simulation techniques based on Real-Time Windows Target. The environment should be set first. The command

```
rtwintgt -install
```

can be used to complete the required specifications.

```
You are going to install the Real-Time Windows Target kernel.
Do you want to proceed? [y] :
```

If error messages are obtained, it may indicate that the current operating systems are not supported. The current version of Real-Time Windows Target only supports operating systems such as 32-bit Windows XP and Vista, while Windows 7 and other 64-bit operating systems are not supported.

As soon as Real-Time Windows Target is installed, it can be used. However, to successfully run Real-Time Windows Target, both Microsoft Visual C++ and the Watcom C compiler should also be installed.

The concepts of "host" and "target" computers will be presented first in this section. The host here is normally the computer running MATLAB and Simulink, while the target computer refers to the computer running the standalone executables generated by Simulink. The RS232 interface and TCP/IP protocol can be used to connect the target with the host computer, and together they can complete real-time simulation tasks. It is not necessary to run the target executables on a different computer, but this application can only be used on the same computer.

Example 9.2   *Consider the Van der Pol equation in Example 4.3. In the Simulink model, an* **XY Scope** *block was used, as shown in Fig. 4.44. For convenience, the model is saved again to c9fvdp1.mdl file, as shown in Fig. 9.3.*

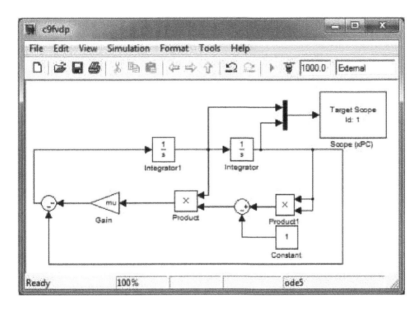

Figure 9.3   Modified Simulink model for the Van der Pol equation (model name: c9fvdp1).

The **Simulation** $\rightarrow$ **Mode** menu supports the following simulation modes.

- **Normal simulation mode**: This is normally used in off-line numerical solutions of systems. This mode is the default one. In the **Simulation** menu, the **Normal** item should be checked.
- **Simulation accelerator mode**: When the menu **Simulation** $\rightarrow$ **Accelerator** is checked, Simulink will automatically translate the Simulink model into C code, and establish an executable *.mexw32 file. This model can be called from Simulink to speed up the simulation process. In this mode, variable step simulation algorithms can be selected, but it is not a standalone program. It has to be executed in the MATLAB environment.

- **External simulation mode**: The menu item **Simulation** → **External** is checked. Simulink models can be executed on the host with no MATLAB installed. In order to get the external function, **Simulation** → **Configuration parameters** or **Tools** → **Real-Time Workshop** → **Option** menus should be selected. In the dialog box, the **System Target File** listbox should be assigned as **Real-Time Windows Target**, as shown in Fig. 9.4. Selecting **Tools** → **Real-Time Workshop** → **Build model** menu, an executable Windows Target code can be generated automatically.

- Simulation also supports hardware-in-the-loop (**HIL**) and processor-in-the-loop (**PIL**) simulation.

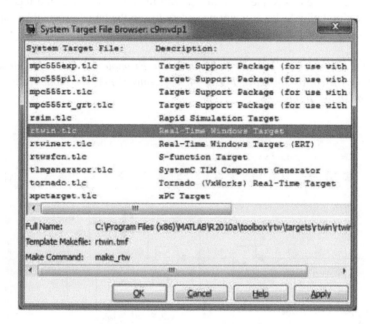

Figure 9.4   Parameter setting dialog box.

From the listbox in Fig. 9.4, it can be seen that, in addition to the ordinary **Real-Time Windows Target** option, a large number of other target formats are also supported.

- **DOS(4GW) Real-Time Target**: The target program can be generated for MS-DOS systems. However, the code can only be executed in the DOS environment, and cannot be executed in a DOS window under the Windows system.
- **Real-Time Workshop Embedded Coder**
- **Rapid Simulation Target**: The rapid simulation code can be generated, mainly used for frequently performed simulation processes, such as the Monte Carlo simulation.
- **Real-Time Windows Target**
- **Tornado (VxWorks) Real-Time Target**

When the compiler is set up, the menu **Tools** → **Real-Time Workshop** → **Build model** can be selected to compile and link the model, such that an executable file c9fvdp1.exe is generated.

Selecting the **Tools** → **External mode control panel** menu item, the dialog box shown in Fig. 9.5 will be opened. Click the **Connect** button, and the executable file can be executed in real-time.

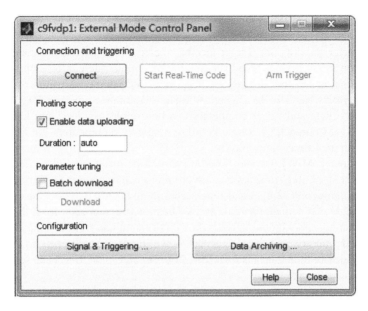

Figure 9.5 Control panel of the external simulation mode.

In a similar way, the target executable file can be generated for MS DOS system. The Watcom C compiler should be used to compile and link the code, and the executable generated can then be run directly under MS DOS.

Real-Time Windows Target also provides its own blockset. The blockset can be opened with either the Simulink model library, or the MATLAB command `rtwinlib`, as shown in Fig. 9.6. Various blocks suitable for real-time simulation are provided.

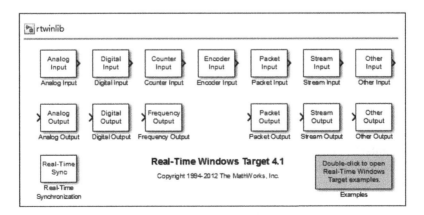

Figure 9.6 Real-Time Windows Target blockset.

9.1.4   Hardware-in-the-loop Simulation with xPC Target

xPC Target is the fast prototype design tool provided by MathWorks. It can be used in real-time testing and development of controllers. Also, C compilers, such as Microsoft Visual C++ or MATCOM C/C++, are required. Through the compatible compilers, an executable file can be generated and used in real-time control.

The major characteristics of xPC Target are that the real-time executable code generated by Simulink can run on the target machine. The maximum sample rate can be as high as 100 kHz. Various standard input and output devices are supported, and interactive parameter tuning on the host or target computers can also be carried out. The communication with RS232 and TCP/IP protocol between the host and target computers can be established. Desktop, laptop computers and PC/104, PC/104+, CompactPCI, single board or single-chip computers can be used as target computers to perform real-time control tasks.

On the host computer, MATLAB and Simulink should be installed, and a C compiler can be used as a development tool. Real-time application programs can be generated. To run the executable program, a disk containing xPC Target real-time kernel should be used to boot the target computer. Once the target computer is booted, the real-time application program can run on that computer.

In the MATLAB command window, execute the `xpclib` command, or click the **xPC Target** icon in the Simulink model library, and the xPC group shown in Fig. 9.7 will open. In the group, there are commonly used device icons such as **A/D**, **D/A**, **Counter** and so on. Users can select appropriate devices as needed.

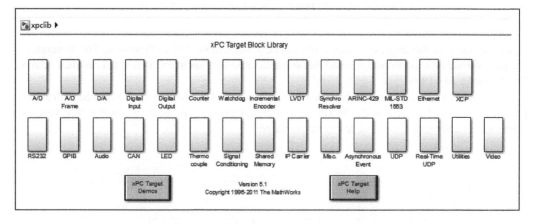

Figure 9.7   xPC Target blockset.

For instance, double click the **A/D** icon, and the A/D converter group will open, as shown in Fig. 9.8, where all the well-known A/D converters are provided. Double click the **Digital Input**, and then the **Advantech** icon, and the Advantech A/D device group shown in Fig. 9.9 is displayed. Suitable A/D devices from the group can be selected, and the controllers can be connected to the control terminals.

Here the Advantech input/output board PCL 1800 is used to introduce the application of the xPC tools. PCL 1800 is an ISA input/output board and it can be used on the main board of the computer. On this board, 16 12-bit analog input channels and two 12-bit D/A converters are provided.

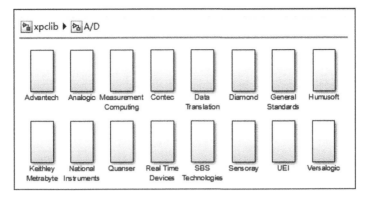

Figure 9.8   A/D converter supported by xPC.

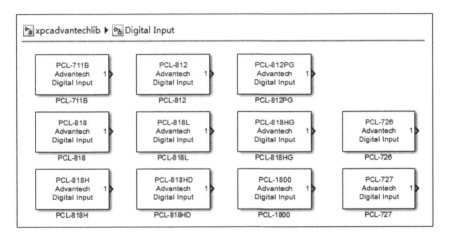

Figure 9.9   AdvanTech A/D converter.

Example 9.3   *Consider the PI control block diagram studied in Example 5.25. In the system, the PI controller can be designed with the mathematical model of the plant. The Simulink block diagram is obtained as shown in Fig. 9.10.*

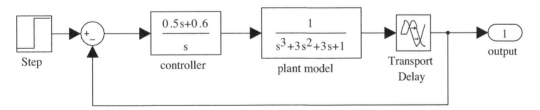

Figure 9.10   Numerical simulation of a control system model (model name: c9mpi).

It has been pointed out that the simulation results are made on the mathematical model of the plant, and the simulation results may not be the same when the actual plant is used in the system. Sometimes the errors can be misleading. In order to test the control behavior of the actual plant, the

mathematical model of the plant must first be deleted from the Simulink model, and then replaced
with the actual plant. This can be done using the D/A and A/D converters of xPC, for instance, with
the Advantech PCL 1800, as shown in Fig. 9.11.

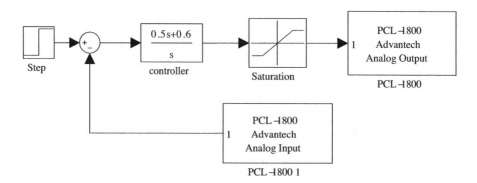

Figure 9.11   Hardware-in-the-loop PI control system (model name: c9mpi1).

With the Simulink model, the actual plant can be connected to the computer through the Advantech
PCL 1800 card. The input terminal of the plant can be connected to the D/A port to accept control
signals. The output terminal of the plant can be connected to the A/D port, which returns the
measured output to the computer. Double click the A/D block and D/A block and the dialog boxes
are respectively shown in Figs 9.12(a) and (b).

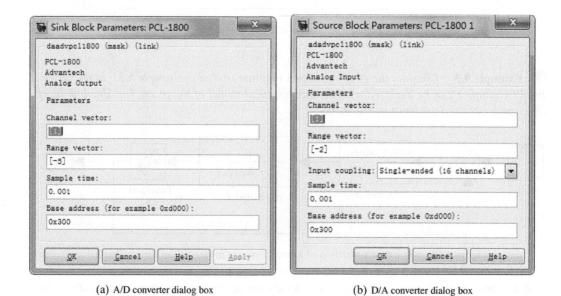

(a) A/D converter dialog box   (b) D/A converter dialog box

Figure 9.12   Dialog boxes of the PCL 1800 AD and DA converters.

- **Channel vector**: If it is connected to a particular channel, the number of it should be entered. For instance, the entry **1** here is the channel number. If many signals are connected to the D/A converters, the channel vector should be specified.

- **Range vector**: Each input and output signal has its own range, which can be expressed by $[v_1, v_2, \cdots]$. If the range of the ith signal is $(-5, +5)$, then it is denoted as $v_i = -5$. However, if the range is $(0, +5)$, it is denoted by $v_i = 5$.

- **Sample time**: This is the sampling interval of the block and it should be the same as the step size of the simulation algorithm.

- **Base address**: The specification is the same as the one in the input and output cards.

📓 **Example 9.4** *It looks as if the system given in Fig. 9.11 is not a closed-loop system; rather it looks like an open-loop structure. In fact, the connection is indeed a closed-loop structure, when the actual plant is considered. Since the plant is part of the simulation system, such a system is referred to as the "hardware-in-the-loop" simulation structure.*

If the input, output and intermediate signals are needed, the simulation model can be redrawn as shown in Fig. 9.13. In such a system, the signal obtained can be saved into data files. The data files can be loaded into the MATLAB workspace, or you can use an x-y scope block, such that real-time display of the results is possible.

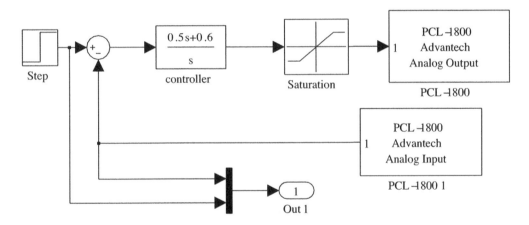

Figure 9.13   Hardware-in-the-loop simulation block diagram with signal detection (c9mpi2).

9.2   Introduction to dSPACE and its Blocks

9.2.1   Introduction to dSPACE

dSPACE (digital Signal Processing And Control Engineering) is the real-time simulation and testing platform, and it can be used along with MATLAB/Simulink. The dSPACE real-time system has high-speed computation capabilities in hardware systems, with processes and I/O ports, and it can easily be used in real-time code generation, downloading and testing [2].

dSPACE real-time control systems have certain advantages. The blocks provided are adequate for constructing real-time control systems, and the real-time facilities are ideal, since the on-board PowerPC processor can be used directly. Seamless connection with MATLAB and Simulink can be used to directly convert the numerical simulation structure to real-time control. dSPACE systems are now widely used in automotive, aerospace, robots, industrial automation and other fields. By using dSPACE systems, product development cycles can be significantly reduced, and the control quality significantly increased.

9.2.2   dSPACE Block Library

Software and hardware environments of the dSPACE real-time simulation system are presented below. The widely used hardware in education and scientific research is the ACE1103 and ACE1104 R&D controller boards. Real-time control software Control Desk, real-time interface RTI and real-time data acquisition interface MTRACE/MLIB are provided in the dSPACE package, and it is very flexible and convenient to use. The DS1104 R&D controller board, which is equipped with a PCI interface and a PowerPC processor, is a cost-effective entry-level control system design product. Here, the DS1104 R&D controller board is used to illustrate the applications in hardware-in-the-loop simulation.

When the hardware and software of dSPACE are installed, there will be a dSPACE group in the Simulink model library; double click its icon, and the group will be opened as shown in Fig. 9.14.

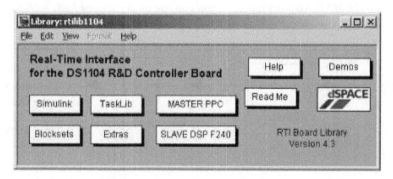

Figure 9.14   dSPACE 1104 blockset.

Double click the **MASTER PPC** block, and the model library shown in Fig. 9.15 will be opened. It can be seen that a lot of components on the board, such as A/D converter, are represented by blocks in the group. Also, double clicking the **Slave DSP F240** icon will open the slave DSP F240 blocks shown in Fig. 9.16. Many practical servo control blocks, such as PWM signal generator and frequency sensor are provided in the group, and they can be dragged to Simulink models. The signals generated by the computer can also be used to drive the actual plant, so that hardware-in-the-loop simulation can be completed.

9.3   Introduction to Quanser and its Blocks

9.3.1   Introduction to Quanser

Quanser products are developed by Quanser Inc., who provides various plants for control education. Quanser has interfaces to MATLAB/Simulink and to LabView of National Instruments. It also provides real-time control software WinCon (now named as QUARC), which is similar to Control Desk from dSPACE. Quanser products are suitable for control education and laboratory research and experimentation, where different control algorithms are being tested. Quanser also offers industrial solutions for hardware-in-the -loop rapid prototyping of real time control systems.

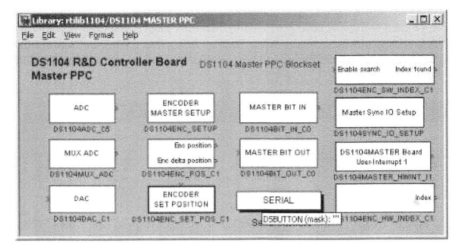

Figure 9.15　**MASTER PPC** group.

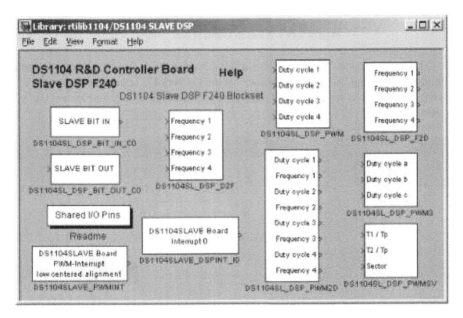

Figure 9.16　**Slave DSP F240** group.

Quanser experimental plants include linear motion control series, rotary series and other special experimental devices. WinCon software enables Simulink modes to directly control the actual plants.

9.3.2　Quanser Block Library

MultiQ3, Q3, Q4, Q8 and other kinds of interface boards are provided in Quanser series products. The boards all provide D/A and A/D converters, and motor encoder input and output ports, such that the actual plants can be connected to the computers to construct closed-loop control structures.

WinCon is a Microsoft Windows based application program implementing real-time control. The program can be used to execute code generated by Simulink models and exchange data with the MultiQ card, to achieve real-time control. When WinCon is installed, a **WinCon Control Box** group will appear in the Simulink model library, as shown in Fig. 9.17, where the subgroups for different Q-series boards are provided.

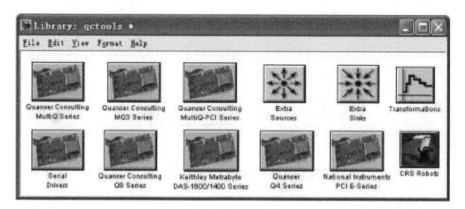

Figure 9.17   **WinCon Control Box** group.

Here the Q4 board is used as an example for further demonstration. Double click the **Quanser Q4 Series** icon in Fig. 9.17 and the Simulink library for the Q4 board is opened as shown in Fig. 9.18. It can be seen that the blocks **Analog Input** and **Analog Output** are used to implement the A/D and D/A converters respectively.

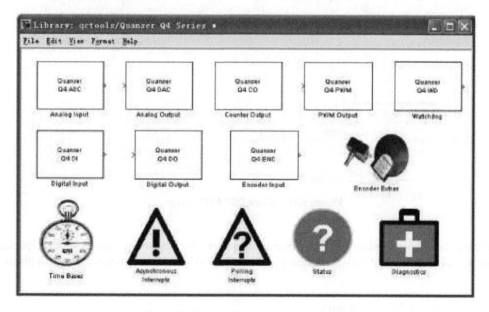

Figure 9.18   Blocks in the MultiQ4 group.

Double clicking the icons of the **Analog Input** and **Analog Output** blocks, the parameter dialog boxes will be opened as shown in Figs 9.19(a) and (b), respectively. The key parameter **Channel** can

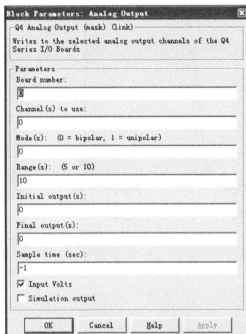

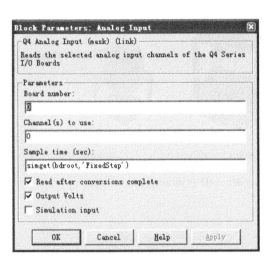

(a) **Analog Input** block setting   (b) **Analog Output** block setting

Figure 9.19   Dialog boxes of the WinCon blocks.

be set in the dialog box, and they should be the same as the one that is actually connected, otherwise the block cannot be used properly.

9.3.3   Plants in Quanser Rotary Series

Quanser experimental plants are mainly the linear motion series and the rotary motion series. The components in the linear series can be used to compose different experiments such as linear speed and position servo control, linear inverted pendulum, double inverted pendulum. In the rotary series, experiments such as rotary inverted pendulum, planar inverted pendulum, ball and beam flexible joint and flexible link can be composed. Other plants such as helicopter attitude control and magnetic levitation systems can also be composed. Some of the plants in the Quanser series system are shown in Fig. 9.20.

9.4   Hardware-in-the-loop Simulation and Real-time Control Examples

9.4.1   Mathematical Descriptions of the Plants

The ball and beam system is one of the experimental systems in the rotary series provided by Quanser. A photograph of this system is shown in Fig. 9.21(a), and the principal structure is shown in Fig. 9.21(b). The principle of the control of the ball and beam system is that the angle θ is

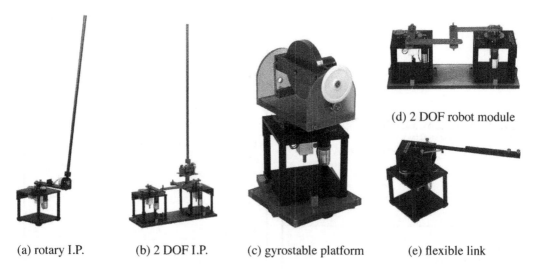

(a) rotary I.P.   (b) 2 DOF I.P.   (c) gyrostable platform   (e) flexible link

(d) 2 DOF robot module

Figure 9.20   Some of the plants in the rotary series (IP is inverted pendulum).

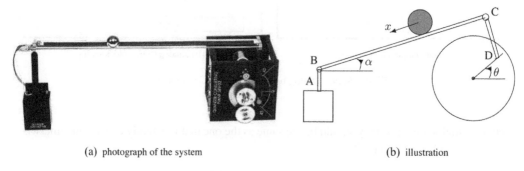

(a) photograph of the system   (b) illustration

Figure 9.21   Ball and beam system.

controlled by the bar, driven by the motor, such that the angle α of the beam BC can be adjusted to the expected position rapidly. The bar AB is a fixed supporting arm.

In the ball and beam system, the position $x(t)$ of the ball is the output signal, and the motor voltage $V_m(t)$ is the control signal. A controller is needed such that the error $e(t)$ between the expected position $c(t)$ and the detected position signal $x(t)$, that is, $e(t) = c(t) - x(t)$, can be used to calculate the control signal $u(t)$. The steel ball on the beam can be used as a variable resistor, such that the position $x(t)$ can be directly detected through the value of the resistor.

1) Mathematical model of the motor drive system

The DC motor model is shown in Fig. 9.22(a). Assume that the electrical efficiency $\eta_m = 0.69$, the equivalent resistance of the motor system is $R_m = 2.6\,\Omega$, viscous damping coefficient is $B_{eq} = 4 \times 10^{-3}\,\text{N} \cdot \text{m} \cdot \text{s/rad}$, transmission ratio $K_g = 70$, EMF constant $K_m = 0.00767\,\text{V} \cdot \text{s/rad}$, torque constant $K_t = 0.00767\,\text{N} \cdot \text{m}$, motor moment of inertia of the equivalent load $J_{eq} = 2 \times 10^{-3}\,\text{kg} \cdot \text{m}^2$, motor moment of inertia $J_m = 3.87 \times 10^{-7}\,\text{kg} \cdot \text{m}^2$, gearbox efficiency $\eta_g = 0.9$.

Constructing a PID controller for the motor, and using derivative action in the feedback loop, the control system structure is as shown in Fig. 9.22(b). A simple tuning method will be given later,

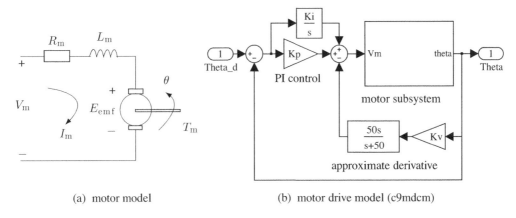

(a) motor model   (b) motor drive model (c9mdcm)

Figure 9.22   Motor and control models.

showing how to design the PID controller parameters to control the angular displacement θ of the model.

In the Quanser experimental system, the Simulink model of the motor system can be constructed as shown in Fig. 9.23. The transfer function of the motor voltage signal $V_m(t)$ to the angle θ can be expressed as [3]

$$G_1(s) = \frac{\theta(s)}{V_m(s)} = \frac{\eta_g \eta_m K_t K_g}{J_{eq} R_m s^2 + (B_{eq} R_m + \eta_g \eta_m K_m K_t K_g^2)s} = \frac{61.54}{s^2 + 35.1\,s}. \tag{9.1}$$

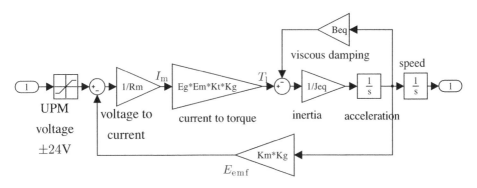

Figure 9.23   Simulink description of motor model.

2) Mathematical model of the ball and beam system

Assume that the length of the beam $l = 42.5$ cm, and the radius of the ball is R. The gravity component on x axis is $F_x = mg\sin\alpha$, $m = 0.064$g, and the moment of inertia of the ball is $J = 2mR^2/5$. The dynamical model of the ball is $\ddot{x} = 5g\sin\alpha/7$.

Since the angle α is usually very small, it can be approximated as $\sin\alpha \approx \alpha$, so the original nonlinear model can be approximated by a linear model. The radius of the eccentric disc is $r = 2.54$ cm. The equivalent arc displacement of beam BC can be expressed as $l\alpha = r\theta$, that is, $\theta = l\alpha/r$. Based on the above formula, the Simulink model shown in Fig. 9.24 can be constructed [4].

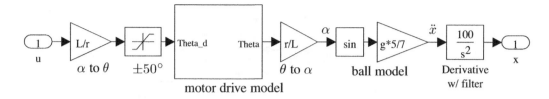

Figure 9.24 Plant model (model name: c9mball).

The Simulink model of the ball and beam control system can be constructed as shown in Fig. 9.25(a), where the whole system is controlled by a PD controller. The parameters of the models and controllers can be entered with the following statements:

```
clear all;   % filename: c9dat_set.m
Beq=4e-3; Km=0.00767; Kt=0.00767; Jm=3.87e-7; Jeq=2e-3; Kg=70;
Eg=0.9; Em=0.69; Rm=2.6; zeta=0.707; Tp=0.200; num=Eg*Em*Kt*Kg;
den=[Jeq*Rm, Beq*Rm+Eg*Em*Km*Kt*Kg^2 0]; Wn=pi/(Tp*sqrt(1-zeta^2));
Kp=Wn^2*den(1)/num(1); Kv=(2*zeta*Wn*den(1)-den(2))/num(1); Ki=2;
L=42.5; r=2.54; g=9.8; zeta_bb=0.707; Tp_bb=1.5;
Wn_bb=pi/(Tp_bb*sqrt(1-zeta_bb^2)); Kp_bb=Wn_bb^2/7;
Kv_bb=2*zeta_bb*Wn_bb/7; Kp_bb=Kp_bb/100; Kv_bb=Kv_bb/100;
```

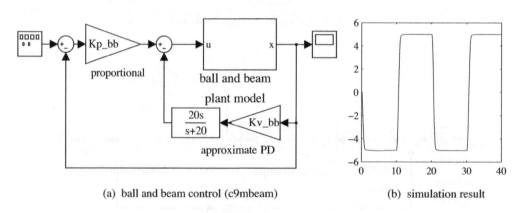

(a) ball and beam control (c9mbeam)   (b) simulation result

Figure 9.25 Control and simulation of the ball and beam system.

Simulating the ball and beam system, the simulation results can be obtained as shown in Fig. 9.25(b). It can be seen that the simulation results are satisfactory. Of course, experimental validation of the controllers can be and should be performed with a hardware-in-the-loop simulation. The hardware-in-the-loop and real-time control of the system are to be demonstrated below.

9.4.2 Quanser Real-time Control Experimentation

It can be seen that the inner PID controller in the system is used to apply control signal V_m to the motor. The position x of the ball and the angle of the motor θ are measured. The control signal can be implemented with the **Analog Output** block, and the position x of the ball can be measured with the **Analog Input** block, and the angle θ of the motor can be measured with the **Encoder Input** block. In real applications, filters can be used, and the real-time control system can be constructed as shown in Fig. 9.26. Note that a fixed step size algorithm is used and the step size is selected as 0.001 s. This can be done by the **Simulation** → **Parameters** menu in the Simulink model, and the **Solver** pane can be set. For instance, the algorithm can be set to **ode4**.

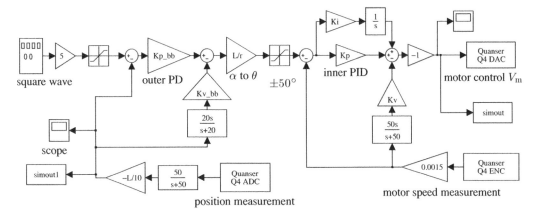

Figure 9.26 Real-time control Simulink model (model name: c9mbbr).

Selecting the **Tools** → **Real-Time Workshop** → **Build Model** menu item, the original Simulink model can be compiled and the dynamic link library file can be generated automatically. The winCon control interface shown in Fig. 9.27 is opened automatically. The plant is controlled directly from the interface. Click the **START** button and the real-time facilities are started. The **Analog Output** block is used to apply the control signal to the motor, and the angle and ball position are measured in real time and fed back to the computer, so that closed-loop control can be achieved.

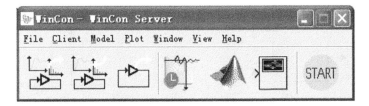

Figure 9.27 WinCon control interface.

The scope button in the interface can be clicked to display the waveforms of the ball positions and control signals, as shown in Fig. 9.28. It should be noted that there are differences between the actual control results and the numerical simulation results. The differences may be caused by modeling error or other minor discrepancies. With the WinCon real-time control interface, the Simulink model runs in the **External** mode. In this mode, if you modify the variables in the MATLAB workspace, real-time control results will also change.

In real-time control systems, the scope data can be saved to the MATLAB workspace with the **File** → **Save** → **Workspace** menu item. The amount of data is determined by the size of the buffer. The buffer size can be selected with the **Buffer** menu; it is set to 40 s in this example.

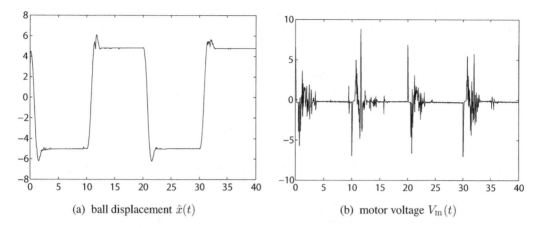

(a) ball displacement $\hat{x}(t)$ (b) motor voltage $V_\mathrm{m}(t)$

Figure 9.28   Real-time control of the ball and beam system.

9.4.3   dSPACE Real-time Control Experimentation

From the Simulink model constructed by the interfaces to the Quanser Q4 components shown in
Fig. 9.26, a new Simulink model can be constructed, with the corresponding components replaced
by dSPACE blocks, as shown in Fig. 9.29. In the dSPACE blockset, the settings of the A/D and
D/A converters are different from those in Quanser. We need to multiply the Quanser A/D converter
figures by 10, and the D/A figures by 0.1, so that the units are unified. Also, since the motor encode
input block of dSPACE is different from that of Quanser, this should be multiplied by 0.006.

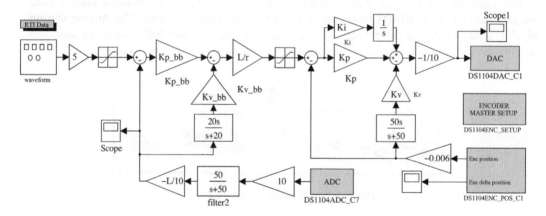

Figure 9.29   Simulink model with dSPACE (model name: c9mdsp).

From the model established, the menu **Tools** → **Real-Time Workshop** → **Build Model** in the
Simulink model window can be selected to compile it, and finally the system description file
c9mdsp.ppc can be generated for Power PC. Open the Control Desk software environment window
[5], and the menu **File** → **Layout** opens a new virtual instrumentation interface. With the controls
on the **Virtual Instruments** toolbar, the control interface can be established. For instance, scroll bars
can be used to access the parameters of the PD controllers. Scopes can be used to display the position
of the ball and the control signals. The final control interface designed is shown in Fig. 9.30.

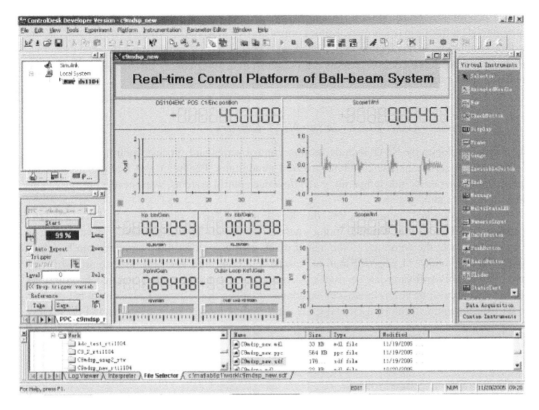

Figure 9.30   Control interface constructed by Control Desk.

Select the **Platform** pane, and the file c9mdsp.ppc saved earlier can be loaded with the file dialog box. The connection between the interface and the *.ppc file generated with Simulink can be established. The parameters in the Simulink model must be associated with the controls in the interface by dragging the Simulink variable names to the relevant controls in the Control Desk.

With the control interface established, real-time control of the actual plant can be performed. The responses of the ball and beam system are as shown in Fig. 9.30. Note that the control signal obtained here is the actual signal written to the **DAC** block from dSPACE, and this should be multiplied by 10 to get the physical signal in the system, which varies in the interval $(-10, 10)$, and it is similar to that obtained with Quanser.

It can be seen that the controller established in Simulink can be used to directly control the practical plant in real time. Also, on-line controller parameter tuning is also allowed. For instance, in the example, the scroll bars can be used to change the parameters of the PD controller, and the control results can be immediately obtained.

The controller and parameters thus created can be downloaded to the actual controllers, such that control can be achieved without MATLAB or dSPACE environments.

9.5   Low Cost Solutions with NIAT

The platform for hardware-in-the-loop simulation and real-time experimentation developed by NIAT is a low-cost solution to experimental education in universities. The Pendubot (double under-actuated inverted pendulum) control system [6] and the water tank control system are ideal research and experimental platforms in teaching real-time control and for MATLAB/Simulink based hardware-in-the-loop simulation. In this section, NIAT blockset is presented first and the modeling and numerical simulation of Pendubot control systems are demonstrated, followed by the hardware-in-the-loop simulation with NIAT.

9.5.1   Commonly Used Blocks in the NIAT Library

The plants developed by NIAT include motion control series and process control series devices. Motion control series include different inverted pendulums, Pendubot and manipulators, while process control series include temperature and flow rate control systems, water-level control systems and double water tank systems.

Universal networked controllers are developed in the NIAT platform with D/A converter **MptSensor**, A/D converter **MptMD**, motor encoder input and output **NpdbENC**. These ports can be used to connect the computer to the actual plant, so that a closed-loop structure can be established. NIAT provides MATLAB/Simulink compatible controller and real-time control software EasyControl. This software can be used to activate the executable code generated by Simulink model, so that real-time control tasks can be achieved. When the software EasyControl is installed, the `niatlib` command can be used to display the Simulink model group, as shown in Fig. 9.31(a).

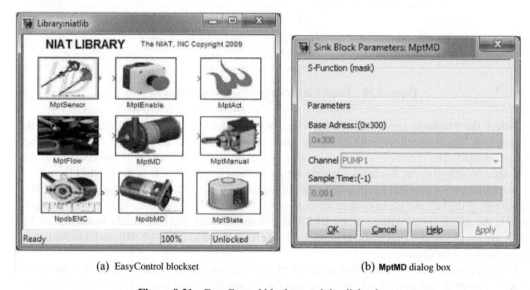

(a) EasyControl blockset   (b) **MptMD** dialog box

Figure 9.31   EasyControl blockset and the dialog box.

Double click the **MptMD** motor block, the parameter dialog box can be opened as shown in Fig. 9.31(b). It can be seen that the **Base Address** and **Sample Time** need to be specified.

9.5.2   Modeling and Simulation of Pendubot Systems

The photograph of the Pendubot experimental device, developed by NIAT, is shown in Fig. 9.32(a). In the photograph, the Pendubot has settled in downward position. The control objective is to swing up the Pendubot arms such that they can maintain an upright position.

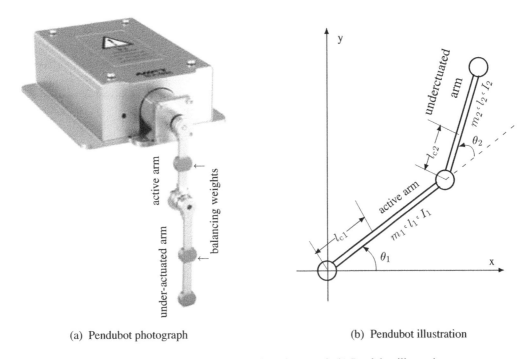

(a) Pendubot photograph (b) Pendubot illustration

Figure 9.32 Pendubot plant.(a) Pendubot photograph.(b) Pendubot illustration.

The Pendubot is a two-degree-of-freedom under-actuated mechanical system provided by NIAT. In the system, there is an active arm driven by a DC torque motor, and an under-actuated arm, without any driving action. Two VLT12 high-precision photoelectric encoders, with a precision of 1250 pulses/cycle. They are installed respectively on the active joint and the under-actuated joint, and these provide the position feedback signals of the arms. The sketch of the Pendubot is given in Fig. 9.32(b), where l_1 is the length of the active arm, l_{c1} is the length from the axle to the center of gravity, l_{c2} is the length from the axle to the center of gravity of the under-actuated arm, m_1 and m_2 are the masses of the two arms and I_1 and I_2 are the moments of inertia to the center of gravity of the two arms. The dynamic equation of the Pendubot system can be written as

$$D(q)\ddot{q} + C(q,\dot{q})\dot{q} + G(q) = \tau, \tag{9.2}$$

where $q = [\theta_1, \theta_2], \tau = [\tau, 0], \tau$ is the torque of the active arm, the coefficient matrices are defined as

$$D(q) = \begin{bmatrix} a_1 + a_2 + 2a_3 \cos\theta_2 & a_2 + a_3 \cos\theta_2 \\ a_2 + a_3 \cos\theta_2 & a_2 \end{bmatrix}, \quad G(q) = \begin{bmatrix} a_4 \cos\theta_1 + a_5 \cos(\theta_1 + \theta_2) \\ a_5 \cos(\theta_1 + \theta_2) \end{bmatrix} g, \tag{9.3}$$

$$C(q,\dot{q}) = \begin{bmatrix} -a_3\dot{\theta}_2 & -a_3(\dot{\theta}_2 + \dot{\theta}_1) \\ a_3\dot{\theta}_1 & 0 \end{bmatrix} \sin\theta_2, \tag{9.4}$$

and $a_1 = m_1 l_{c1}^2 + m_2 l_1^2 + I_1$, $a_2 = m_2 l_{c2}^2 + I_2$, $a_3 = m_2 l_1 l_{c2}$, $a_4 = m_1 l_{c1} + m_2 l_1$, $a_5 = m_2 l_{c2}$. The parameters were identified as $a_1 = 0.0106$ kg·m^2, $a_2 = 0.00597$ kg·m^2, $a_3 = 0.00509$ kg·m^2, $a_4 = 0.0751$ kg·m, $a_5 = 0.0367$ kg·m under the current balancing weights. The objective of the balancing control is that when θ_1 approaches the set-point θ_{1d}, the under-actuated arm maintains an upright position, and θ_2 settles down at the given balancing point θ_{2d} such that $\theta_{1d} + \theta_{2d} = 90°$.

Since the Pendubot system is driven by a DC motor, in the mathematical model of the actuator and DC motor, the control torque $\tau(t)$ of the active arm can be obtained from

$$\tau(t) = K_1 K_2 u(t), \tag{9.5}$$

where K_1 is the torque constant of the motor and K_2 is the gain of the actuator. In this control system, these two parameters are 0.4125 N·m/A and 2.68 A/V respectively. The signal $u(t)$ is the input voltage of the actuator, generated directly by the controller.

It can be seen from (9.2) that when the substate vectors are selected as $x_1 = q, x_2 = \dot{q}$, the original system can be written as

$$\begin{bmatrix} \dot{x}_1 \\ \dot{x}_2 \end{bmatrix} = \begin{bmatrix} x_2 \\ D^{-1}(x_1)[\tau - C(x_1, x_2)x_2 - G(x_1)] \end{bmatrix}. \tag{9.6}$$

More specifically, the state variables can be selected as $x_1 = \theta_1$, $x_2 = \theta_2$, $x_3 = \dot{\theta}_1$, $x_4 = \dot{\theta}_2$. Thus the state space model of the system can be written as

$$\begin{cases} \dot{x}(t) = F(t, x(t), u(t)) \\ y(t) = x(t). \end{cases} \tag{9.7}$$

It can be seen that, the state space model is suitable to be described by an S-function. In the model, there are four continuous state variables and no discrete states. There is one input signal $u(t)$ and four output signals, that is, the four states. There are also four additional variable vectors, the coefficient $a = [a_1, \cdots, a_5]$, the initial state vector x_0 and constants K_1 and K_2. The system model can be described by the following S-function

```
function [sys,x0,str,ts]=pendubot(t,x,u,flag,a,x0 s,K1,K2)
switch flag,
case 0
    sizes = simsizes; sizes.NumContStates=4; sizes.NumDiscStates=0;
    sizes.NumOutputs=4; sizes.NumInputs=1;
    sizes.DirFeedthrough=0; sizes.NumSampleTimes=1;
    sys=simsizes(sizes); x0=x0 s; str=[]; ts=[0 0];
case 1
    c1=cos(x(1)); c2=cos(x(2)); c12=cos(x(1)+x(2)); x2=x(3:4);
    D=[a(1)+a(2)+2*a(3)*c2, a(2)+a(3)*c2; a(2)+a(3)*c2, a(2)];
    G=[a(4)*c1+a(5)*c12; a(5)*c12]*9.81;
    C=[-a(3)*x(4), -a(3)*(x(3)+x(4)); a(3)*x(3), 0]*sin(x(2));
    sys=[x2; inv(D)*([K1*K2*u; 0]-C*x2-G)];
case 3, sys=x;
case {2, 4, 9}, sys = [];
otherwise, error(['Unhandled flag=',num2str(flag)]);
end
```

The four additional variables can be assigned with the following statements

```
>> a0=[0.0106, 0.00597, 0.00509, 0.0751, 0.0367];
   x0=[1.3; 0.3; 0; 0]; K1=0.4125; K2=2.68;
```

The plant model can be linearized, and a controller can be designed from such a model. To get the linearized model, a masked Simulink model can be constructed, as shown in Fig. 9.33(a), where the dialog box of the Pendubot model is given in Fig. 9.33(b). For the equilibrium point $(90°, 0°)$ with zero input, that is, $u_0 = 0$, the linearized model can be obtained with the following MATLAB statements

```
>> G=linearize('c9mniat3',[pi/2; 0; 0; 0],0)
```

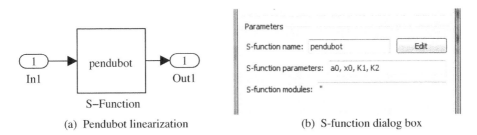

(a) Pendubot linearization   (b) S-function dialog box

Figure 9.33   Simulink model of the Pendubot system (model name: c9mniat3).

The linear state expression can be obtained as

$$A = \begin{bmatrix} 0 & 0 & 1 & 0 \\ 0 & 0 & 0 & 1 \\ 68.651 & -49.033 & 0 & 0 \\ -66.876 & 151.14 & 0 & 0 \end{bmatrix}, \quad B = \begin{bmatrix} 0 \\ 0 \\ 176.59 \\ -327.15 \end{bmatrix}, \quad C = I_{4\times4}, \, D = 0_{4\times1}.$$

Assume that, at initial time, the Pendubot system is settled around its equilibrium point $(90°, 0°)$. Linearization can be performed on the original system, and based on the results, an optimum linear quadratic controller can be designed, with state feedback vector $K = [-22.3589, -19.8736, -4.0484, -2.4394]$. The Simulink model with state feedback control is constructed as shown in Fig. 9.34(a), with the set-point input vector given by $x_d = [\pi/2, 0, 0, 0]$.

Assume that the control objective is to keep the two arms in an upright position, that is, $\theta_{1d} = 90°$ and $\theta_{2d} = 0°$. The following MATLAB statements can be executed and the simulation results are obtained as shown in Fig. 9.34(b).

```
>> vec_in=[pi/2 0 0 0]; K=[-22.3589,-19.8736,-4.0484,-2.4394];
   [t,x,y]=sim('c9mpendsim',[0,5]); plot(t,y(:,1:2))
```

It can be seen that when the initial conditions are near the upright position, that is, $\theta_{10} = 1.3$ rad$= 74.5°$, $\theta_{20} = 0.3$ rad$= 17.2°$, within a very short time the system settles down at its equilibrium point, such that $\theta_2 = 0.5\pi$ rad $= 90°$, $\theta_2 = 0°$.

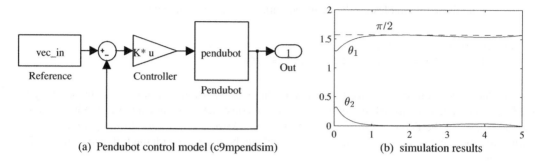

(a) Pendubot control model (c9mpendsim) (b) simulation results

Figure 9.34 Pendubot control system simulation.

If the control object is to have $\theta_{1d} = 1$ rad and $\theta_{2d} = (\pi/2 - \theta_{1d})$ rad, the following statements can be used and the simulation results are then as shown in Fig. 9.35.

```
>> vec_in=[1 pi/2-1 0 0]; [t,x,y]=sim('c9mpendsim',[0,5]);
   plot(t,y(:,1:2),t,y(:,1)+y(:,2))
```

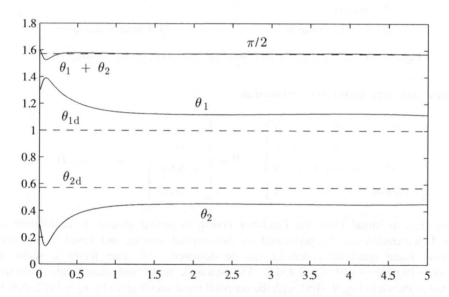

Figure 9.35 Simulation results when the control objectives are changed.

An optimum linear quadratic regulator alone cannot be used to achieve the expected x_d equilibrium point. If such a result is to be achieved, gravity compensation needs to be introduced so that $u(t) = K(x_d - x(t)) + u_0$, where the gravity compensation can be obtained from

$$u_0 = \frac{g}{K_1 K_2}\Big[a_4 \cos\theta_{1d} + a_5 \cos(\theta_{1d} + \theta_{2d}) \Big]. \tag{9.8}$$

With the gravity compensation technique, the new Simulink model can be constructed as shown in Fig. 9.36(a). The calculation of the gravity compensation constant u_0 can be embedded in the **PreLoadFcn** property of the model. For this system, $u_0 = 0.3601$ V. Simulation results under the compensation can be obtained as shown in Fig. 9.36(b). In this way, the set-point of the angle θ_{1d} can be arbitrarily chosen. From (9.8), if $\theta_{1d} = 90°$, $\theta_{2d} = 0°$, the compensation constant is $u_0 = 0$, which means that the compensation method also works for the original upright control problem.

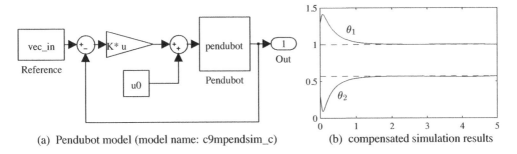

(a) Pendubot model (model name: c9mpendsim_c)   (b) compensated simulation results

Figure 9.36   Pendubot system with gravity compensation.

9.5.3   Hardware-in-the-loop Simulation Experiment of Pendubot Systems

With the mechanism of hardware-in-the-loop simulation, the actual Pendubot system can be used to substitute the mathematical model, such that the new hardware-in-the-loop Simulink model shown in Fig. 9.37 can be constructed, where the two angles θ_1 and θ_2 can be measured with the **NpdbENC** encoder blocks, and the gain of $2\pi/(4 \times 1250)$ can be used to convert it to radians. The derivatives of the angles can be obtained with the approximate derivative block with $80s/(s+80)$. The control signal u can be calculated with the **MptMD** block, and fed to the driving motor. The signal can be applied to the driving motor through the A/D converter, to form the closed-loop control system. The model can be used to control directly the real Pendubot system in real-time. The control results shown in Fig. 9.38 can be achieved.

The optimum linear quadratic controller can be used to move the two arms of the Pendubot relatively quickly into the upright position from their initial states, and can remain in that position. This means that the system has clear robustness.

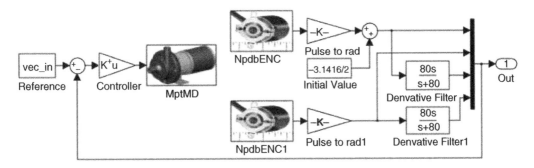

Figure 9.37   Pendubot hardware-in-the-loop model (model file: balancing_LQG_top).

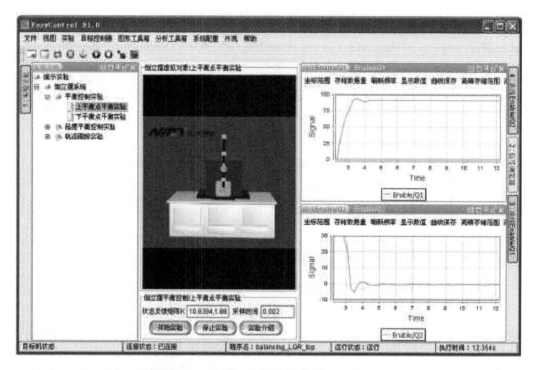

Figure 9.38   Real-time control result in Pendubot system.

9.6   HIL Solutions with Even Lower Costs

It can be seen from previous sections that HIL and rapid prototyping solutions can be obtained by products such as those from Quanser and NIAT. In this section, we will introduce another HIL solution using low-cost hardware based on Arduino™ for hardware-in-the-loop real-time control rapid prototyping. This solution is not only low cost but also very portable and it can be taken home for doing real-time closed-loop systems control experiments. This is particularly useful in a mobile age.

9.6.1   Arduino Interface Installation and Settings

In the latest versions of MATLAB such as R2012b, Simulink provides support packages for third-party products such as Arduino, LEGO Mindstorms™ and PandaBoard™.

This feature can be accessed in any Simulink model window through the **Tools** → **Run on Target Hardware** → **Install/Update Support Package...** menu item, as shown in Fig. 9.39, or simply by typing `targetinstaller` in the MATLAB command window. Alternatively, in older versions of MATLAB, users can go to the **Tools** menu in a Simulink model window and click **Add-Ons** → **Get Hardware Support Packages** menu item.

After the above action, users will be directed to the **Target Installer** in which the target support packages are listed. For a complete list of the supported products, see [7]. On completing the installation procedure described previously, the "Simulink Arduinolib" is automatically embedded into the Simulink Library Browser and is ready to use. It includes the blocks shown in Fig. 9.40.

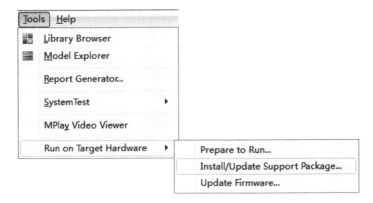

Figure 9.39 Additional package support menu.

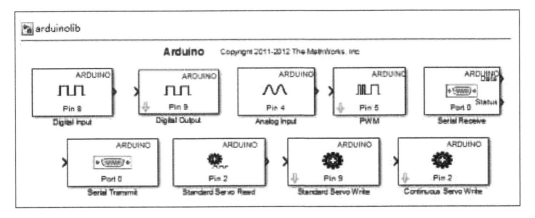

Figure 9.40 Simulink Arduinolib blockset.

The Simulink model created in this way is set up to run on the target, which is the Arduino board in this case. Hence, pre-settings need to be performed prior to execution. Again, refering to Fig. 9.39, clicking the **Prepare to Run** menu item will pop up the standard Simulink configuration dialog box. From the dialog box, if the **Run on Target Hardware** is selected, the **Arduino Uno** item can be specified from the **Target hardware** list box.

Follow the steps of specifying the serial communication port and Baud rate, then the popup menu appears as in Fig. 9.39. Now, the Simulink model can run on the Arduino target as a stand-alone application.

9.6.2 Applications of Arduino Control

There are two main ways of interfacing with the open-source hardware Arduino:
1) interfacing with it as a normal target through code generation
2) "virtual machine style" host-target communication.

The block library given earlier can be regarded as the first approach, while the Simulink model can be executed directly with the connected hardware.

Here, the second approach is presented as a case study on Arduino Uno. This interfacing approach does not support code generation, meaning that it does not generate code to download to the Arduino target. Instead, an I/O description file needs to be fed into the Arduino in advance as if it were firmware. Afterwards, all the I/Os on the Arduino board can be accessed either through Simulink or through MATLAB commands using serial communication. This is similar to the mechanism of a "Java virtual machine", which manages the bottom layer hardware while providing users with generic APIs.

To use this approach, a third-party Arduino support package is needed which is freely available from MATLAB Central [8], and it can be initiated with `arduino_io_lib`. The blocks in the package are shown in Fig. 9.41.

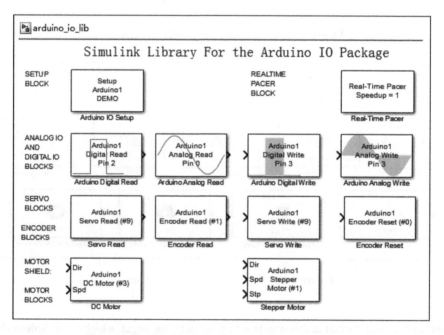

Figure 9.41   The blocks in the Arduino support package.

Example 9.5   *This example shows a simple closed-loop control system developed on the Simulink Arduino support package. The hardware set-up is shown in Fig. 9.42 (a). The control objective is to make the ball track the height set-point and, after settling down, maintain its height against disturbances.*

*A MATLAB GUI, shown in Fig. 9.42 (b), is designed to visualize the control and response. Users can adjust the height set-point by sliding the bar in the GUI. The Simulink model is displayed in Fig. 9.43, in which the **Arduino analog read Pin 0** is connected to an infrared (IR) sensor on top of the tube, and **Arduino analog write Pin 3** is connected to a motor-driven fan to blow the ball. The IR sensor is used to detect the height of the ball within the tube.*

It is worth mentioning that, during the running of such an application, the serial communication between the host PC and the Arduino target cannot be interrupted because the signal processing is in fact being carried out on the PC and only the sensing and actuation is performed on the Arduino.

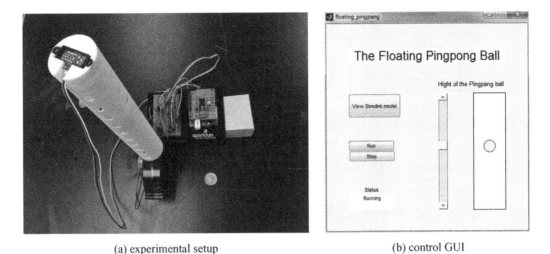

(a) experimental setup   (b) control GUI

Figure 9.42   The ping-pong ball floating experiment.

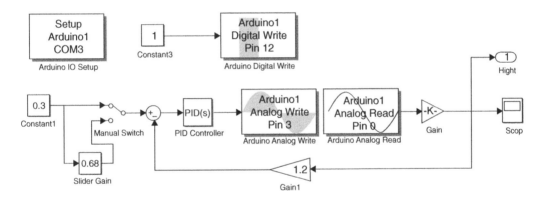

Figure 9.43   The Simulink model (model name: pingpang_sim).

9.6.3   The MESABox

An educational experimental toolbox, named MESABox™, based on the Arduino Uno core and peripheral hardware components, was developed in the MESA (Mechatronics, Embedded Systems and Automation) Laboratory at University of California, Merced, for the purpose of making the traditionally cumbersome mechatronics laboratories into a small-sized portable handset [9]. The cost of the key components in MESABox is approximately $180. The assembled box is shown in Fig. 9.44.

In MATLAB R2012b, varieties of "Apps" such as the "Floating ball" and "Fan-and-Plate" have been developed for the MESABox to support its educational usage in the MESA LAB. They are seamlessly integrated into MATLAB as a Toolbox/App.

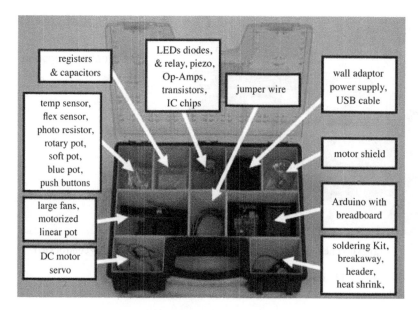

Figure 9.44   MESABox for take-home mechatronics and control labs.

Exercises

9.1   To help understand the concepts and applications of different simulation modes in Simulink with practical examples, the modes **Normal, Accelerator** and **External** should be tested and suitable applications of different modes explored.

9.2   Consider the four-bar mechanism studied in Chapter 7, with the model name c7mmech4.mdl. Convert the model into a standalone executable file, and run the program in the DOS command box environment and compare the simulation results.

9.3   Select one of the Simulink models created earlier. Use two methods to simulate the model in real time. The executable simulation program can be executed on host and target computers. Also try to control the target computer through internet and complete the simulation process.

9.4   Design a virtual instrumentation interface that acquires a signal from an actual A/D converter, and display the signal with gauges. Use the tracker-differentiator to display the original signal and its derivative. Select suitable parameters for the tracker-differentiator and compare the results. Note that in real-time simulation, S-functions written in MATLAB are not supported. An S-function written in C should be used instead.

9.5   If the reader has access to tools such as Quanser, dSPACE, NIAT or Arduino Uno, try to construct a nonlinear plant model, and design a controller for it.

9.6   In practical control systems, the derivatives of some signals cannot be measured easily. Alternative indirect methods should be used instead, such as the filters $50s/(s + 50)$, $80s/(s + 80)$. Of course, the tracker-differentiator discussed in Section 6.3.2 can be used. Construct a simulation model for the Pendubot system with the tracker-differentiator and find suitable parameters.

9.7   Consider the mathematical model of the Pendubot system. Try to design a switching controller such that, starting at an initial downward position, it will swing up the Pendubot, and when the arms enter the equilibrium point, the controller switches to an optimum linear quadratic controller.

References

[1] The MathWorks Inc. Real-time windows target user's guide, 2010

[2] dSPACE Inc. DS1104 R&D controller board installation and configuration guide, 2001

[3] Quanser Inc. SRV02 – Series rotary experiment # 1: Position control, 2002

[4] Quanser Inc. SRV02 – Series rotary experiment # 3: Ball & beam, 2002

[5] dSPACE Inc. Control Desk – experiment guide, Release 3.4, 2002

[6] Mechatronic Systems Inc. Pendubot model P-2 user's manual, 1998

[7] MathWorks. Hardware for project-based learning. `http://www.mathworks.com/academia/hardware-resources/index.html`, 2012

[8] MathWorks Classroom Resources Team. MATLAB Support Package for Arduino (aka ArduinoIO Package). MATLAB Central # 32374, 2011

[9] B Stark, Z Li, B Smith, and Y Q Chen. Take-home mechatronics control labs: a low-cost personal solution and educational assessment, Proceedings of the ASME 2013 International Design Engineering Technical Conferences & Computers and Information in Engineering Conference, Portland, Oregon, USA, 2013

Appendix

Functions and Models

Bold page numbers indicate where to find the syntax explanation of the function. The entries labelled by * are those developed by the authors. The entries labelled mdl are Simulink models, the slx files are also available.

Index

System Simulation Techniques with MATLAB® and Simulink®, First Edition. Dingyü Xue and YangQuan Chen.
© 2014 John Wiley & Sons, Ltd. Published 2014 by John Wiley & Sons, Ltd.

Printed and bound by CPI Group (UK) Ltd, Croydon, CR0 4YY

16/04/2025

14658561-0005